MATLAB and Simulink

FOR ENGINEERS

AGAM KUMAR TYAGI

Assistant Professor
Electrical Engineering Department
University of Petroleum and Energy Studies
Dehradun

OXFORD

UNIVERSITY PRESS

OXFORD
UNIVERSITY PRESS

Oxford University Press is a department of the University of Oxford.
It furthers the University's objective of excellence in research, scholarship,
and education by publishing worldwide. Oxford is a registered trademark of
Oxford University Press in the UK and in certain other countries

Published in India by
Oxford University Press
YMCA Library Building, 1 Jai Singh Road, New Delhi 110001, India

© Oxford University Press 2012

The moral rights of the author have been asserted

First published in 2012
Third impression 2013

MATLAB® and Simulink® are registered trademarks of The Math Works, Inc.

ISBN-13: 978-0-19-807244-7
ISBN-10: 0-19-807244-9

Typeset in Times New Roman
by Country Caramels, Ghaziabad
Printed in India by Multicolour Services, New Delhi 110 020

Third-Party website addresses mentioned in this book are provided
by Oxford University Press in good faith and for information only.
Oxford University Press disclaims any responsibility for the material contained therein.

*Humbly dedicated to
the lotus feet of
my beloved teacher and most revered
Prof. Prem Saran Satsangi Sahab*

PREFACE

MATLAB® is a high-performance technical computing language. It has an incredibly rich variety of functions and is often referred to as the Bible due to its vast programming capabilities. In MATLAB, computation, visualization, and programming are integrated such that the data can be expressed in a familiar mathematical notation. Evolved over many years with constant inputs from various users, it is now being widely used as a programming language for scientific and technical computation. One can perform powerful operations in MATLAB by using simple commands. Hence, writing programs in MATLAB is easier compared to other high-level languages such as FORTRAN or C. Users can even build their own set of functions for a particular application. It is an interactive system in which the basic data element is an array.

MATLAB is used as a standard tool in various introductory and advanced courses of almost all streams of engineering. It is used as a tool for analysis, modeling and simulation, and for high productive research and development activities. It can be used for various functions such as mathematical computation, algorithm development, data acquisition, modeling, simulation, prototyping, data analysis, exploration, visualization, and engineering graphics development as well as building graphical user interface.

Simulink is a software package for modeling, simulating, and analysing dynamic systems. MATLAB and Simulink are integrated and one can simulate, analyse, or revise the models in either environment. MATLAB also features a family of add-on application-specific features called toolboxes in Simulink. These toolboxes can be used for modeling and simulation of specialized technology in a real-time environment.

This book attempts to train engineering students of different streams to use the functions and toolboxes of MATLAB and Simulink for the study, design, and analysis of different electrical circuits and systems. All these toolboxes can be used to build a real-time prototype of the system.

ABOUT THE BOOK

This book gives an overview of the MATLAB and Simulink environment from the scratch. An engineer totally unknown to this software can become proficient in using this software after going through this book.

Salient Features of the Book

- Uses the latest version, 2010a, of MATLAB.

This book uses the latest version of MATLAB which was launched in 2010. This ensures that students get to learn the latest version of the software.

- Emphasis on circuit simulations and designing using simulations

A number of programs and simulations of electrical engineering circuits and systems are presented along with the theoretical explanation of commands and functions in detail. The book emphasizes on those functions or toolboxes which are useful for an engineer. A number of solved practical examples are illustrated in each chapter so that one can easily learn the programming skills. Simulink models are also illustrated in much detail in most of the chapters in such a manner that one can easily construct and analyse models.

- Includes a few color plates in the book with MATLAB graphs in color

Chapters 2 and 9 illustrate some graphs which are color specific and hence separate color plates have been appended in this book to ensure that students are able to view these graphs in color.

- Supplements any course that uses MATLAB by presenting projects from electrical, computer, and mechanical engineering.

Students will find this book very useful for their projects. Since MATLAB works as a laboratory, one can experiment, analyse, and simulate any experiment in it. Various simulations of rectifiers, inverters, choppers, projects, etc. are elaborated for better understanding of these topics.

- Includes a CD comprising user interactive programs and projects discussed in the book.

All the solved examples and projects are provided in the CD accompanying with this book so that the user can utilize them for their experimentation and learning. The CD can also be used for detailed study and analysis of the various problems and projects discussed and elaborated in the book.

CONTENT AND COVERAGE

The first three chapters of the book are designed to serve as a material in order to teach the basics of programming and simulation. The remaining six chapters focuses on applications of the various engineering areas like basic electrical engineering, power electronics, power system, communication systems, control systems, mechanical engineering, computer science, and other advanced topics in the end related to electrical engineering.

This book consists of eleven chapters and two appendices.

Chapter 1 provides an introduction to MATLAB. The concept of creating a new M-file and executing it is explained in this chapter. It also provides some basic concepts of Simulink with the help of a simple model. This chapter is useful for a novice user.

Chapter 2 focuses on the programming concepts of MATLAB. It discusses about variables, arrays, arithmetic, logical and relational operators, 2-D and 3-D plots, branching and looping, string, input and output functions with the help of various examples.

Chapter 3 focuses on Simulink and its utilization as a tool for design, modeling, and simulations of practical systems. An introduction to Simulink is provided along with related examples. Instead of going deep inside this tool, this chapter focuses on the application part. More stress is given on modeling and simulation aspect so as to provide a clear view about this tool.

Chapter 4 deals with MATLAB applications in basic electrical engineering. In this chapter Ohm's law and Kirchhoff's laws were verified in MATLAB. Analysis of R-L, R-C, and R-L-C circuits is done with the help of Simulink models. In addition to this, series and parallel resonance circuits are simulated along with network theorems like Thevenin's, Norton's, superposition, reciprocity, and maximum power transfer. In the end various types of electrical powers are described along with the electrical transformers.

Chapter 5 discusses design and analysis of different types of rectifiers along with their performance parameters. A basic understanding of different types of power electronic switches and their functioning as well as Fourier series and Laplace transform is a pre-requisite. A brief coverage of various power electronics switches is provided. Models of various single phase and three phase controlled and uncontrolled rectifiers and their output waveforms under different loading conditions are provided. One can further analyse these output waveforms by experimenting the models provided in the CD. It gives a practical insight about the design and working of the rectifiers. Four simulation projects are also discussed in the end of the chapter.

Chapter 6 discusses various types of inverters. Single phase half-wave and full-wave inverters are simulated and their performance parameters are analysed. A brief insight is provided about current source inverters and three-phase inverters. In the end, dual converters are discussed. Mc Murray Bedford full bridge inverter is also analysed in project 2 of this chapter.

Chapter 7 contains simulation models of various types of choppers and cycloconverters. This explains the principle of chopper and cycloconverter operations in a very lucid manner.

Chapter 8 deals with power system engineering. A brief introduction about power system and power quality is provided. Issues related to power quality and power harmonics are discussed. Power definitions under non-sinusoidal conditions are defined in time and frequency domain which are very important for a power system engineer. Load flow studies are done by Gauss Seidel and Newton Raphson methods. Power system stability and load frequency control are also discussed. Three simulation projects are also undertaken in order to provide a better understanding of the field.

Chapter 9 covers control system engineering and electrical machines. In control system engineering, time response analysis of first- and second-order systems and in-frequency response analysis Bode, Nyquist and Nichols chart are discussed. In electrical machines section, various types of DC machines, three-phase induction and synchronous motors are analysed.

Chapter 10 contains analog and digital communication systems, computer animations, artificial neural networks, fuzzy logic, and mechanical engineering applications.

Chapter 11 aims to discuss about the recent trends in the area of electrical engineering. The main focus is on matrix converters, PWM rectifiers in terms of their performances and technical issues. This chapter does not assume any special knowledge in these devices, only elementary knowledge about power electronic converters is sufficient. The design and analysis of these converters is performed by using MATLAB/simulink and the results obtained are further analysed and elaborated.

Appendix A contains Fourier series of some well known waveforms and Laplace transforms. Appendix B incorporates the proof of maximum power transfer theorem, derivation of EMF equation of transformer, and EMF equation of a DC machine.

As this book deals with applications of MATLAB in engineering, suggested readings are given at the end of each chapter so that further technical knowledge can be acquired on that topic with the help of these resources.

ACKNOWLEDGEMENTS

I am grateful to Dr Kamal Bansal, Head of the Department, Electrical, Electronics and Instrumentation Engineering, UPES, for his support, motivation, and continuous inspiration in bringing out this book. I would also like to thank my colleagues in UPES—Rajesh Singh, Vineet Mediratta, Jayadeep Chakravorty, Vivek Kondal, N.B. Soni, B. Bagchi, Madhu Sharma, Adesh Kumar, M.K. Gupta, and Manish H. Bilgaye of BITS Pilani—for providing comments on the manuscript in their respective area of specialization.

I appreciate the insights, suggestions, and comments received from the reviewers who reviewed my manuscript and helped enhance the quality of the book. I would especially like to thank Dr Abraham T. Mathew of NIT Calicut and Prof. Udhaya Kumar of Anna University, Chennai for their valuable suggestions.

I am also grateful to my father Prof. Guru Sewak Tyagi, Dayalbagh Educational Institute, Agra and grandfather Prof. (Retired) Gurudas Singh Tyagi from whom I acquired skills. I wish to thank my wife Karuna and daughter Agrima for their motivation, patience, understanding, encouragement, and support in preparing this book. They are a constant source of inspiration for me. My sincere regards to my mother Ms Saroj Devi and my younger brother Mr Aman Tyagi for their support.

Lastly, I am grateful to the editorial team of Oxford University Press for providing the required support and cooperation for completing this book.

The aim of writing this book is to make learning a pleasant experience for students by presenting the matter in a clear, understandable, and well-organized manner. Every effort is made to provide an insight into the subject in a practical manner rather than just explaining results theoretically. Even a fraction of success achieved in meeting this goal will make all the pains taken to write this book worthwhile. If readers want to share any suggestions or comments for improving the book, they can write to me at tyagiagam@rediffmail.com.

Agam Kumar Tyagi

BRIEF CONTENTS

DETAILED CONTENTS

LIST OF PLATES

1

INTRODUCTION TO MATLAB PROGRAMMING

1.1 INTRODUCTION

This chapter focuses on some basic concepts of MATLAB programming along with some application-oriented exercises. After going through the chapter, we will have a basic understanding of MATLAB and will be able to write simple programs in it. A brief introduction to Simulink is also demonstrated by the simulation of a simple circuit. This gives us an idea about Simulink and the manner of performing simple simulations.

1.2 WHAT IS MATLAB?

The name MATLAB is an abbreviation for MATrix LABoratory. Cleve Moler, a numerical analyst, wrote the first version of MATLAB in 1970. Since then, it became a successful computational and commercial software. As matrix mathematics became more common, owing to its potential and simplicity, and as the engineering systems became more flexible and complex, the task of computation became tedious. Thus, MATLAB gained popularity due to its computational power, user-friendliness, and its modifications according to the changing scenario. In most of the engineering colleges and universities, MATLAB has replaced the principal computational tools such as FORTRAN and C++ as this software takes care of many computational problems and one can devote more time on experimentation. The results in MATLAB are accurate and reliable as highly robust algorithms are used in this software.

Owing to its vast programming capabilities, MATLAB has an incredibly rich variety of functions and is often referred to as the Bible of computation. Many toolboxes are also available in Simulink for modeling and simulation. By merely using one or two commands, many powerful operations can be performed in MATLAB. A set of functions can be built for an application, if required. This book makes an attempt to aid the electrical engineers in the study and analysis for different electrical circuits and systems by using these functions and toolboxes. So, the main focus of this book is on practical applications related to electrical engineering.

1.3 GETTING STARTED—STEP 1

MATLAB can be started by the double clicking on the MATLAB icon on the desktop or by selecting MATLAB from the Windows Start menu. A technical person can help in the installation of MATLAB software on the system. Also, one should be sure that the system is compatible and has sufficient memory for installing and running the MATLAB. Normally, the system becomes slower after the installation.

The default configuration of MATLAB window is shown in Fig. 1.1. It has many options available for files, variables, functions, help, and demos within the MATLAB environment. The latest MATLAB version 7.10.0.499 (R2010a) has been used by the author while writing this book.

The first window accessible to the user after opening the MATLAB is the command window which is shown in Fig. 1.1. Check the options available on the toolbar. By clicking on the File button, one can open or save a new MATLAB file, close any window, print a file, import data from a file, and set path for directory. By selecting the Edit option, one can undo, redo, cut, copy, paste, select/deselect, and clear command and command history windows. This option is normally used for editing the MATLAB files. Debug can be used for debugging the error at any step of the program. Debug is available in the edit window, i.e., when a new file is opened, it has the 'run' option to run the M-file. M-files can also be run by pressing function key F5 from the keyboard. The Desktop button on the toolbar is used for organizing the desktop. Different windows, i.e., command window, command history, current directory, and workspace, can be viewed by selecting the Desktop option. Command window as the name indicates is used for

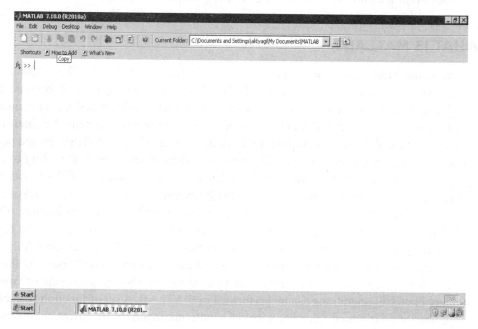

Fig. 1.1 MATLAB desktop (The exact appearance might vary depending upon the version and the type of platform used)

executing the commands. It also shows the output of a program and various variables can also be defined and used in this window. Suppose a variable, say A = 2, has to be assigned, then just type >> A = 2 after the command prompt, it will take 'A is equal to two'. If the value of a variable has to be visualized, then just type that variable, say A, and press Enter on the command window. Simple programs can be easily written and run on the command window itself. In order to clear the command window, type 'clc' after the command prompt and then press Enter.

All the variables are stored in the workspace with their name, value, and class. Workspace can be seen through the desktop button on the toolbar. Workspace can be utilized to view all the variables stored in an array form defined in the programs. Command history contains all the information regarding the commands executed in the MATLAB. Besides these, edit window and figure window are also present in the MATLAB. An edit window is used to create new M-files or to edit the existing ones, while a figure window is a separate window for displaying the graphics.

The last icon on the tool bar is Help. It contains information about all functions, commands, and toolboxes and explains how to use them. Online help is also available along with the available Web resources. Help can also be used to find out about the usage of a particular command. Figure 1.2 clearly depicts the usage of the INT (integrate) command along with the various other options available for that command. It also gives examples elaborating the usage of all the commands/toolboxes so that the user can choose the most suitable one and then utilize

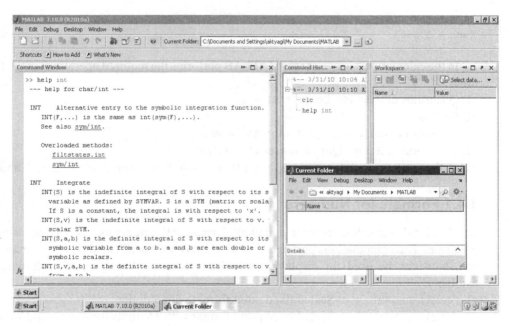

Fig. 1.2   Desktop showing command window, current directory, and command history. How to utilize help for using int command in MATLAB is also shown in the command window

it conveniently. Similarly, Help can be taken for any other command by just typing that command after help on the command window after the command prompt. In a similar fashion, Lookfor command can also be used. The difference between Lookfor and help commands is that the former gives a summary related to the function or the command mentioned, while the latter directly goes for the exact function itself.

Some shortcuts, namely new file, open file, cut, copy, paste, undo, redo, Simulink, guide, and help, are available on the toolbar. These shortcuts are convenient to use and are also time-saving. If required, more shortcut keys can also be added by using 'how to add' button as per the convenience of the programmer. At the bottom of the MATLAB desktop, Start button is provided for utilizing various options. This button can also be used for directly starting a file, for a demo, for help, etc.

1.4 GETTING STARTED—STEP 2

The basic arithmetic operators that can be conveniently used in MATLAB are +, −, *, and / and these are used in conjunction with brackets: ( ). The symbol $^\wedge$ is used to get exponents (powers): $3^\wedge4 = 81$, symbol \ for left division, and symbol c for complex conjugate transpose. Operators i and j both have value $\sqrt{-1}$ by default. We should type the commands after the prompt: ».

```
» 3+2/(4*6)
ans = 3.0833
```

In the calculation 3 + 2 / (4*6), MATLAB performs the required operations in the following order:
1. Quantities in brackets, i.e. (4 * 6) = 24
2. Powers
3. * /, working left to right {2 / (4 * 6)= 0.0833}
4. + −, working left to right (3 + 0.0833 = 3.0833)
Thus, the earlier calculation for 3 + 2 / (4*6) was done according to points 1, 3, and 4.

1.4.1 Formats for Numbers

In MATLAB variables, there is a default format. All calculations are done using double precision. The computations performed in MATLAB are not affected by the formats. Formats can, however, be used for displaying a variable or a number using the following options available:
1. FORMAT SHORT: Scaled fixed point format with 5 digits.
2. FORMAT LONG: Scaled fixed point format with 15 digits for double and 7 digits for single.
3. FORMAT SHORT E: Floating point format with 5 digits.
4. FORMAT LONG E: Floating point format with 15 digits for double and 7 digits for single.
5. FORMAT SHORT G: Best of fixed or floating point format with 5 digits.

6. FORMAT LONG G :   Best of fixed or floating point format with 15 digits for double and 7 digits for single.

7. FORMAT HEX: Hexadecimal format.

8. FORMAT + :   The symbols +, −, and blank are printed for positive, negative, and zero elements respectively. Imaginary parts are ignored.

9. FORMAT BANK: Fixed format for dollars and cents.

10. FORMAT RAT: Approximation by ratio of small integers.

The following formats may be used to affect the spacing in the display of all variables as follows:

1. FORMAT COMPACT: Suppresses extra line-feeds.

2. FORMAT LOOSE: Puts the extra line-feeds back in.

For instance, the following commands are executed on the command window:

```
>> format short, pi
ans = 3.1416
>> format long, pi
ans = 3.14159265358979
>> format short e, pi
ans = 3.1416e+000
>> format long e, pi
ans = 3.141592653589793e+000
>> format short g, pi
ans = 3.1416
>> format long g, pi
ans = 3.14159265358979
```

1.5 GETTING STARTED—STEP 3

A user new to MATLAB must know some useful commands. As such, there are numerous commands in MATLAB but the list of commands shown in Table 1.1 would be quite useful for a novice user.

Table 1.1   General purpose commands

S. No.	Command	Description
1.	simulink	Opens Simulink library browser. Simulink is used for modeling and simulation of a system and has various inbuilt blocks and functions
2.	demos	Demonstrates about the capabilities of MATLAB. Inbuilt demos can be run for better understanding
3.	ver	Gives information about MATLAB, Simulink version
4.	clc	Clears the command windows
5.	clf	Clears the current figure
6.	clear all	Removes all functions, variables, and global variables from the MATLAB

(Continued)

Table 1.1 (*Continued*)

S. No.	Command	Description
7.	quit	Exits MATLAB
8.	who	Lists the variables currently in the memory
9.	inmem	Lists the functions in MATLAB memory
10.	java	Uses JAVA within MATLAB environment
11.	recycle	Moves deleted files to recycle bin
12.	pack	Saves all variables on the disk after clearing the memory and then reloads the variables
13.	%	Writes comments as MATLAB ignores whatever is written after this symbol

In general, it is a good idea to begin programs with the `clc` and `clear all` commands to be sure that the memory has been cleared along with the workspace window.

Table 1.2 represents some predefined symbols frequently encountered in MATLAB.

Table 1.2 Predefined symbols in MATLAB

S. No.	Symbol	Description
1.	pi	Contains value of π up to 15 significant numbers
2.	i, j	$\sqrt{-1}$, imaginary number
3.	Inf	Infinity. Represents a very large or a very small number
4.	NaN	Not a Number. Signifies an output that is very large
5.	clock	Represents an array containing six elements—year, month, day, hour, minute, and second
6.	date	Contains the current date like 24-Jul-2010
7.	eps	Represents the smallest difference between two numbers
8.	ans	Stores the result of an output by default

1.6 GETTING STARTED—STEP 4

Simulink is a blockset-based programming tool developed by M/s Math Works Inc. for modeling, simulating, and analyzing multi-domain dynamic systems. It is basically a graphical extension of MATLAB in which system models are constructed on screen by using various blocks. It offers tight integration with rest of the MATLAB environment such that it can either drive MATLAB or be scripted from it. A number of blocksets, or say toolboxes, such as voltage source, current source, oscilloscopes, and function generators are available. It is widely used by researchers and engineers in the fields of power system, control system, power electronics, etc. for multi-domain analysis and design.

Simulink can be opened by clicking on the shortcut key or on the start button. Different toolboxes for designing and analyzing a system can be found in the library browser of Simulink. It has a standard library which has several blocks for different functions as shown in Fig. 1.3.

The toolbar has File, Edit, View, and Help buttons which have the functions according to their names. Through the file option on the toolbar or the shortcut key below the 'file' button, a new file can be opened. A Simulink file has an . mdl extension while a MATLAB file has an .m extension. By default, untitled name is assigned to a new file which a user can change later on. Blocks from the library browser can be taken by dragging or by right clicking the mouse and then adding them to the Simulink file. These blocks cannot be added to MATLAB files. If a block is required from the library, it can be searched by typing the name on the Simulink toolbar and then pressing the 'Enter' key from the keyboard. The browser will search the entire library for that block name. Some blocks—Unit Delay, Demux, Switch, Constant, Bus Selector, Terminator, and others—and their symbols are given in Fig. 1.3.

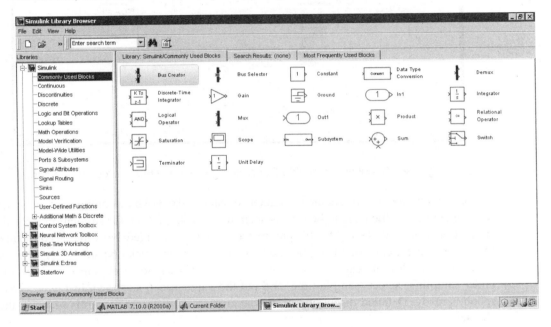

Fig. 1.3 Window showing Simulink library browser

These blocks function as per their nomenclature, e.g., an AC voltage source provides an AC voltage of a particular magnitude, frequency, phase, and so on. The above-mentioned blocks can be found in 'SimPowerSystems' blockset of the Simulink library. This blockset is mostly used by electrical engineers.

A Simulink file is made up of assembly of these blocks. A Simulink file shown in Fig. 1.4 contains two sine wave sources and a scope. In this file, there is a small demonstration of using blocks. To construct this file, drag a sine wave block from Simulink/sources sub-blockset and a scope from Simulink/sinks sub-blockset and drag it to the file 'untitled'. After this, right click the sine wave source and drag it to make another sine wave source in the same file. Connect the two sine wave sources to the scope by simply holding the left click of mouse first to the output of sine wave block and then join it to the scope input. After connecting these blocks, the circuit will look like the circuit in Fig. 1.4.

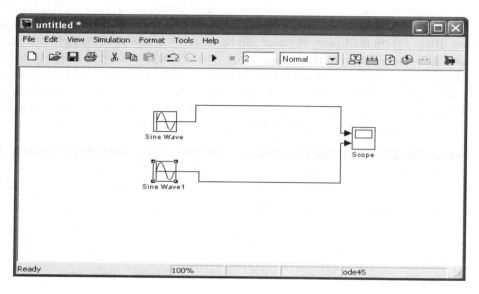

Fig. 1.4 Simulink file 'untitled' depicting two sine wave sources and a scope

Before running this circuit, the parameters of each and every block along with the simulation parameters for the file have to be set. As seen in Fig. 1.4, the time for simulation is 2 s and the simulation is normal. Rest of the file parameters remain unchanged. Now there is only one input for scope. Double click on the scope and then on the toolbar, click on parameters. The 'Scope' parameters window appears as shown in Fig. 1.5. In this window, set number of axes to 2 and under sampling, select 'sample time' and put the value 10e–6 as shown in Fig. 1.5. Now the scope is ready. Detailed discussion on the parameters of different blocks is given in the later chapters.

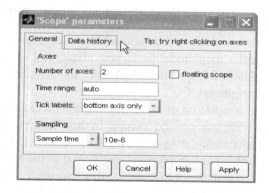

Fig. 1.5 Scope parameters

Similarly, by double clicking on the sine wave block, a new window showing the parameters of the block will open as shown in Fig. 1.6. This window also explains to the user about the various properties of a sine wave. The various values taken are shown in Fig. 1.6. After entering these values, click the 'ok' button. In a similar fashion, set the values for the second sine wave block.

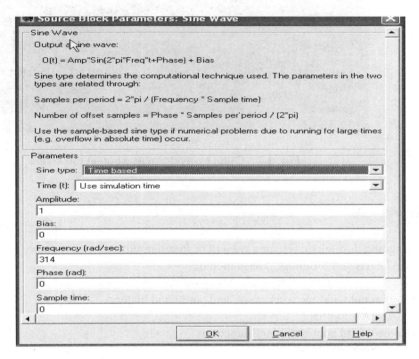

Fig. 1.6 Window showing different parameters of sine wave source

Now the simulation of the file can be started. For running the simulation, click the start simulation key just below the 'help' key or click on simulation/start. Observe the file while the simulation is running. This simulation can be run for different time periods, say 4 s, 7 s, and so on. After running the simulations, double click on the scope. On the scope, the sine waves which are given as inputs can be seen. The output of the scope is shown in Fig. 1.7. Click on the auto scale button on the toolbar for viewing the wave uniformly. In this figure, magnitude of the sine wave is shown on the x-axis and the time of the sine wave is shown on the y-axis. This result can also be printed, if required.

This exercise might help in understanding the Simulink. Similarly, different circuits can be tried by picking up blocks from the library browser. For different kinds of sources, 'sources' sub-block can be explored and for different display/output devices, 'sinks' sub-block can be explored. Help for a specific block can be quickly accessed by right clicking the block and selecting help option. This will also demonstrate about using that block for a model.

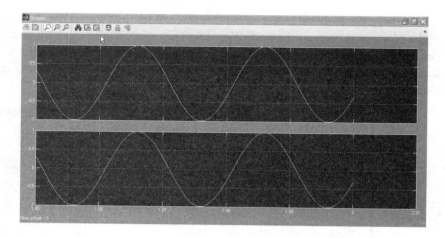

Fig. 1.7 Output waveform as seen in the scope

SUMMARY

After going through this chapter we will be familiar with some elementary operations performed in MATLAB and Simulink. These include

- Starting and exiting the MATLAB/Simulink
- Command window, workspace, and command history
- MATLAB toolbar
- Some useful commands and functions
- Managing variables
- Simulink toolboxes
- Simulation of elementary circuits
- Making Simulink files and viewing plots

REVIEW QUESTIONS

1. What is MATLAB? Elaborate its applications in the field of electrical engineering.
2. Mention the functions of command window, edit window, and figure window in the MATLAB.
3. Which command is used to clear the contents of the workspace?
4. What is Simulink? Elaborate its applications in electrical engineering.
5. Explain the various ways of taking help about a function in MATLAB.
6. Mention the various kinds of 'sources' and 'sinks' in the Simulink sub-block along with their uses.
7. Explain the different parameters of a sine wave block.

8. Use `lookfor` command and find out various integration functions available in MATLAB.
9. Find help on MATLAB command window for `exp` function.
10. Utilize `help` function and find out how to use `plot` function for plotting graphs.
11. Utilize `help` function and find how to use `sin`, `cos`, `tan`, `atan`, and `tand` functions.
12. Find the value of `pi` stored in MATLAB.
13. Use the signal generator source to obtain a cosine wave of frequency 100 rad/s and plot it.
14. List all the blocks under SimPowerSystems blockset in Simulink.
15. Find the applications of clock and date functions defined in MATLAB.
16. Elaborate on the applications of MATLAB in the fields of power system, power electronics, and electrical machines.
17. Search for various solvers present in Simulink for simulations and find their uses.
18. How to use `diff` function by using `help`?
19. Explore ways to receive data in MATLAB from external device and state its advantages.
20. What are the disadvantages of MATLAB programming software?
21. How many versions of MATLAB have been released till date and what are their features?

EXERCISES

1. Perform the following calculations in MATLAB, taking variables x = 100 and y = 50:
   (a) z = x/y
   (b) u = x * y
   (c) v = x + y
   (d) r = √x/y
   (e) t = √x * y
   (f) q = x sin(5y)
2. Taking x = 2.10101 and y = 4.3457, estimate the following expressions in MATLAB:
   (a) $5x/3y$
   (b) $45x^{-8}/y^{-7}$
   (c) $x^{-3} - y^{-4}$
   (d) $x^{6.7}/(4x + y^{-3.1})^2$
3. Calculate the following using MATLAB:
   (a) $e^{(-1.3)3} + 16 \sin(40\pi)$
   (b) $\pi \log 21 + \sqrt[5]{291}$
   (c) $\sin^2 16\pi/40 + \cos^2 16\pi/40$
   (d) $\tan (4\pi/5)^2$
   (e) $\tan^{-1}0.5^2$
4. Consider variables x = 6 + 7i and y = -5 + 3i. Estimate the following with the help of MATLAB:
   (a) x + y
   (b) xy
   (c) x/y
   (d) $x^2 + y^2$
5. Construct a circuit for analyzing the waveform of a sine wave source using a scope in Simulink. Take the source frequency as 50 Hz, 60 Hz, 80 Hz,

and 100 Hz and verify it on the scope by viewing its time period.
6. Estimate the rms, average, and peak values of the waveforms analyzed in 5 using Simulink blocks.
7. Obtain a sawtooth waveform of 100 Hz, 54 V and view this waveform on a scope.
8. Construct the circuit of a half-wave rectifier and determine the rms and average values of the output voltage waveform if the input source is 50 Hz, 200 V sinusoidal.
9. Design a three-phase star-connected balanced power voltage source of 220 V rms, 50 Hz and view it using a scope.
10. Write codes on the command window to convert angle in degrees to radians.
11. Perform the following operations on command window and analyze the result:
    (a) 0/0
    (b) 1/0
    (c) inf/inf
    (d) cos(pi)
    (e) 0/100
    (f) log(10)
    (g) 10^pi

2

FUNDAMENTALS OF
MATLAB PROGRAMMING

2.1 INTRODUCTION

This chapter focuses on the fundamental concepts of programming in MATLAB. It introduces arrays, plots, matrices, looping functions, logical functions, and various other functions along with their applicability and utility in engineering problems. After going through this chapter, we will be able to use these functions as programming applications for various purposes.

2.2 VARIABLES

A variable is a symbolic name used to address the storage of data. It is linked with a value. The linked value can be changed during the course of running the program. In MATLAB, assigning values to the variables is simple. Values can result from computation, from the output of a function, or from an external device. There is no need for the programmer to declare their type, size, etc. If no variable is assigned to the output value, *ans* is taken as a variable by default. For instance,

```
>> 4 + 5
ans =
9
>> ans*2
ans=
18
```

Hence, the output of the first calculation, i.e., 4 + 5, is labeled as *ans* and the same default variable *ans* is used in the second calculation also, where its value has changed to 18. Variables can also be assigned for a calculation. For instance,

```
>> x = 4 + 5
x =
9
>> y = x*2
y=
18
```

Here the value of variable x is 9 and that of y is 18. These variables can be used for further calculations as they are now saved in the memory area which is the MATLAB workspace.

In MATLAB, variables are case-sensitive, i.e., variable x is not same as X, so this should always be borne in mind before using assigned variables. Also, the name of the variable should start from a letter but it can be followed by a number, digits, or underscores. For example, the following variables are acceptable:

```
Net_current, Voltage2, V2, I1, Irms, V_out, P, out
```

The following variables are *not* acceptable:

```
Net-current, 2votage, %V, 1I, @2, $E
```

Also, built-in MATLAB function names cannot be assigned to variable names.

Suppose that two variables of a single-phase voltage source with peak amplitude of 220 V and frequency of 50 Hz have to be defined. This can be done by

```
>> A = 220; f = 50
```

where variable A is the amplitude and f the supply frequency. The units of these variables (volts and hertz) should not be written on the command line as MATLAB will not understand them. Now, after these parameters are defined, they can be used for computational work later on. Table 2.1 lists some commands useful while operating with variables

Table 2.1   Useful commands to be known while working with variables

S. No.	Command	Description
1.	Clear	Removes all variables from the workspace
2.	Load	Loads variables in workspace from the hard disk
3.	Save	Saves workspace variables to hard disk
4.	Who	Lists current variables
5.	Whos	Lists more information about variables
6.	exist	Checks whether a function or variable is defined
7.	isglobal	True if the variable is global otherwise false

(Continued)

Table 2.1 (*Continued*)

S. No.	Command	Description
8.	symvar	Searches for the symbolic variable in an expression
9.	export2wsdlg	Exports variables to the workspace
10.	flow	A simple function of three variables
11.	peaks	A simple function of two variables
12.	finihsav	Saves current variables in MAT-file
13.	Bluetooth_init	Initializes variables for Bluetooth voice demo
14.	qrydata	Links data variables for interactive queries
15.	sym	Constructs symbolic numbers, variables, and objects
16.	issame	Tests whether two variables are similar
17.	findym	Finds symbolic variables in a symbolic expression or matrix
18.	updatenames	Updates the names of the variables in the formula from the current names of the input value object
19.	resetvariables	Resets the values of variables

2.3 ARRAYS

An array is the fundamental unit of data in a MATLAB program. All data processed in MATLAB—as an input data, output data, or for computational work—is in the form of arrays. This is a major attraction for electrical engineers. Normally, a vector or a one-dimensional matrix ($N*1$ or $1*N$) is referred to as an array. Suppose the current in an AC circuit is I = (4 + $j5$) amperes, we can represent this current as I = [4, 5]. Similarly, a three-dimensional vector say A = 2i+3j+5k can be represented as A = [2, 3, 5] or A = [2 3 5]. The elements of a row vector A are enclosed in brackets and are separated by a comma or simply space as shown in the following.

```
>> A = [2 3 4]
A =
2 3 4
>> a = [4, 5, 6]
a =
4 5 6
```

If the vector is a column vector, the elements of the vector are separated by a semicolon. For example, it is represented as V = [6; 8; 9]. Thus, it should be understood that in MATLAB, comma or space is the column separator while semicolon is a row separator. Suppose A is a row vector or array and V be a column vector or array and we want to compute C which is, C = A*V.

C can be computed by writing the following program in a MATLAB file:

```
>> A = [3, 6, 9]; % Row vector or Array
>> V = [1; 3; 9]  % Column vector or Array
>> C = A*V         % Multiply A and V vector
```

The output of this program at the command window will be as follows:

```
C =
102
```

In order to get the transpose of vector V, add another program line to the file as

```
>> F = V' % Find transpose of array V
```

It will display F as

```
F =
1    3    9
```

An array can also be defined by the following syntax:

```
Initial: increment: terminator
```

For example, an array X = [1 3 5 7 9] can be defined as

```
>> X = 1:2:10
X =
1    3    5    7    9
```

In this case, '1' is the initial value, '2' is the incremental value, and '10' is the termination value. So the elements values for array X are 1, 1 + 2 = 3, 3 + 2 = 5, and so on till 9. This can further be elucidated by Example 2.1.

Example 2.1

```
t = 0:10e-6:.04;        % Time is taken from 0 sec to 0.04 sec in steps of 10
                        micro sec.
f = 50;                 % Frequency is taken as 50 Hz.
w = 2*pi*f;             % Angular frequency in rad/sec.
A = 220;                % Peak amplitude 220 V
V_out = A*sin(w*t);     %V_out is the output voltage waveform from 0 sec to 0.04
                        sec.
plot(t,V_out)
```

This program plots the waveform (two cycles) of a voltage source of 220 V peak amplitude, 50 Hz frequency (time period = 0.02 s) from time 0 s to 0.04 s, i.e., two complete cycles. Hence in the first line we define the time array, in the second line variable f, in third variable w, in fourth A, and in fifth plot V_out. The output of this program is shown in Fig. 2.1.

Table 2.2 lists some useful commands to be known while working with arrays. These commands will further help you for the computational work involving arrays.

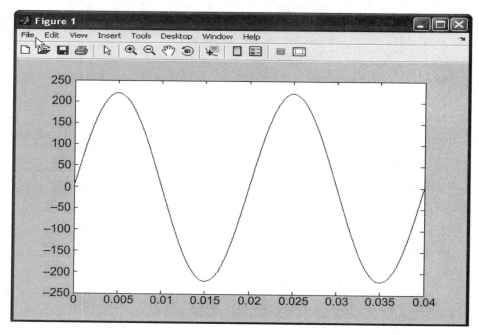

Fig. 2.1 Plot for t versus V_out

Table 2.2 Commands to be known while working with arrays

S. No.	Command	Description
1.	Cat	Concatenate arrays
2.	Isequal	True if arrays are numerically equal
3.	isequalwithequalnans	True if arrays are equal numerically
4.	meshgrid	X and Y arrays for 3-D plots
5.	ndgrid	Generation of arrays for N-D functions and interpolation
6.	printtables	Methods to create cell arrays of data on drivers and options
7.	isfloat	True for floating point arrays both single and double
8.	isinteger	True for arrays of integer data type
9.	Isjava	True for JAVA object arrays
10.	isnumeric	True for numeric arrays
11.	Instget	Retrieves data arrays from an instrument variable
12.	Nbinrnd	Random arrays from the negative binomial distribution
13.	Geornd	Random arrays from geometric distribution

(Continued)

Table 2.2 (*Continued*)

S. No.	Command	Description
14.	Random	Generates random arrays from a named distribution
15.	Sort	Sorts for cell arrays of strings
16.	checkarraysize	Checks and harmonizes array dimensions
17.	Fixgeocoords	Replaces from value with to value in the coordinate arrays

2.4 MATRICES

A matrix (plural matrices or matrixes) is a rectangular array of numbers. An item in a matrix is called an entry or an element. The horizontal and vertical lines in a matrix are termed as rows and columns respectively. To specify a matrix's size, a matrix with *m* rows and *n* columns is called an *m-by-n* matrix, where *m* and *n* are its dimensions. Entering matrices in MATLAB is similar as in the case of arrays. Consider a 3-by-3 matrix A that has to be entered. This can be done on the command window as follows:

```
>> A = [1 2 3; 3 4 5; 2 3 4]
A =
1    2    3
3    4    5
2    3    4
```

A matrix can be entered in this manner and then can be used later on in the program. For instance, consider the following example.

```
A = [1 2 3; 3 4 5; 2 3 4]; % Defines matrix A
B = [1 1 1; 2 2 2; 3 3 3]; % Defines matrix B
C = A+B;                    % Add A & B
```

The output of this program can be found by checking the values of the variables C, D, E, and F on the command prompt. The output will be as follows:

```
D = C';              % Take transpose of C
E = RANK(B);         % Find rank of matrix B
F = A*B;             % Multiply A & B
C =
2    3    4
5    6    7
5    6    7
D =
2    5    5
3    6    6
4    7    7
E =
1
F =
```

```
14    14    14
26    26    26
20    20    20
```

These variables can also be viewed at the workspace. By double clicking these variables, we can get their values in the matrix form on the right-hand side window titled 'array editor'.

Matrices, in general, find vast applications in electrical engineering problems. Consider an electrical network described by the equation V = IR, where V is a 3*1 voltage matrix, I is a 3*3 current matrix, and R is a 3*1 resistance matrix. Currents and resistances in the network are known and we have to calculate the voltages. The solution to this problem can be elucidated by Example 2.2.

Example 2.2

```
% This program demonstrates the multiplication of two vectors V & I
% Where,[P]= [I]*[V]and
% P1 = I1*V1+I2*V2+I3*V3, P2 = I4*V1+I5*V2+I6*V3, P3 = I7*V1+I8*V2+I9*V3
V = [1 2 3; 4 5 6; 7 8 9];    % Matrix for voltage
I = [6 7 8];                   % current matrix
P = I * V;                     % Power matrix
```

The output voltage matrix can be seen on the command window as follows:

```
>> P
P =
90    111    132
```

Table 2.3 provides some useful commands which can be utilized for computations involving matrices.

Table 2.3 Some useful commands for matrix operations

S. No.	Command	Description
1.	mldivide \	Divides the left matrix
2.	mpower ^	Power of the matrix
3.	Mrdivide /	Divides the right matrix
4.	mtimes *	Multiplies the matrices
5.	company	Companion matrix
6.	Diag	Diagonal matrices and diagonals of a matrix
7.	Eye	To get identity matrix
8.	Expm	To get exponential of a matrix
9.	Inv	Inverse of a matrix
10.	Logm	Logarithm of a matrix to the base e

(Continued)

Table 2.3 (*Continued*)

S. No.	Command	Description
11.	Norm	Norm of a matrix
12.	Rank	Rank of a matrix
13.	Sqrtm	Square root of a matrix
14.	Cov	Covariance of a matrix
15.	Nnz	Number of non-zero elements in the matrix
16.	mean2	To compute means of all the elements of matrix
17.	std2	To compute the standard deviation of the elements of matrix
18.	Nncopy	To copy a matrix or a cell array
19.	Normc	To normalize columns of a matrix
20.	Normr	To normalize rows of a matrix
21.	Nnsumr	Sums each row of a matrix
22.	findmax2	Interpolates the maxima in a matrix of data
23.	Ctrb	To compute the controllability matrix
24.	Obsv	To compute the observability matrix
25.	Ones	Creates matrix of elements '1'
26.	Zeros	Creates matrix of elements '0'

2.5 MATLAB OPERATORS

Different types of MATLAB operators are defined below.

2.5.1 Arithmetic Operators

Arithmetic operators are used for numerical computational work involving numerals, arrays, and matrices. Table 2.4 gives a brief description of the arithmetic operators.

Table 2.4 Arithmetic operators

S. No.	Command	Description
1.	plus +	For performing addition
2.	uplus +	For unary addition
3.	minus −	For performing subtraction
4.	uminus −	For unary subtraction
5.	mtimes *	For matrix multiplication

(*Continued*)

Table 2.4 (*Continued*)

S. No.	Command	Description
6.	times .*	For array multiplication
7.	mpower ^	For power of a matrix
8.	power .^	For power of an array
9.	mrdivide /(slash)	For dividing right-hand side matrix
10.	mldivide \(backlash)	For dividing left-hand side matrix
11.	ldivide .\	For dividing left-hand side array
12.	rdivide ./	For dividing right-hand side array
13.	Kron	For Kronecker tensor product

Example 2.3 demonstrates some of the arithmetic operators.

Example 2.3

```
% Program to demonstrate arithmetic operators
 R = [3 7 9]; % Different radii
Area = pi * R .^2; % Areas
Volume = (1/3)* pi *R .^3; % Volumes
```

The output of this program will be as follows:

```
>> Area
Area =
28.2743   153.9380   254.4690
>> Volume
Volume =
28.2743   359.1888   763.4070
```

2.5.2 Relational Operators

Relational operators are used for comparing the operands quantitatively for conditional branching of execution. Table 2.5 gives a description of the relational operators available in MATLAB.

Table 2.5   Relational operators

S. No.	Command/Operator	Description
1.	eq ==	Right-hand side operand is equal to the left-hand side operand of the operator
2.	ne ~=	Right-hand side operand is not equal to the left-hand side operand of the operator

(*Continued*)

Table 2.5 (*Continued*)

S. No.	Command/Operator	Description
3.	lt <	Left-hand side operand is less than the right-hand side operand of the operator
4.	gt >	Left-hand side operand is greater than the right-hand side operand of the operator
5.	le <=	Left-hand side operand is less than or equal to the right-hand side operand of the operator
6.	ge >=	Left-hand side operand is greater or equal to the right-hand side operand of the operator

Example 2.4 demonstrates some of the relational operators.

Example 2.4

```
% Program to demonstrate few relational operators
A = [1 2 3; 4 5 6; 7 8 9]; % Defines matrix A
B = [1 1 1; 5 5 5; 9 9 9]; % Defines matrix B
A == B % Equal to operator
```

The output after execution will be as follows:

```
ans =
1    0    0
0    1    0
0    0    1
```

If instead of A == B we put A ~= B, the output will be

```
ans =
0    1    1
1    0    1
1    1    0
```

Here, what MATLAB does is, it tries to answer the question 'Is A equal to B?' or 'Is A not equal to B?' The answer 'yes' is given as '1' and answer 'no' is given as '0'.

2.5.3 Logical Operators

In MATLAB, the following three kinds of logical operators are available for performing logical functions:
1. Element-wise operators—operate on elements corresponding to the logical arrays as shown in Table 2.6.
2. Bit-wise operators—operate on corresponding bits of the arrays as shown in Table 2.7.
3. Short-circuit operators—operate on scalar or logical operations as shown in Table 2.8.

Element-wise Operators

The following logical operators perform element-wise logical operations on their operands:

Table 2.6 Element-wise operators

S. No.	Command/Operator	Description
1.	and &	Returns 1 for every element location that is true in both arrays, otherwise returns 0, e.g., A&B or and(A,B)
2.	Or |	Returns 1 for every element location that is true in either one or the other array, otherwise returns 0, e.g., A|B or or(A,B)
3.	not ~	Returns complement of each element of the input array, e.g., ~A or not(A)
4.	xor	Returns 1 for every element location that is true in only one array, otherwise returns 0, e.g., xor (A,B)
5.	all	True, returns 1 if all the elements of vector are non-zero, e.g., all(A)

Example 2.5 demonstrates some of the relational operators.

Example 2.5

```
% Program to demonstrate element-wise AND operator
A = [1 0 0 0 1 1 1 0 0 0 0 1]; % Binary number A
B = [0 1 0 1 0 1 1 0 1 0 1 1]; % Binary number B
C = and(A,B);   % AND operator
D = A|B;        % OR operator
```

The arrays C and D will be

```
C =
0    0    0    0    0    1    1    0    0    0    0    1
D =
1    1    0    1    1    1    1    0    1    0    1    1
```

Bit-wise Operators

The following operators perform bit-wise logical (binary) operations on non-negative integers. If the input is an array, these operators produce same-sized output.

Table 2.7 Bit-wise operators

S. No.	Command/Operator	Description
1.	bitand	Returns the bit wise AND of two non-negative integers or arguments, e.g., bitand (A,B)
2.	bitor	Returns the bit-wise AND of two non-negative integers or arguments, e.g., bitor (A,B)

(Continued)

Table 2.7 (*Continued*)

S. No.	Command/Operator	Description
3.	bitcmp	Returns the bit-wise complement as an *n*-bit number, where *n* is the second input argument to the command, e.g., bitcmp (A,7)
4.	bitxor	Returns the bit-wise exclusive OR of two non-negative integer or arguments, e.g., bitxor (A,B)
5.	bitmax	Returns the maximum unsigned double precision floating point integer. It is the value when all the bits in the mantissa are set, e.g., bitmax
6.	bitset	Sets the bit position in the argument to 1, where argument must be an unsigned integer and 'bit' must be a number between 1 and the last bit of the unsigned integer, e.g., bitset (A,4)
7.	bitget	Returns the value of the bit at the mentioned position in the argument, where argument must be an unsigned integer and 'bit' must be a number between 1 and the last bit of the unsigned integer, e.g., b = bitget (A,4)
8.	bitshift	Returns the value of the argument shifted by *n* number of bits. The argument must be an unsigned integer. Shifting by *n* is the same as multiplication by 2^n

For instance, Example 2.6 demonstrates some of the bit-wise operators.

Example 2.6

```
A = 29; % binary 11101
B = 20; % binary 10100
C = bitand (A,B); % bitand operator
D = bitor (A,B); % bitor operator
```

The values of C and D are as follows:

```
>> C
C = 20
>> D
D = 29
```

Short-circuit Operators

The following operators perform AND and OR operations on logical expressions of scalar quantities. These are short-circuit operators as they evaluate the second operand only when the result is not fully determined by the first operand.

Table 2.8   Short-circuit operators

S. No.	Command/Operator	Description
1.	&&	Returns logical 1 (true) if both the operands come to be 1 (true), in other cases it is 0 (false), e.g., A&&B
2.	||	Returns logical 1 (true) if either of the operand or both operands are 1 (true), otherwise it returns logical 0 (false), e.g., A||B

Example 2.7 demonstrates some of the short-circuit operators.

Example 2.7

```
% Program to demonstrate short circuit operators
A = 29;% binary 11101
B = 20;% binary  10100
C = A&&B; %short circuit and operator
D = A||B; %short circuit or operator
```

The values of C and D are as follows:

```
>> C
C =
1
>> D
D =
1
```

2.5.4 Operator Precedence

The expressions are normally evaluated from left to right. The precedence rules for the operators in MATLAB are as follows:

1. Parentheses ( );
2. Transpose (.'), power (.^), complex conjugate transpose ('), matrix power (^);
3. Unary plus (+), unary minus (−), logical negation (~);
4. Multiplication (.*), right division (./), left division (.\ ), matrix multiplication (*), matrix right division (/), matrix left division (\ );
5. Addition (+), subtraction (−);
6. Colon operator (:);
7. Less than (<), less than or equal to (<=), greater than (>), greater than or equal (>=), equal to (==), not equal to (~=);
8. Element-wise AND (&);
9. Element-wise OR (|);
10. Short circuit AND (&&);
11. Short circuit OR (||).

It must be remembered that MATLAB always gives the ampersand (&) operator precedence over the OR (|) operator. That is, it evaluates the expressions from left to right but the expression a|b&c will be evaluated as (b&c)|a.

2.6 MATLAB GRAPHICS

In the MATLAB environment, a vast number of techniques are available for plotting and displaying data. A variety of graphical functions can be utilized to create and manipulate graphs according to the data. Graphs or figures can be easily printed, exported to other media, or can be used in the presentations. Graphs are displayed in a special window called figure window. To construct a graph, a coordinate system needs to be defined. Each grap has its associated axes. Actual visual representations of the data can be achieved with graphical objects like lines or surfaces. The actual data is stored as properties of the graphics objects that can modify every attribute of the graph using handle graphics. Various types of graphs can easily be constructed in this programming language.

2.6.1 Plots

Two-dimensional plot can be obtained using the function plot. As illustrated earlier, this will plot the graph between the two vectors. For example, the code in Example 2.8 plots the following figure of the cosine function.

Example 2.8

```
x = 0:pi/200:4*pi;
y = cos(x);
plot(x,y)
```

Now, if desired, the color, marker, or plotting style of the waveform can also be changed. Blue is taken as default color, none as default marker, and solid line as default plot style (Fig. 2.2).

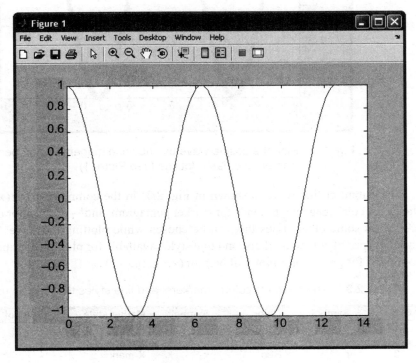

Fig. 2.2 Two-dimensional plot of a cosine function

For plotting a cosine wave of magenta color, pentagram as a marker, and dash-dot as line, the program in Example 2.9 is to be written.

Example 2.9

```
x =0:pi/200:4*pi;
y = cos(x);
plot(x,y,'mp-.')
```

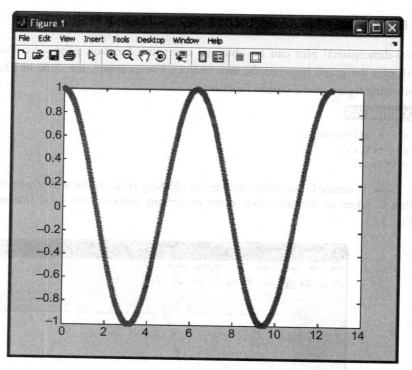

Fig. 2.3 Plot of a cosine wave by choosing magenta color, pentagram marker, and dash-dot line (see Plate 1)

The output cosine wave is shown in Fig. 2.3. In the command plot(x, y,'mp-.'), m stands for color magenta, p stands for marker pentagram, and —. stands for dash-dot line. Table 2.9 shows some other styles that can be chosen while plotting the wave. This table lists the various types of colors, markers, and line styles available for plotting a graph. For example, the command for green color plot will be plot(x, y,'gp-.').

Table 2.9 List of various colors, markers, and line styles that can be used for plotting

S. No.	Symbol	Color	Symbol	Marker style	Symbol	Line style
1.	b	Blue	.	Point	-	Solid
2.	g	Green	o	Circle	:	Dotted
3.	r	Red	x	X-mark	-.	Dash-Dot
4.	c	Cyan	+	Plus	--	Dashed

(Continued)

Table 2.3 (*Continued*)

S. No.	Symbol	Color	Symbol	Marker style	Symbol	Line style
5.	m	Magenta	*	Star	(none)	No line
6.	y	Yellow	s	Square		
7.	k	Black	d	Diamond		
8.			v	Triangle (down)		
9.			^	Triangle (up)		
10.			<	Triangle (left)		
11.			>	Triangle (right)		
12.			p	Pentagram		
13.			h	Hexagram		
14.			(none)	No line		

The function `legend (string1, string2, …)` puts a legend on the current plot using the specified strings as labels. By using this function, one can label any colored graph or surface object. The command `legend (string1, string2,…, position)` will place the legend according to the `position` specified by the string `position`. The font size and font name for the legend strings matches the axes font size and font name. The command `legend OFF` removes the legend from the current axes and deletes the legend handle. Commands `legend HIDE` and `legend SHOW` makes the legend invisible and visible respectively.

The function `gtext ('string')` displays the figure window and puts up a cross-hair. Then it waits for a mouse button or keyboard key to be pressed so that it can write the `'string'` on the figure window. The cross-hair can be positioned with the mouse. For example, the command `gtext ('This book is for Engineers')` will place the string `'This book is for Engineers'` on the figure window.

The commands `xlabel`, `ylabel`, and `zlabel` can be used to label the x-axis, y-axis, and z-axis of the graph. For example, the command `ylabel ('This is y-axis')` adds the text `'This is y-axis'` above the y-axis on the current axis.

The function `title ('Graph Title')` adds a title at the top of the current axis of the graph.

2.6.2 Subplots

If we want to plot more than one graph in a figure window, we can use subplots. The subplot function divides the figure window into a matrix of addressable sub-windows. This command has the following syntax:

`subplot(m, n, p)`: It breaks the figure window into small axes of dimension $m*n$ and selects the pth axes for the current plot.

`subplot(m, n, p,'replace')`: It replaces the already existing axis and creates a new one.

`subplot(m, n, p,'align')`: It aligns the axes such that they do not overlap.

`subplot('position', [left bottom width height])`: It creates an axis at the specified position in normalized coordinates.

`subplot(m,n,P,PROP1,VALUE1,PROP2,VALUE2,………)`: It sets the specified property-value pairs on the subplot axes.

Axis scaling and appearance can also be set similarly by the following syntax:

```
axis([xmin xmax ymin ymax])
axis([xmin xmax ymin ymax zmin zmax])
```

The `axis` is applicable for `plot` function also.

Creating subplots in the figure and scaling of the axes can be demonstrated by the program given in Example 2.10:

Example 2.10

```
clc
clear all
clf
x = 0:pi/400:4*pi;
y = cos(x);
z= sin(x);
Subplot(2,1,1),plot(x,y,'r.:')
Axis([0 12 -1 1])
Subplot(2, 1, 2),plot(x,z,'kd--')
Axis([0 12 -1 1])
```

The first and second line clears the workspace window and variables. Then the array x is defined which gives the angle in radians. Cosine and sine of these angles are calculated in lines 4 and 5, and plotted. The subplots are made as per syntax. Axis is also defined as per the syntax. The output of this program is shown in Fig. 2.4.

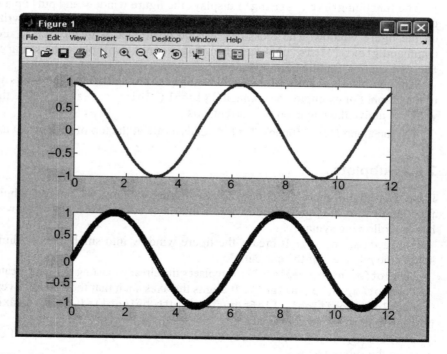

Fig. 2.4 Subplots of cosine and sine waves in the figure window (see Plate 1)

We can also edit and save the figure window. The settings of the figure can also be saved as M-file by 'generate M-file'. Various other options such as print setup, page setup, save as, preferences, etc. are also available. Figure window toolbar has many features needed for modifying the graph. We can label the *x*-axis and *y*-axis and assign a title to the graph in each subplot of a sub-window. By using the plotting tools, we can set the property of a graph, select a graph type, and drag and drop data into the graph. There are several other options available to modify the graph which can be easily learnt by just playing with these options. Figure 2.5 illustrates some of the modifications. In this figure, the sine and cosine waves are rotated in 3-D with the help of a mouse in order to get a different view. *x*-Axis is labeled 'angle (radian)' and *y*-axis 'amplitude'. The title of the graph and figure are labeled 'cosine & sine waves' and 'cos & sin waves' respectively.

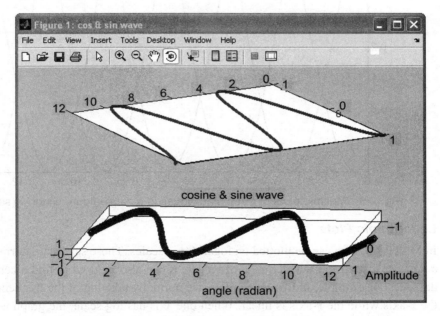

Fig. 2.5 Edited subplots of cosine and sine waves in 3-D (see Plate 2)

2.6.3 Other Types of Plots

Multiple Plots in a Graph

Many plots can be plotted in a graph having one common axis. This allows us to create multiple graphs in a single window by using a single plot command. For example, the program in Example 2.11 plots three sine waves on a single window as shown in Fig. 2.6. The three waves can be identified by their colors.

Example 2.11

```
x = 0:pi/200:4*pi;
y = sin(x);
r = sin(x-(2*pi/3));
b = sin(x+(2*pi/3));
plot(x,y,x,r,x,b)
```

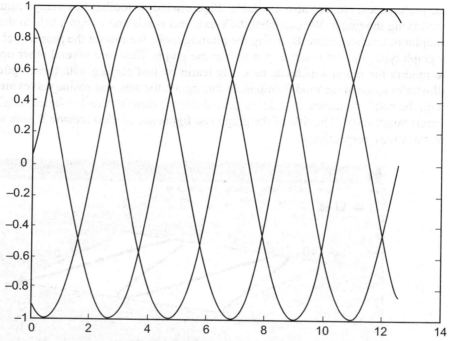

Fig. 2.6 Multiple plots of three sine waves in a single figure window (see Plate 2)

Logarithmic Plots

In MATLAB, data can be plotted on logarithmic scales. These scales are normally used to plot a data set that covers a wide range of values. It can also be used to find a certain trend in the data set. For example, in frequency plots, we normally take log of the frequency, i.e., log w on the x-axis while the y-axis is linear. When one axis has log scale, the graph is called semi-log graph (Fig. 2.7). In case both the axes having log scales, it is called log–log graph (Fig. 2.7). For these graphs, log to the base 10 scale is used. The following commands are available in MATLAB for logarithmic plots:

`Semilogx(x,y):` This command produces a semi-log graph of `log` x versus y.

`Semilogy(x,y):` This command produces a semi-log graph of `log` y versus x.

`Loglog(x,y):` This command produces a semi-log graph of `log` x versus `log` y.

The set of instructions in Example 2.12 plots semi-log graph of the function that can be seen in Fig. 2.7.

Example 2.12

```
G1 = 1/(1 + 0.1s)
% Program to demonstrate semilog function
w = 1:10:10000; %Angular Frequency
G1 = (1./(1+i*0.1*w)); % Define function
```

(Continued)

Example 2.12 (*Continued*)

```
mag = abs(G1); % Magnitude of the function
semilogx (mag,w);
title('\bf Semilog plot of G1');
xlabel('\bf Magnitude');
ylabel('\bf Log w');
```

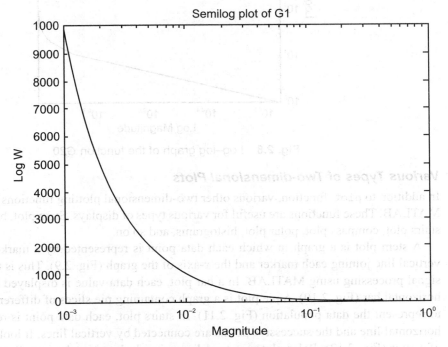

Fig. 2.7 Semi-log graph of the function G1

The set of instructions in Example 2.13 plots a log–log graph of the following function as can be seen in Fig. 2.8.

Example 2.13

```
G2 = (1 + s)/(1 + 0.1s)
% Program to demonstrate loglog function
w = 1:10:10000; %Angular frequency
G2 = ((1+i*w)./(1+i*0.1*w)); % Define function
mag = abs(G2); % Magnitude of the function
loglog(mag,w);
title('\bf loglog plot of G2');
xlabel('\bf Log magnitude');
ylabel('\bf Log w');
```

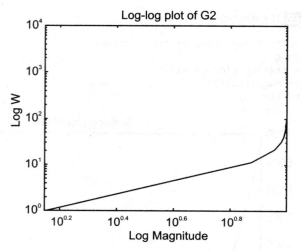

Fig. 2.8 Log–log graph of the function G20

Various Types of Two-dimensional Plots

In addition to `plot` function, various other two-dimensional plotting functions are available in MATLAB. These functions are useful for various types of displays. Stem plot, bar plot, pie plot, stairs plot, compass plot, polar plot, histograms, and so on.

A stem plot is a graph in which each data point is represented by a marker along with a vertical line joining each marker and the x-axis of the graph (Fig. 2.9). This is useful in digital signal processing using MATLAB. In a bar plot, each data value is displayed by a vertical or horizontal bar (Fig. 2.10). A pie plot is a graph containing pie slices of different sizes in order to represent the data population (Fig. 2.11). In stairs plot, each data point is represented by a horizontal line and the successive points are connected by vertical lines. It looks like the steps of a stair (Fig. 2.12). Polar plots are two-dimensional plots in polar coordinates. If the polar coordinates are (θ, r), where θ is the angular coordinate and r the radial coordinate of a point, the function `polar(θ,r)` will construct a polar graph (Fig. 2.13). Compass plot is a graph which represents each data value by an arrow whose length is proportional to the data value (Fig. 2.14).

Consider the following example:

Plotting a sine wave by two-dimensional functions.

For a clear and better understanding of the two-dimensional functions defined in this section, let us plot a sine wave with the help of two-dimensional functions. For the instruction in Example 2.14, lines will create a stem plot of a sine wave as seen in Fig. 2.9.

Example 2.14

```
t = 0:10e-4:.04;      % Time is taken from 0 sec to 0.04 sec in steps of 10 micro sec.
f = 50;               % Frequency is taken as 50 Hz.
t = 0:10e-4:.04;      % Time is taken from 0 sec to 0.04 sec in steps of 10 micro
                        sec.
```

(Continued)

Example 2.14 (*Continued*)

```
f = 50;                 % Frequency is taken as 50 Hz.
w = 2*pi*f;             % Angular frequency in rad/s.
A = 220;                % Peak amplitude 220 V
V_out = A*sin(w*t);     % V_out is the output voltage waveform from 0 sec to 0.04 sec.
stem(t,V_out)
xlabel('\bftime in seconds')
ylabel('\bfAmplitude')
```

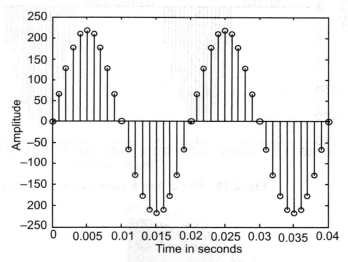

Fig. 2.9   Plot of a sine wave by stem plot function

For getting the bar plot, follow the underline instructions give in Example 2.15. The output can be seen in Fig. 2.10.

Example 2.15

```
t = 0:10e-4:.04;        % Time is taken from 0 sec to 0.04 sec in steps of 10 micro
                          sec.
f = 50;                 % Frequency is taken as 50 Hz.
w = 2*pi*f;             % Angular frequency in rad/s.
A = 220;                % Peak amplitude 220 V
V_out = A*cos(w*t);     % V_out is the output voltage waveform from 0 sec to 0.04
                          sec.
bar(t,V_out,0.5)        % 0.5 is the width of the bar
xlabel('\bftime in seconds')
ylabel('\bfAmplitude')
```

Similarly, for a pie plot use pie(t,V_out), for a stairs plot use stairs(t,V_out), for a polar plot use polar(t,V_out), and for a compass plot use compass(t,V_out) commands and we will obtain the plots shown in Figs 2.11–2.14 respectively.

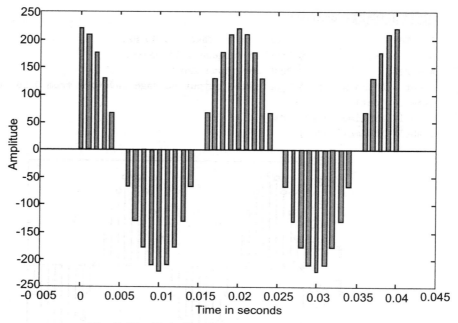

Fig. 2.10   Plot of a sine wave by bar plot function

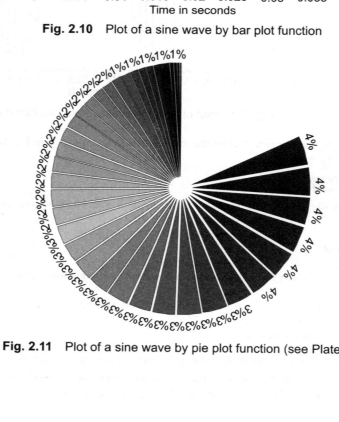

Fig. 2.11   Plot of a sine wave by pie plot function (see Plate 3)

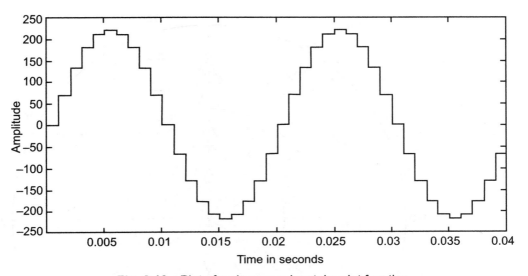

Fig. 2.12   Plot of a sine wave by stairs plot function

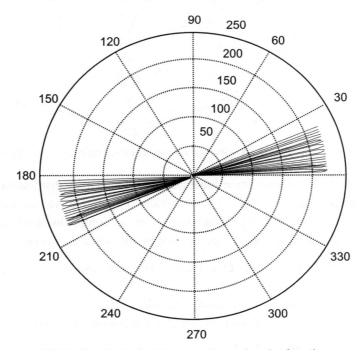

Fig. 2.13   Plot of a sine wave by polar plot function

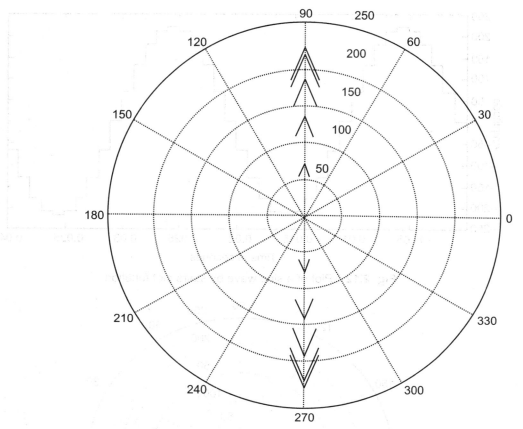

Fig. 2.14 Plot of a sine wave by compass plot function

A histogram is a two-dimensional graph used to display distribution of values in a data set (the frequency of the values). In a histogram, the range of values present in the data set are divided into equally spaced bins and then the values coming in each bin are determined in order to plot the count. This function can be used by the following syntax:

HIST(Y): The ,elements of the data set Y are binned into 10 equally spaced bins and plotted.

HIST(Y,N): This function uses N bins instead of 10.

HIST(Y,X): Y is distributed according to the bin centers as specified by vector X.

For instance, the following instructions create a histogram as seen in Fig. 2.15. The commands in Example 2.16 create a data set of 1,000 normally distributed numbers and create a histogram of the data by using 50 equally spaced bins. One can define the number of bins, color, etc.

Example 2.16

```
N = randn(1000,1); % Generates normally distributed random numbers
n = 50;            % defines the interval or bin
hist(N,n)
```

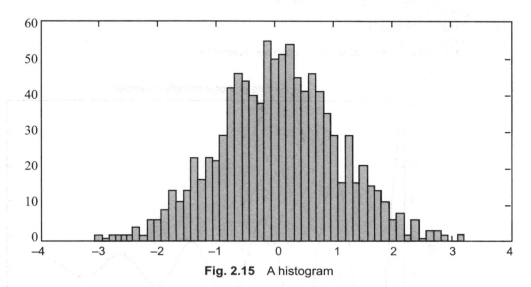

Fig. 2.15 A histogram

Plotting Complex Numbers

Complex numbers are numbers with a real part and an imaginary part. Electrical engineers often encounter complex numbers. For example, the steady-state values of the electrical quantities like current, voltage, and impedance are complex quantities in the alternating current (AC) analysis. A complex has the form

$$Z = R + iX$$

or

$$Z = R + jX$$

where R and X are the real numbers and Z is a complex number. This can also be expressed in polar coordinates as

$$Z = z/\theta$$

where $z = \sqrt{(r^2 + x^2)}$ and $\theta = \tan^{-1}(X/R)$. MATLAB uses rectangular coordinates in order to represent complex numbers. Plotting complex numbers is different from plotting real data. For example, consider a function

$$v(t) = e^{-0.01t}(10 \cos t + i\, 10 \sin t)$$

If this function is plotted by using only the conventional plot function, then the real data will be plotted along with a warning on the command window that imaginary part of complex x and y arguments are ignored. The instructions in Example 2.17 plots the function v(t) as can be seen in Fig. 2.16.

Example 2.17

```
t = 0:pi/50:10*pi;
v = exp(-0.1*t).*(10*cos(t)+i*10*sin(t));
t = 0:pi/50:10*pi;
v = exp(-0.1*t).*(10*cos(t)+i*10*sin(t));
plot(t,v)
xlabel('\bftime');
ylabel('\bfv(t)');
title('\bf Simple plot of a complex number')
```

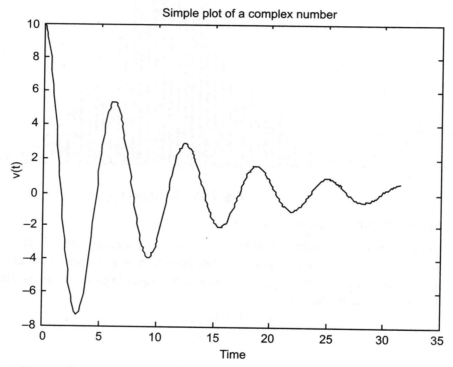

Fig. 2.16 Plot of $v(t) = e^{-0.01t}$ ($10\cos t + i\,10\sin t$) by function `plot(t, v)`

If both the real and imaginary parts are required to be plotted as shown in Fig. 2.17, then others options are available. The instructions in Example 2.18 plot both the parts on the same axis.

Example 2.18

```
t = 0:pi/50:10*pi;
v = exp(-0.1*t).*(10*cos(t)+i*10*sin(t));
plot(t,real(v),t,imag(v))
xlabel('\bftime');
ylabel('\bfv(t)');
title('\bf plot of a complex number')
```

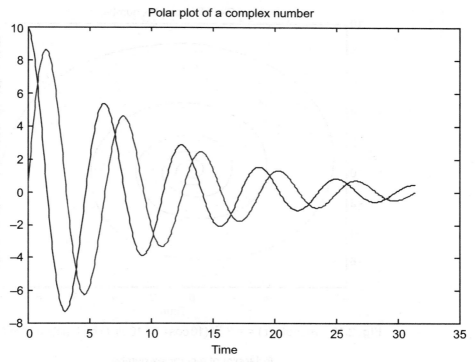

Fig. 2.17 Plot of $v(t) = e^{-0.01t} (10\cos t + i\, 10\sin t)$ by function `plot(t, real(v), t, imag(v))`

Similarly, the real part can also be plotted with respect to the imaginary part as shown in Fig. 2.18.

Example 2.19

```
t= 0:pi/50:10*pi;
v = exp(-0.1*t).*(10*cos(t)+i*10*sin(t));
plot(v)
xlabel('\bftime');
ylabel('\bfv(t)');
title('\bf plot of a complex number')')
```

If required, complex function can also be plotted as a polar plot displaying magnitude versus angle as shown in Fig. 2.19.

Example 2.20

```
t = 0:pi/50:10*pi;
v = exp(-0.1*t).*(10*cos(t)+i*10*sin(t));
polar(abs(v),angle(v))
title('\bf Polar plot of a complex number')
```

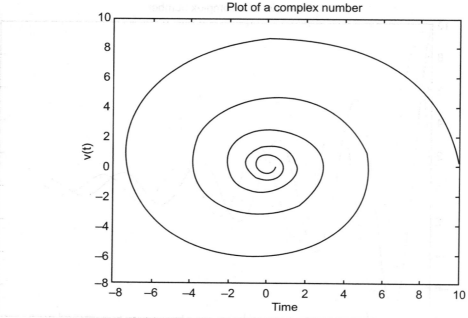

Fig. 2.18 Plot of $v(t) = e^{-0.01t}(10\cos t + i\,10\sin t)$ by function `plot(v)`

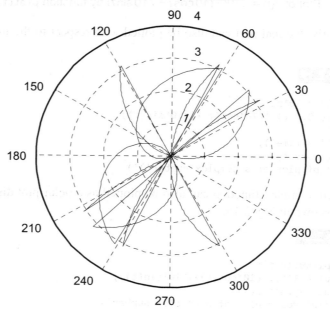

Fig. 2.19 Plot of $v(t) = e^{-0.01t}(10\cos t + i\,10\sin t)$ by function `plot(v)`

Three-dimensional Plots

MATLAB has a rich variety of three-dimensional plots. These plots can be used for displaying certain types of data and are more effective when we want to display two functions which depend on a common independent variable or a function that depends on two independent variables. These plots are more informative and realistic in nature. The following commands can be used for three-dimensional plots:

Plot3(x,y,z): This command creates a three-dimensional plot where x, y, and z are equal-sized arrays containing the locations of data points to be plotted.

Mesh(x,y,z): This command produces wireframe surfaces in three dimensions that color only the lines connecting the defining points. Here x, y, and z are two-dimensional arrays containing the values of the data to be displayed.

Surf(x,y,z): This command creates three-dimensional plot that colors both the connecting lines and the faces of the surface. Here x, y, and z are two-dimensional arrays containing the values of the data to be displayed.

Contour(x,y,z): This commands constructs a three-dimensional contour plot. The counter is normally based on the current color map. Here vectors x, y, and z have the same meaning as for a surf plot.

Consider the example—plotting a function in 3-D

In this example, let us take the output of a system to be (exp(-0.2*t)) + (sin(3*x))^2. This output is a function of two variables x and t. Let t be the time and x the input to the system. Let us evaluate this system output for certain values of x and t, say from 0 to 5. For defining these values, it is convenient to use meshgrid function with the following syntax:

```
[t,x] = meshgrid(t_start_point : t_increment : x_end_point, x_start_point : x_
increment : x_end_point);
```

It defines the values of t, x to be included in the grids. With the help of this function we create the arrays of t, x values in order to plot the output function for each pair of these values. For plotting the output function, any of the earlier commands, i.e., plot3, surf, mesh, and contour, can be used.

The colormap function can be used to view the color bar in the figure window. An HSV colormap varies the hue component of the hue–saturation–value color model. The colors begin with red, pass through yellow, green, cyan, blue, and magenta, and return to red as shown in Figs 2.20–2.22. The map is particularly useful for displaying periodic functions.

The statements in Example 2.21 will create a plot using plot3 function as shown in Fig. 2.20. The x-axis in the graph is labeled as time, y-axis as input, and z-axis as amplitude.

Example 2.21

```
clear all;                        % clear all variables
clc;                              % clears the command window
clf                               % clears the figure window
[t,x] = meshgrid(0:0.1:5,;        % specifies values of t,x
-0:0.1:5)                           to be included in the grid
y = (exp(-0.2*t))+(sin(3*x))^2;   % Defines an output function
```

(Continued)

Example 2.21 *(Continued)*

```
plot3(t,x,y)              % to create surface plot
xlabel('\bftime');        % x axis label as time
ylabel('\bfinput');       % y axis label as input
zlabel('\bfAmplitude')    % z label  as Amplitude
colormap hsv
colorbar
```

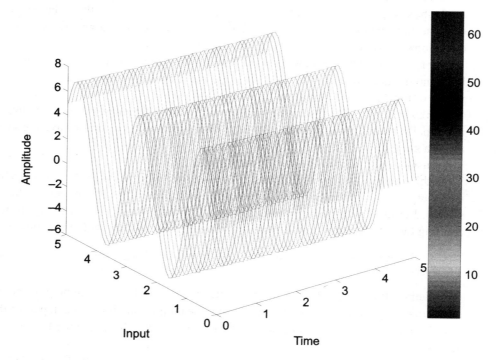

Fig. 2.20 Plot of `(exp (-0.2*t))+(sin(3*x))^2` by using `plot3` function (see Plate 3)

In order to plot the output function y by using **mesh** command, the following statements can be written. The statements in this program are same as were in case of `plot3` program except that `mesh (t,x,y)` command is used for plotting the graph. Figure 2.21 shows the graph created by the program in Example 2.22.

Example 2.22

```
clear all;                % clear all variables
clc;                      % clears the command window
[t,x] = meshgrid          % specifies values of t,x to be
```

(Continued)

Example 2.22 (*Continued*)

```
(0:0.1:5, -0:0.1:5);              included in the grid
y = (exp(-0.2*t))+(sin(3*x))^2;  % Defines an output function
mesh(t,x,y)                      % to create surface plot
xlabel('\bftime');               % x axis label as time
ylabel('\bfinput');              % y axis label as input
zlabel('\bfAmplitude') % z label as Amplitude
colormap hsv
colorbar
```

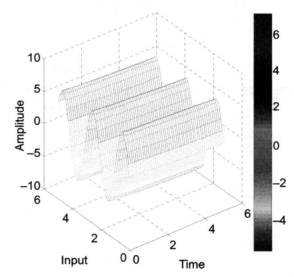

Fig. 2.21   Plot of `(exp(-0.2*t)) + (sin(3*x))^2` by using mesh function (see Plate 4)

Similarly, the graph can be created by using surf function. The statements in Example 2.23 create the graph as shown in Fig. 2.22.

Example 2.23

```
clear all; % clear all variables
clc; % clears the command window
[t,x] = meshgrid(0:0.1:5, -0:0.1:5);% specifies values of t,x to be included in
the grid
y = (exp(-0.2*t))+(sin(3*x))^2; % Defines an output function
surf(t, x, y) % to create surface plot
xlabel('\bftime'); % x axis label as time
ylabel('\bfinput'); % y axis label as input
zlabel('\bfAmplitude') % z label as Amplitude
colormap hsv
colorbar
```

Similarly, in order to plot a contour plot, contour `(t, x, y)` function can be used. The contour graph of the output function `y` is shown in Fig. 2.23.

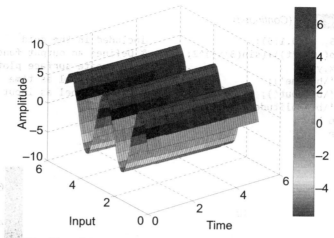

Fig. 2.22 Plot of `(exp(-0.2*t)) + (sin(3*x))^2` by using surf function (see Plate 4)

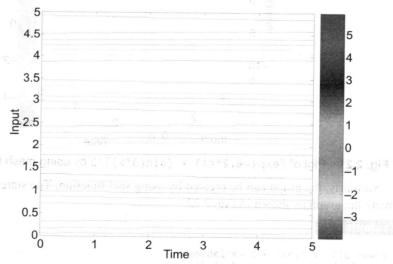

Fig. 2.23 Plot of `(exp(-0.2*t)) + (sin(3*x))^2` by using contour function (see Plate 5)

Other Three-dimensional Plots

In addition to the three-dimensional plots already seen, MATLAB supports other specialized plots also. This section introduces some of these plotting functions which can be used to display data in 3-D. For example, `meshz` function is same as `mesh`, but a curtain or reference plane is also drawn beneath the plot as shown in Fig. 2.24. The vectors X and Y define the axes. Functions specified for a rectangular grid plots well by this command. The instructions in Example 2.24 illustrate the application of `meshz` command. The graph produced by these instructions is shown in Fig. 2.24. Similarly, `meshc` command can also be used for creating three-dimensional graphs. This command plots a contour plot beneath the mesh plot in the graph as can be seen in Fig. 2.25. It is used only for rectangular grids. The graph created by using this command, i.e., `meshc(X, Y, f)`, is shown in Fig. 2.25. `Waterfall` command is also used for three-dimensional plots. This command creates same graph as by `mesh` command except that the column lines of the mesh are not drawn as can be seen in Fig. 2.26. For column line orientation, use `Waterfall(X', Y', f')` command.

Example 2.24

```
clear all;
clc;
[X, Y] = meshgrid (-10:0.25:10, -10:0.25:10);
f = sinh(sqrt((X/pi).^2+(Y/pi).^2));
meshz(X, Y, f)
xlabel('\bfx'); % x axis label as time
ylabel('\bfy'); % y axis label as input
zlabel('\bfAmplitude') % z label as Amplitude
colormap hsv
colorbar
```

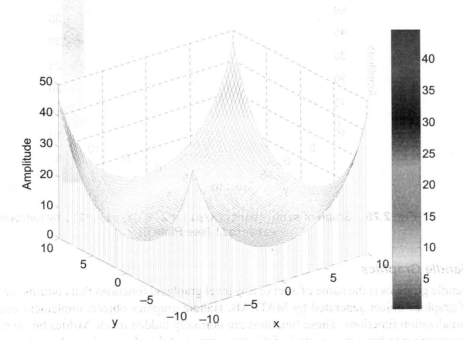

Fig. 2.24 Graph of `sinh (sqrt((X/pi).^2 + (Y/pi).^2))` by function meshz (see Plate 5)

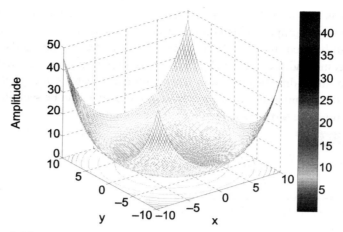

Fig. 2.25 Graph of sinh (sqrt((X/pi).^2 + (Y/pi).^2)) by function meshc (see Plate 6)

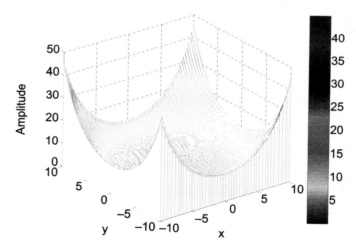

Fig. 2.26 Graph of sinh (sqrt((X/pi).^2 + (Y/pi).^2)) by function waterfall (see Plate 6)

Handle Graphics

Handle graphics is the name of a set of low level graphical functions that controls the properties of graphics object generated by MATLAB. Handle graphics objects implement graphing and visualization functions. These functions are normally hidden inside M-files but they allow the programmer to have fine control of the appearance of the plots and graphs that they generate. When a MATLAB plotting function is called, it creates the graph using various graphics objects, such as a figure window, axes, lines, text, and so on. One can set the value of each property of a graph. For example, the statement:

```
figure('color','red','Toolbar','none')
```

creates a figure with a red background color and without a toolbar.

Objects Handles   Whenever MATLAB creates a graphics object, it assigns an identifier called a handle to the object. The handle is a unique identifier called a handle to the object. The handle is a unique integer or real number that is used by MATLAB to identify object. One can use this handle to access the object's characteristics with the help of `set` and `get` functions. A handle is automatically returned by any function that creates a graphics object. For example, the following statements create a new graph and return the handle of that graph in variable `hndl`.

```
x = 1 : 5;
y = x.^2;
hndl = plot(x, y);
```

By typing `hndl` on the command window, we get the value as follows:

```
>> hndl
hndl =
139.0016
```

Now, the handle `hndl` can be used to set the properties of the line series object. For example, the color property can be set as follows:

```
set(hndl, 'color', 'yellow')
```

If we query the line series property,

```
get(hndl, 'LineWidth')
```

we obtain the result as

```
ans =
0.50000
```

We can also use the handle to view what properties a particular object contains. For example,

```
>> get(hndl)
   Color: [1 1 0]
   EraseMode: 'normal'
   LineStyle: '-'
LineWidth: 0.5000
 Marker: 'none'
   MarkerSize: 6
MarkerEdgeColor: 'auto'
MarkerFaceColor: 'none'
   XData: [1 2 3 4 5 6 7 8 9 10]
   YData: [1 4 9 16 25 36 49 64 81 100]
   ZData: [1x0 double]
BeingDeleted: 'off'
ButtonDownFcn: []
Children: [0x1 double]
   Clipping: 'on'
```

```
            CreateFcn: []
            DeleteFcn: []
           BusyAction: 'queue'
     HandleVisibility: 'on'
              HitTest: 'on'
        Interruptible: 'on'
             Selected: 'off'
    SelectionHighlight: 'on'
                  Tag: ''
                 Type: 'line'
        UIContextMenu: []
             UserData: []
              Visible: 'on'
               Parent: 138.0012
          DisplayName: ''
            XDataMode: 'manual'
          XDataSource: ''
          YDataSource: ''
          ZDataSource: ''
```

By using set function, we can set any number of properties of a graph. The properties can be changed by calling the set function as

```
>> set(hndl, 'property1', propvar1, 'property2', propvar2,.....)
```

where property1, property2, and so on are the properties of the graph and propvar1, propvar2, and so on are set values of these properties.

2.7 BRANCHING AND LOOPING FUNCTIONS

Until now we have developed and executed several programs. All these programs were very simple and instructions written are executed step-by-step starting from the first instruction. These types of programs are often known as sequential programs. A sequential program consists of several series of instructions to be executed one after another. In these programs, there is no possibility of repeating some or all instructions based upon a particular condition. They are simple to develop and understand. In engineering problems, we often require programs to control a process based on a condition or to repeat a set of instructions until the desired conditions are met. There are two types of instructions required to perform these actions, branching instructions and looping instructions. Branching instructions specify the section of instructions to be executed while looping instructions make a section of instructions to be repeated a number of times. These instructions though, on one hand, provide more controllability, but on the other hand, they increase the complexity of the program. We will discuss about these functions in the following section.

2.7.1 Branching Functions

The branching constructs allow structured programming using MATLAB. These functions allow us to select and execute a particular set of instructions while skipping the rest. These

Plate 1

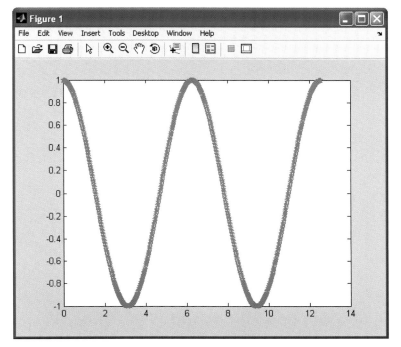

Plot of a cosine wave by choosing magenta color, pentagram marker, and dash-dot line (*Chapter 2, page 26*)

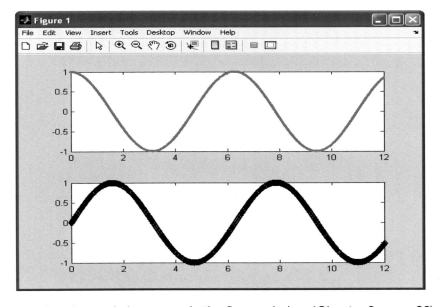

Subplots of cosine and sine waves in the figure window (*Chapter 2, page 28*)

Plate 2

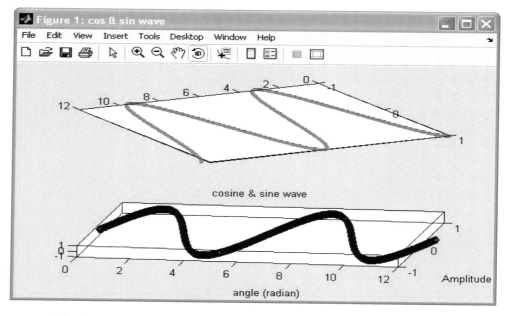

Edited subplots of cosine and sine waves in 3-D (*Chapter 2, page 29*)

Multiple plots of three sine waves in a single figure window (*Chapter 2, page 30*)

Plate 3

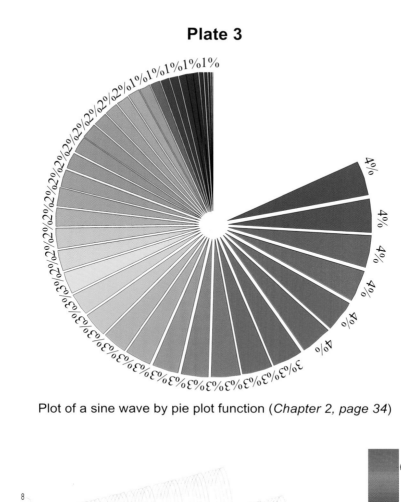

Plot of a sine wave by pie plot function (*Chapter 2, page 34*)

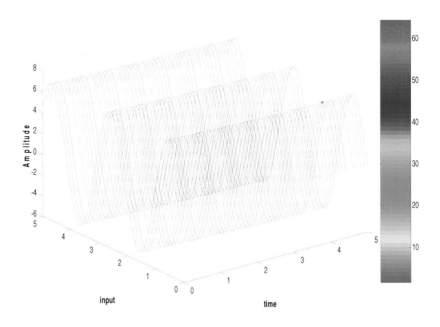

Plot of (exp (-0.2*t))+(sin(3*x))^2 by using plot3 function (*Chapter 2, page 42*)

Plate 4

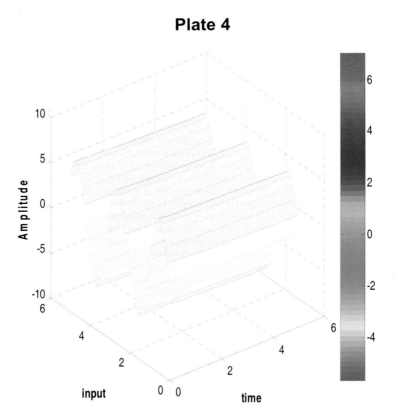

Plot of (exp (-0.2*t)) + (sin(3*x))^2 by using mesh function (*Chapter 2, page 43*)

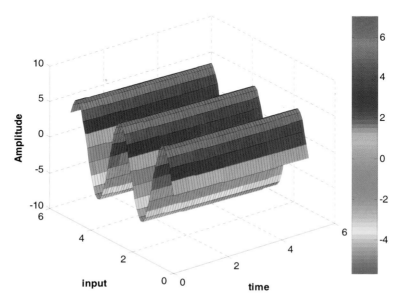

Plot of (exp (-0.2*t)) + (sin (3*x))^2 by using surf function (*Chapter 2, page 44*)

Plate 5

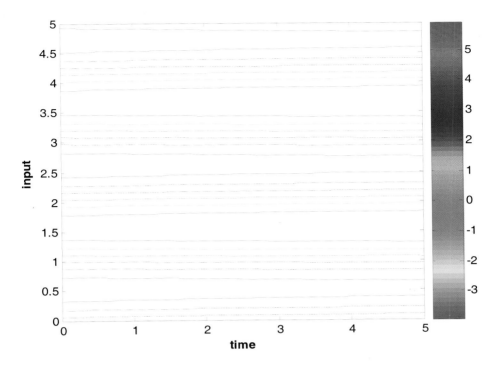

Plot of (exp (-0.2*t)) + (sin (3*x))^2 by using contour function (*Chapter 2, page 44*)

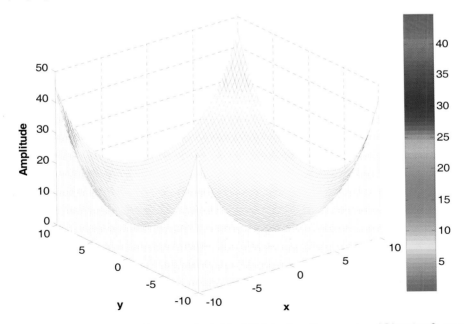

Graph of sinh (sqrt((X/pi).^2 + (Y/pi).^2)) by function meshz (*Chapter 2, page 45*)

Plate 6

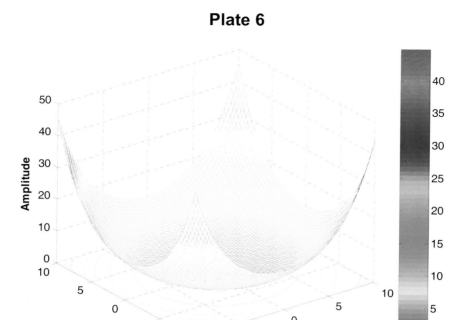

Graph of sinh (sqrt((X/pi).^2 + (Y/pi).^2)) by function meshc (*Chapter 2, page 46*)

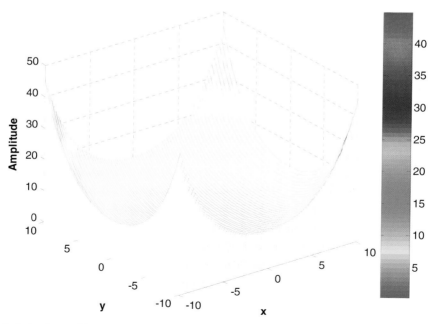

Graph of sinh (sqrt((X/pi).^2 + (Y/pi).^2)) by function waterfall (*Chapter 2, page 46*)

Plate 7

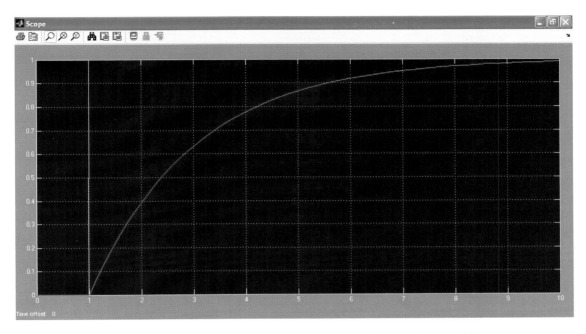

Unit step response of the first-order system (*Chapter 9, page 306*)

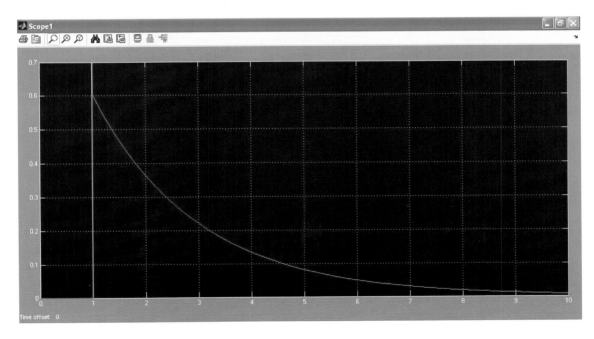

Impulse response of the first-order system (*Chapter 9, page 307*)

Plate 8

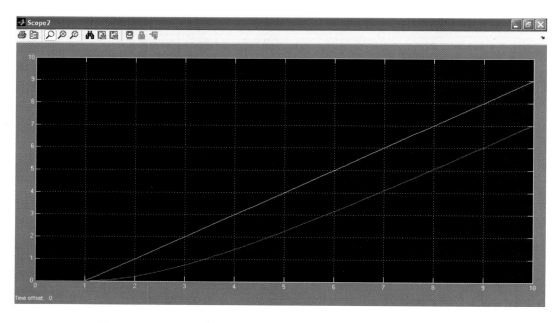

Ramp response of the first-order system (*Chapter 9, page 308*)

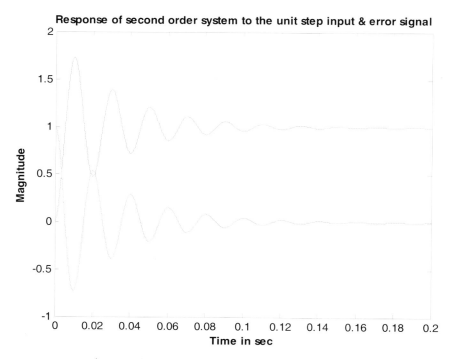

Step response and error signal of **the second**-order system (*Chapter 9, page 309*)

functions include if, switch, try/catch, and so on. The program execution branches route according to the decision and execution continues.

if *Function*

The if function has the following syntax:

```
if condition 1
        instructions
elseif condition 2
        instructions
eleseif condition 3
        instructions
else condition 4
        instructions
end
```

The instructions after the if command are executed only if condition 1 is true, i.e., real part is non-zero, after that it skips the other conditions and goes to end. Similarly, instructions after eleseif command are executed only if condition 2 is true and so on. In this function, we can use any number of elseif commands (zero also), but at most one else command can be used. The control expression in each clause will be tested only if the control expressions in every clause before it are false, i.e., 0. If all the control expressions given by if or elseif statements are false, then the program executes the instructions in the else clause. If there is no else clause, the execution continues after the end statement without executing any part of the if function. For instance, consider Example 2.25.

Example 2.25

```
% Program to demonstrate if-elseif-else function
V = 1:1:20            % Defines a voltage source
if V < 5              % If voltage is less than 5 V
out_1 = 0             %Output is zero
elseif V == 5         % If voltage is equal to 5 V
    out_1 = 0.5       % Output is o.5
else
    out_1 = 1         % In other cases output is 1
end
```

Switch Function

This is another form of branching which switches among cases based on the conditions mentioned in the expression. This function permits a programmer to select a particular code block to execute based on the value of a single integer, character, or a logical expression. The general form of a switch function is as follows:

```
switch expression
    case expression,
        statement_1, statement_2,..................,statement_n
```

```
     case {expression_1, expression_2,............., expression_n}
         statement_1, statement_2,.......................,statement_n
   ...
     otherwise
         statement_1, statement_2,.......................,statement_n
   end
```

The expressions can be a scalar or a string. The instructions or statements following the first case command where the *expression* after `switch` command matches the *expression* after `case` command are executed. If there are number of expressions after the `case` command, then if any of the *expression* matches the `switch` *expression*, the statements following that case are executed. Only the *statements* between the matching `case` and the next `case`, `otherwise`, or `end` are executed. For instance, consider Example 2.26.

Example 2.26

```
% Program to demonstrate switch-case instruction
switch 5
    case{1,2,3,4}
        out_1 = 0;      %Output voltage is zero
    case{5}
        out_1 = 0.5     %Output voltage is 0.5
    otherwise
        out_1 = 1       %Output voltage is 1
end
```

Try/Catch Function

This function has the following syntax:

```
try,
   statement_1,
     .
     .
     .
   statement_n,

catch,
   statement_1,
     .
     .
     .
   statement_n,
end
```

The `try/catch` is another type of branching instruction designed for trapping errors. The usage is similar to that in JAVA. Statements between the `try` and `catch` instructions are normally executed. If any error occurs while executing statements between `try` and `catch` instructions, that error is captured and the statements between `catch` and `end` instructions are executed. If there is an error in `catch` statements, the execution stops unless caught again by a `try/catch` instruction. For instance, consider Example 2.27.

Example 2.27

```
%Program to demonstrate try/catch instruction
V = 1:1:20;
try
    if V < 5
        out_1 = 0
    elseif V == 5
        out_1 = 0.5
        otherwise
            out_1 = 1
    end
catch
    if V > 10
        disp('overvoltage')
    else
        disp('ok')
    end
end
```

Error Function

The syntax for this command is as follows:

```
error('message')
```

This instruction displays the error message contained in the string *message* and causes an error exit from the currently executing M-file to the keyboard.

2.7.2 Looping Functions

Looping instructions permit us to execute a sequence of instruction a number of times. This is also a part of the structured programming constructs. There are two types of looping functions in MATLAB, `for` loop and `while` loop.

for Function

This instruction repeats the statements for a particular number of times. The syntax of `for` instruction is as follows:

```
for variable = expression,
    statement_1,
    .
    .
    .
    statement_n
end
```

The statements between the `for` and the `end` instructions are executed repeatedly during pass of `for` loop. For instance, consider Example 2.28.

Example 2.28

```
% Program to demonstrate for instruction
x = 1;
y = 1;
for i = 1:5% i is a complex number
    x = x + i
    y = y + 1/i
end
```

While Function

This instruction repeats the statements indefinite number of times. The syntax for while instruction is as follows:

```
While expression
    Statement_1,
        .
        .
        .
        .
    Statement_n
end
```

The statements executed till the real part of the *expression* has all non-zero elements. In other words, the statements between the while and the end instructions continue to be executed till the condition in the *expression* remains true. If the condition in the *expression* is false, the loop ends. For instance, consider Example 2.29.

Example 2.29

```
%Program to demonstrate while loop
current = 20;
n = 0;
while current < 200
    n = n + 2;
    current = current*1.23 + 0.9;
end
current
n
```

Break and Continue Functions

The break instruction terminates the execution of while or for loops. In the case of nested loops, it exists from the innermost loop only if it is defined inside a loop. The continue instruction passes control to the next iteration of for or while loop which it appears and skips the remaining statements present in the for or while loops.

2.8 MISCELLANEOUS FUNCTIONS

This section deals with string and input/output MATLAB functions. String functions are required while working with strings and input/output functions are required for data input or output operation.

2.8.1 String Functions

A character or string can also be assigned to a variable in MATLAB. For example, the statement

```
Name = 'Albert Einstein'
```

will store the string *Albert Einstein* in the variable Name. It is also possible to create two-dimensional character arrays but each row must be of exactly the same length. For example, the statement

```
Name = ['Albert Einstein'; 'Is a scientist']
```

will produce an error message. An easier way is to use a char function. This function automatically adjusts the length of all the strings to the largest input string. For example, the statement

```
name = char('Albert Einstein','Is a scientist')
```

will produce

```
name =
Albert Einstein
Is a scientist
```

There are many other operations possible on strings. Table 2.10 lists some functions available in MATLAB for performing various operations on strings.

Table 2.10 Selected string functions

S. No.	Function	Description
1.	strcat	Concatenates two or more strings horizontally ignoring any trailing blanks but preserves the blanks within strings
2.	strcmp	Determines whether the two strings are identical or not
3.	strfind	Finds a string in another string
4.	isletter	For a string it is 1 for letters having alphabets else 0
5.	isspace	It is true for whitespace letter
6.	strmatch	Searches for possible match of a string
7.	regexp	Matches for regular expression

2.8.2 Input/Output Functions

MATLAB has a variety of input–output functions available. These functions can be used for data input from the standard input device like a keyboard and for data output from a device like a monitor. Data can also be sent or received from an external device like DSP. In addition to these functions, powerful file handling functions are also available in it.

Input Functions

Input functions help in getting the input data for the program or simulation. For example, the input command is used for taking input data from the user through keyboard during runtime. The statement

```
N = input('How many students are in the class')
```

gives the user the prompt in the text string and waits for the input from the keyboard. Suppose the user enters input as 20, it will take $N = 20$. Another instruction keyboard when placed in the program stops executing the program and gives the control to the user keyboard. The changed status is indicated by k before the command prompt. All the instructions and commands are valid and can be changed by the user. We can exit from this mode by the return command. Table 2.11 lists some input functions available in MATLAB.

Table 2.11 Input functions

S. No.	Function	Description
1.	Menu	Displays a number of choices in the menu for the user. For instance, C = menu('choose a profession', 'Doctor', 'Engineer') asks for the choice Engineer or Doctor
2.	Pause	Pauses the execution and waits for user signal to restart. For instance, pause(10) pauses the program for 10 s
3.	Textread	Reads formatted text from the file. For instance, T = textread('untitled') reads numeric data from the file untitled
4.	Textscan	Reads the complete file in blocks. For instance, v = textscan(untitled,'Type of format',N) reads the data from file untitled into array v
5.	Strread	Reads data from a formatted string. For instance, S = strread(name,'format') reads numeric data into from name to S
6.	Xlsread	Reads data and text from a spread sheet in a Excel workbook. For instance, numeric = xlsread(untitled) reads numeric data from the excel sheet untitled
7.	Dlmread	Reads data from ASCII delimited file. For instance, roll_num = dlmread(roll_list) reads numeric data from the ASCII delimited file roll list
8.	Fread	Reads binary data from a file. For instance, B = fread(untitled) reads binary data from the file untitled and writes it into matrix B

(Continued)

Table 2.11 (*Continued*)

S. No.	Function	Description
9.	Fgets	Reads line from the file and no additional characters are read after the line terminator. For instance, line = fgets(file,char) returns at the most char characters of the next line
10.	Fgetl	Reads line from the file, the line terminator. Line = fgetl(file) reads the line text of the file
11.	Fscanf	Reads data from the specified file. For instance, A = fscanf(file) reads data from the file and returns it to matrix A

Output Functions

Output functions are used for sending the processed data for display or to an external device. For example, the function disp displays an array without printing its name. The statements

```
A = [1 2 3]; disp(A)
```

will display 1 2 3 at the command window. Table 2.12 lists some output functions available in MATLAB.

Table 2.12   List of output functions

S. No.	Function	Description
1.	Fwrite	Writes binary data to the file. For instance, C = fwrite(file,B) writes the elements of matrix B to the file. The data is written in the column order. C is the number of elements successfully written
2.	Diary	Saves text of MATLAB session. For instance, dairy untitled causes a copy of all subsequent command window input and most of the resulting command window output to be appended to the untitled file
3.	Sprintf	Writes formatted data to string. For instance, C = sprint(format, var) formats the data in array var and returns it to the variable C
4.	Dlmwrite	Writes in ASCII delimited file. For instance, dlmwrite('file', A) writes matrix A into file using ',' as the delimiter to separator matrix elements
5.	Fprintf	Writes formatted data to file. For instance, V = fprintf(file, format, A) formats the data in array A and writes it in file

Formatted Input/Output Functions

The following input/output functions are discussed in this section.

fscanf function

The fprintf function is used to write formatted data in a user-specified formatted file. This function has the following form:

```
count = fprintf(fid, format, A, ......)
fprintf(format, A,......)
```

It formats the data in the real part of matrix A under the control of the specified format string. Then it writes it to the file associated with the file identifier `fid`. `fprintf` and returns a count of the number of bytes written. If `fid` is missing, the data is written to the standard output device, i.e., the command window.

The format string specifies the conversion specification. A conversion specification controls the notation, alignment, significant digits, field width, and other aspects of output format. The format string can contain escape characters to represent non-printing characters and tabs. Conversion specifications begin with the % character and contain these optional and required elements:

1. Flags (optional)
2. Width and precision fields (optional)
3. A sub-type specifier (optional)
4. Conversion character (required)

We can specify these elements in the following order:

```
% - 12.5 e
```

Here, (%) is the conversion specification starter, (–) is the flag, (12) is the field width, (5) the precision, and (e) the conversion number. MATLAB help can be referred to for further details regarding this function,

`sprintf` function: The `sprintf` function is used to write formatted data to a character string of a file. This function has the following form:

```
string = sprintf(format, A, …..)
```

The `sprintf` formats the data in the real part of array A under control of the specified `format` string. Then it returns it in the string variable `string`. The `sprintf` function is the same as `fprintf` except that it returns the data in a MATLAB string variable rather than writing it to a file.

`fscanf` function: The `fscanf` function reads formatted data in a user-defined form from a file. It is of the form,

```
[Array, cnt] = fscanf (fid, format, size)
```

where `fid` is the field id of a file from which the data will be read, `format` is the string controlling how data is read, and `array` is the array that receives the data. The output argument `cnt` returns the number of values read from the file. The `size` is optional as it limits the number of elements that can be read from the file. If the size is not specified, the entire file is considered. The valid entries for size are as follows:

n: read at most n elements into a column vector,

inf: read at most to the end of the file,

[m, n]: read at most $m \times n$ elements filling at least an m by n matrix, in a column order. The value of n can be `inf` but not m.

textscan function: This function reads ASCII files that are formatted into columns of data, where each column can be of a different type. It stores the contents into the columns of a cell array. This function is very useful for importing data tables of other applications. This function has following form:

```
b = textscan (fid, 'format')
b = textscan (fid, 'format', N)
b = textscan (fid, 'format', parameter, value, ………)
b = textscan (fid, 'format', N, parameter, valuie, ……)
```

where fid is the file id of a file that has already been opened by fopen function, format is a string containing a description of the type of data in each column, and N is the number of times to use the format specifier. The format string contains the same types of format descriptors as function fprintf.

2.9 PROGRAMMING A THREE-PHASE VOLTAGE SOURCE

A three-phase voltage source is defined as a voltage source which has three sources displaced by angle 120° in phase. The voltage equations for a three-phase voltage source are as follows:

$$V_A = V_m \sin (w * t),$$
$$V_B = V_m \sin (w * t - 120°), \text{ and}$$
$$V_C = V_m \sin (w * t + 120°)$$

where

V_A = Voltage of the first phase,
V_B = Voltage of the second phase,
V_C = Voltage of the third phase,
V_m = Peak voltage or maximum voltage,
w = Angular frequency in rad/s,
t = Time in seconds.

This voltage source is programmed in Example 2.30. The maximum voltage (V_m), frequency of the supply (f) and time up to which the voltage is required are asked from the user. After receiving these inputs, the desired three-phase voltages are plotted as shown in Fig. 2.27.

Example 2.30

```
% Program to develop user defined Three Phase voltage source
clc;
clear all;
Vm = input('Enter Peak Magnitude Required in volts');
f = input('Enter Supply frequency in Hz');
w = 2 * pi * f; % Angular frequency in rad/s
time = input('Enter time in sec upto which output is required');
t = 0:10e-6:time;
First_phase = Vm * sin(w * t);
Second_phase = Vm * sin(w * t + 120);
Third_phase = Vm * sin(w * t + 240);
```

(Continued)

Example 2.30 (*Continued*)

```
subplot(3,1,1),plot(t,First_phase,'r-')
axis([0 time -Vm Vm])
title('\bfFirst Phase Voltage');
xlabel('\bfTime in sec');
ylabel('\bfAmplitude in volts');
subplot(3,1,2),plot(t,Second_phase,'y-')
axis([0 time -Vm Vm])
xlabel('\bfTime in sec');
ylabel('\bfAmplitude in volts');
title('\bfSecond Phase voltage');
subplot(3,1,3),plot(t,Third_phase,'b-')
axis([0 time -Vm Vm])
xlabel('\bfTime in sec');
ylabel('\bfAmplitude in volts');
title('\bfThird Phase Voltage ');
```

When this program is run, MATLAB asks for the following parameters of the three-phase voltage source as given in the following:

```
Enter Peak Magnitude Required in volts 200
Enter Supply frequency in Hz 50
Enter time in sec up to which output is required 0.04
```

The voltage entered here is 200, supply frequency 50, and time 0.04. For this data, the voltage of the three phases are plotted as shown in Fig. 2.27. These types of usages allow the user to run a number of trials without actually modifying the code. This becomes convenient in simulation of circuits, systems, etc.

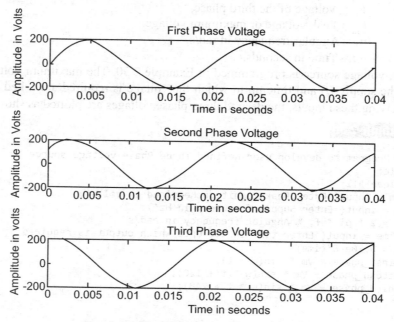

Fig. 2.27 A three-phase voltage source

ADDITIONAL PROGRAMS

1. To develop a program to check if a given number is greater than, less than, or is equal to 1.

```
% Program to check whether a number is greater than 1 or equal to 1 or less than
1
number = input('Enter a number:');
if number > 1,
    disp('This number is greater than 1');
elseif number == 1,
    disp('This number is equal to one');
else
disp('End of the program');
```

The result of this program for given three numbers 0.5, 1, and 24 is

```
Enter a number:.5
This number is less than 1
End of the program
Enter a number:1
This number is equal to one
End of the program
Enter a number:24
This number is greater than 1
End of the program
```

2. To develop a program to evaluate the resonant frequency and quality factor of a parallel resonance RLC circuit for given values of R, L, and C.

```
% Program to find the resonant frequency and Quality factor of
% a parallel resonance RLC circuit
R = input('Enter the value of resistance in ohms:');
L = input('Enter the value of inductance in Henry:');
C = input('Enter the value of capacitance in Farads:');
F = {(1/2 * pi)*((1 /( L * C))-(R^2 / L^2))^0.5};
    % Resonance Frequency
Q = {R/(2 * pi * L)}; % Quality factor
disp('Resonance frequency of the supply is:' );F
disp('Quality factor of the supply is:');Q
```

The result of this program for R = 1, L = 10e–3, and C = 200e–6 is

```
Enter the value of resistance in ohms:1
Enter the value of inductance in Henry:10e-3
Enter the value of capacitance in Farads:200e-6
Resonance frequency of the supply is:
F = [1.0996e+003]
Quality factor of the supply is:
Q = [15.9155]
```

3. To develop a program to plot the current flowing in a series RLC circuit I vs f frequency of the supply. The data required should be taken from the user.

```
% Program to plot current vs frequency semilog plot of
% a series RLC circuit
R = input('Enter value of resistance in ohms:');
L = input('Enter value of inductance in Henry:');
C = input('Enter value of capacitance in Farads:');
V = input('Supply voltage in volts:')
f = 0 : 1: 1000000;
XL = 2 * pi .* f * L;
XC = 1./(2 * pi .* f * C);
Z = (R^2-(XL-XC).^2).^0.5;
I = V ./ Z;
semilogx(f,real(I));
title('\bfCurrent Vs Log frequency plot' );
xlabel('\bfLogf');
ylabel('\bfCurrent in A');
```

The result of this program for the following set of inputs can be seen in Fig. 2.28.

```
Enter value of resistance in ohms:10
Enter value of inductance in Henary:23e-3
Enter value of capacitance in Farads:234e-6
Supply voltage in volts:340
V = 340
```

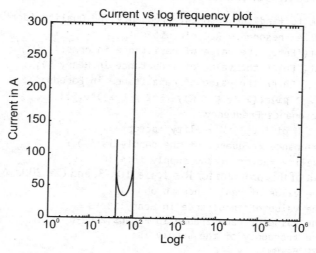

Fig 2.28 Plot of log *f* vs *I* for a series RLC circuit

SUMMARY

After going through the chapter we will be familiar with a number of MATLAB functions and operations. All these functions are the backbone of MATLAB programming. In short, they include

- Functions and operations using variables
- Functions and operations on arrays
- Functions and operations on matrices
- Various arithmetic operations
- Various relational and logical operators
- Various two-dimensional plots and subplots
- Three-dimensional plotting
- Program control functions
- Input, output, and string operators

We will also be familiar about forming branches and loops in the program for given conditions. Also, plotting different data or functions in different dimensions is elaborated with the help of various figures.

EXERCISES

1. Construct a plot for the function $f(x) = e^{-x^2} \sin \pi x^2$ over the interval of $[-2,2]$ of x.

2. Develop a program to compute 103rd prime number.

3. Plot the following functions on a single figure window:
   (a) $\sin(\omega * t)$ (b) $\cos(\omega * t)$
   (c) $\sin^2(\omega * t)$ (d) $\cos^2(\omega * t)$
   Take $w = 314$ and $t = [0, 4]$.

4. Some closed curves on a sphere are defined as
   $x = \cos(mt) \cos(nt)$; $y = \sin(mt) \sin(nt)$; and $z = \sin(nt)$
   for t from $-\pi$ to π. Construct plots for values of $m = 5, 6, 7$ and $n = 9, 11, 13$. Take sufficient samples in order to get a smooth curve and construct the plots.

5. Construct following subplots in a figure window by using suitable graphical functions.
   (a) $e^{-t} \cos t$
   (b) $\cos t + \sin t$
   (c) $e^{-t} \sin t$
   (d) Show the compound plot of e^{-t}, $\sin t$, and $\cos t$
   Define various arrays for t and observe the results.

6. Construct an epicycloids defined by
   $X(t) = (a + b) \cos t - b \cos (a/b + 1)t$
   $Y(t) = (a + b) \sin t - b \sin (a/b + 1)t$
   Take $a = 12$, $b = 5$, and $t = [0; 10\pi]$.

7. The output of a second-order system for step input is given by
   $c(t) = 1 - [(e^{-\xi \, wn \, t})/\sqrt{(1 - \xi^2)} \sin \{(\omega_n \sqrt{(1 - \xi^2)} \, t)$
   $+ \tan^{-1} (\sqrt{(1 - \xi^2)}/\xi)\}]$
   where $c(t)$ is the output of the system. For $w_n = 100$ rad/s, plot the output for different values of ξ, where ξ lies between 0, 1.

8. The current through a p-n junction diode can be expressed by
   $I_D = I_0 (e^{(qvd/kT)} - 1)$
   where

I_D = Current flowing through the diode in amps

I_0 = Leakage current of the diode in amps

q = Charge on an electron, i.e., $1.6 * 10^{-16}$ C

v_d = Voltage across the diode in volts

k = Boltzmann's constant, i.e., $1.38*10^{-23}$ J/K

T = temperature in kelvin

Now, suppose the leakage current of the diode is 15 μA. Develop a program to evaluate the current flowing through this diode for all voltages ranging from –1.5 V to 5 V in the steps of 0.05 V. Repeat this process for the following temperatures: 10°C, 25°C, 40°C, and 55°C. Create a plot of the current as a function of applied voltage V_d, with the curves for the different temperatures with different colors.

9. The mechanical power output of a rotating motor is given by

$$P_M = \tau_d w_m$$

where τ_d is the torque induced on the shaft of the motor in N m, w_m is the angular rotation of the shaft in rad/s, and P_M is the mechanical power in watts. Now, taking the rotational speed of a particular motor shaft to be given by

$w_m = 188.5 (1 - e^{(-02t)})$ rad/s

and the induced torque on the shaft given by

$\tau_d = 10e^{-0.2t}$ N m,

construct a plot for torque, speed, and power versus time from 0 s to 20 s. Label the graph properly with appropriate symbols. Construct two plots, one with the power displayed on a linear scale and another with the output power displayed on a logarithmic scale. Time should be taken on a linear scale.

10. A character array R is defined by

R = 'perception is quite different from reality'

Develop a program to change the content of R to: 'our perception is quite different from reality', without retyping any of the words in R. Hint: Use function isprime and find the length of array R.

11. Write a program to calculate factorial of a number, i.e., $N!$. Take care of 0! and report error if the number is negative or not an integer.

12. Given that arrays,

$A = [1\ 2\ 3]$ and $B = [2; 4; 8]$

Find

(a) $A * B$

(b) $B * A$

(c) $A^T * B^T$

(d) Arrange the elements of A and B in descending order

13. Given that the matrix,

$D = [4\ -6\ 8;\ 7\ 10\ -12,\ 13\ 3\ 4]$

Write the instructions to get

(a) All the elements of second row but second column

(b) All the elements of third row but all columns

(c) Elements of first row and first column.

14. Consider a matrix F given by

$F = [1\ 2\ 4;\ 6\ 8\ 36;\ 7\ 9\ 49]$

Apply the following commands on this matrix and see the output.

(a) Diag(F)   (b) Poly(F)

(c) Diag(F, n)   (d) Eig(F)

15. Evaluate the product of $(s + 1)$, $(s + 2)$, and $(s + 3)$.

16. Integrate the polynomial $y(s) = 6s^3 + 4s^2 + 3s + 1$ with respect to s. Assume constant of integration to be 10.

17. Write a program to generate the first 100 prime numbers.

18. Write a program to subplot sin θ, cos θ, and sin $(\pi - θ)$, for θ = 0 to 2π, using subplot command with each plot on separate sub-windows.

19. Write a program to check whether a number entered by the user is greater than, equal to, or less than 100. Display the result at the command window.

20. Develop a program that returns 1 if its input from the keyboard is a prime number and returns 0 otherwise. Test this for numbers less than 300.

21. Find the Laplace transform of the following functions:
    (a) $u(t)$ (b) t
    (c) $\sin t$ (d) $\cos t$
    (e) e^{-t} (f) $e^{-t} \sin 3t$

22. Find the solution for the following differential equations, for the interval $t = 0$ to $t = 2$.
    (a) $dy/dx = y (e^{-x} - 2)$, with $y(0) = 0$
    (b) $dy/dx = y (e^{-x} - 2)$, with $y(0) = 10$
    (c) $dy/dx = e^{-x} \sin x$, with $y(0) = 0$
    (d) $dy/dx = x e^{-x}$, with $y(0) = 1$

23. Develop a program to count the even and odd numbers from a given set of integers. Display the result at the command window.

24. Plot the bar graph for the following data:
    (a) $x = [0\ 1\ 2\ 3\ 4\ 5\ 6\ 7\ 8\ 9\ 10]$
    (b) $y = [5\ 7\ 8\ 10\ 12\ 12\ 14\ 21\ 16\ 15\ 12]$

25. For $x = 3e^{-2t} \sin 7t$ and $y = \cos t$, obtain a 3-D plot for $0 \le t \ge 10$.

26. Find the roots of the following equations:
    (a) $s^2 + 4s + 2 = 0$
    (b) $s^3 + s^2 + s + 1 = 0$
    (c) $7s + 8 = 0$
    (d) $-s^3 + 7s^2 - 10s + 100 = 0$

27. The root mean square (rms) value of a series of numbers is the square root of the arithmetic mean of the squares of the series of numbers.
    $$\text{rms} = \{1/N \Sigma_{i=1 \text{ to } N} X_i^2\}^{0.5}$$
    where $X_1, X_2, \ldots, X_N$ are the N numbers. Develop a set of instructions in MATLAB that accepts N numbers from the user and then display their rms value at the output window.

28. The EMF equations for a single-phase transformer are given by
    $$E_1 = 4.44 f \Phi_m N_1, \qquad E_2 = 4.44 f \Phi_m N_2$$
    where
    E_1 = Primary winding EMF in volts,
    E_2 = Secondary winding EMF in volts,
    f = Supply frequency in hertz,
    Φ_m = Maximum flux in webers,
    N_1 = Number of primary winding turns, and

N_2 = Number of secondary winding turns.
Plot the primary and secondary winding EMFs vs supply frequency graph for $\Phi_m = 10$ m Wb, $N_1 = 100$, $N_2 = 2,000$, and take supply frequency f as $0 \le f \le 1$ kHz.

29. For a series RLC resonance circuit, the resonance frequency and the circuit current are given by
    $f_r = 1/2\pi \{1/(LC)\}^{0.5}$ Hz
    $I = V/R$, at resonance
    Develop a program to calculate the resonant frequency and current flowing in the circuit for different input values of R, L, and C.

30. For a series RLC circuit, the current flowing in the circuit is given by
    $$I = V/Z, Z = \{R^2 + (X_L - X_C)^2\}^{0.5}$$
    where
    X_L = $2\pi f L$ Ω,
    X_C = $1/(2\pi f C)$ Ω,
    L = 10 mH,
    C = 10 μF,
    R = 10 Ω, and
    V = 250 V.
    Plot a semi-log graph of the circuit current I vs supply frequency f for $0 \le f \le 10$ MHz.

31. Total harmonic distortion (THD) in a system is given by
    $$\text{THD} = \{(I_1/I)^2 - 1\}$$
    Develop a MATLAB program which takes input values of I_1, i.e., first harmonic current, and I, i.e., total current, and then evaluates THD in the system.

32. The output voltage of a buck converter is given by
    $$V_{out} = D V_{in}, \qquad D = T_{ON}/T$$
    where
    V_{out} = Output voltage in volts,
    V_{in} = Input voltage in volts,
    D = Duty cycle,
    T_{ON} = On time, and
    T = Total time
    Plot V_{out} vs T_{ON} graph for $T = 30$ ms and T_{ON} from $0 \le T_{ON} \ge T$.

33. In a DC generator, the EMF generated due to the rotation of the armature is given by

$E = (n * P * \Phi * Z)/(60 * A)$

where

E = EMF generated by the DC generator in volts,

n = Speed of rotation of armature in rpm,

P = Total number of poles,

Φ = Flux per pole in webers,

Z = Total number of conductors in the armature, and

A = number of parallel paths.

For lap winding $A = P$ and for wave winding $A = 2$.

Write a program in MATLAB to plot EMF vs speed (n) graph for a lap connected generator. Take $\Phi = 27$ m Wb.

3

FUNDAMENTALS
OF SIMULINK

3.1 INTRODUCTION

This chapter focuses on Simulink and its utilization as a tool for design, modeling, and simulations of practical systems, and gives an introduction to Simulink along with related examples. Rather than elaborating on the tool, this chapter focuses on the application-oriented perspective. The aspects of modeling and simulation are emphasized so as to provide a straightforward view about the tool. After going through this chapter, we will be able to efficiently utilize this tool for the simulation of models.

Simulink is a commercial tool developed by the Math Works Inc. under the MATLAB® family for modeling, simulating, and analyzing multi-domain dynamic systems. It is basically a graphical extension of MATLAB in which the systems are constructed on screen by using blocks. It offers tight integration with rest of the MATLAB environment such that it can either drive MATLAB or be scripted from it. A number of block diagrams, or say blocks, such as voltage source, current source, oscilloscopes, and function generators are available. It is widely used by the researchers and engineers in the fields of power system, control system, power electronics, etc. for multi-domain simulation and design. Simulink gives the programming environment for block set-based programming.

With Simulink, we can move one step forward beyond the idealized linear models to explore more realistic nonlinear models by factoring in leakage resistance, leakage magnetic flux, device ON time, OFF time, and many other things related to real-world phenomena. It simply transforms the computer into a laboratory for modeling and simulating systems that is simply not possible or practical otherwise. It supports the linear and nonlinear systems modeled in continuous or discrete time, or a combination of the two. Systems can also be simulated at a multi-rate, i.e., various parts of the system are sampled and updated at different rates.

For modeling a system, various building blocks known as blocks are available along with the graphical user interface (GUI). Models can be constructed from the block diagrams by simple click and drag mouse operations. Models constructed are hierarchical models. A hierarchical model is similar to a network model except that it displays a collection of blocks in trees, rather than displaying arbitrary. This approach gives a much deeper insight about the organization and mutual interaction of the various components of the model. The blocks available for the display also aid in analyzing the output from different viewpoints. Models can easily be simulated by different solvers available like discrete, ode45 (Dormand–Prince), ode23 (Bogacki–Shampine) and many others by choosing desired step sizes. The menus are particularly convenient for interactive work, while the command line approach is quiet fruitful for executing a batch of simulations. Simulink platform encourages trying step by step and then observing the results. Models can be easily built and changes can also be made in the existing model. As these models are user interactive, the modeling parameters can be easily set and reset. In this platform, the system in the simulation time can be studied taking the practical factors into account.

As Simulink software is started, Simulink library browser pops up displaying different block sets. Each of these block sets contains a number of blocks that belong to a particular category. In order to view all the blocks available in a particular block set library, it has to be expanded by clicking on it. If a particular block from the library has to be searched, type the name on the toolbar and click on find block button. A list of some frequently required blocks along with their functions are given in Table 3.1.

Table 3.1 Some frequently used Simulink blocks

S. No.	Block Name	Function
1.	Derivative	Provides derivative of the continuous input signal as output
	Abs	The output of this block is the absolute value of the input
2.	Integrator	The output is the integral of continuous input signal at the current time step
3.	Display	Displays the numeric value of the input, it can be used like a digital meter
4.	Scope	Displays the input waveform on its screen, it is like a CRO
5.	Terminate	Terminates unwanted output signals
6.	Ground	Grounds signals
7.	Pulse Generator	Generates output pulses of desired amplitude, width, and frequency
8.	Signal Generator	Generates sine, square, sawtooth, and random waveforms of desired amplitude and frequency
9.	Sine Wave	Generates a sine wave of desired amplitude, phase, frequency, and phase

(Continued)

Table 3.1 (*Continued*)

S. No.	Block Name	Function
10.	AC Voltage Source	Generates AC (power) voltage sinusoidal waveform of desired amplitude and frequency
11.	AC Current Source	Generates AC (power) current sinusoidal waveform of desired amplitude and frequency
12.	Three-phase Source	Generates three-phase sinusoidal voltage waveform of desired rms value and frequency
13.	Ground	Grounds connection for power elements
14.	Linear Transformer	Provides three winding linear windings
15.	Parallel RLC Branch	Provides R, L, and C elements in parallel
16.	Series RLC Branch	Provides R, L, and C elements in series
17.	DC Machine	Provides DC machine of a desired rating
18.	Permanent Magnet Synchronous Machine	Provides a three-phase permanent magnet synchronous machine with sinusoidal flux distribution
19.	Current Measurement	Measures power current
20.	Impedance Measurement	Measures impedance between two points as a function of frequency
21.	Voltage Measurement	Measures power voltage
22.	Detailed Thyristor	Provides detailed model of thyristor, with internal resistance, inductance, and series R-C snubber circuit
23.	Diode	Provides diode with parallel snubber circuit, the model has an internal resistance R_{ON}, and an internal inductance L_{ON}
24.	Ideal Switch	Provides a model of an ideal switch controlled by a gate pulse, the ON resistance of this switch is R_{ON} and the OFF resistance is infinite
25.	IGBT	Provides IGBT model with R_{ON}, L_{ON}, and parallel snubber circuit
26.	MOSFET	Provides MOSFET model with R_{ON}, L_{ON}, and parallel snubber circuit
27.	Thyristor	Provides thyristor model with R_{ON}, L_{ON}, and parallel snubber circuit

3.2 COMMONLY USED BLOCKS

This section gives an introduction about the commonly used blocks library. In this section the functioning of some blocks included in the library are described, and simulations are performed to illustrate their applications. The frequently required blocks in the commonly used blocks library are shown in Fig. 3.1.

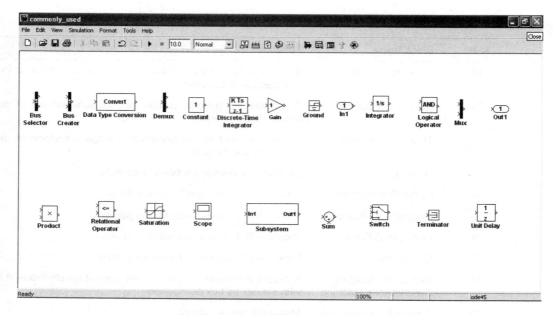

Fig. 3.1 Frequently required blocks

3.2.1 Bus Selector, Bus Creator, and Scope Blocks and Configuration Parameters

Consider a bus consisting of several bundled wires held together. Each bus graphically represents a composite signal made up of several individual signals. Bus Creator block is used to create signal buses and Bus Selector block is used to access the components of a bus. The names of the Bus Creator and Bus Selector blocks are hidden, when they are sent to a Simulink model but can be made visible by clicking on *Show name* on *Format* menu. The font, font style, and font size of the block name can be modified by selecting the block and then clicking on the *Font* option on the *Format* menu and then selecting the appropriate choices. In Fig. 3.1, the Arial font, Bold, and size 12 are taken for displaying the names of the blocks. The Bus Creator and Bus Selector are shown by heavy and black vertical line.

Example 3.1

The model in Fig. 3.2 simulates the Sine Wave, Step, and Pulse Generator signals by combining them into a bus by using Bus Creator block and then separates them by Bus Selector block.

The Sine Wave, Step, and Pulse Generator can be taken from Sources library and then dragged into a new Simulink file. Scope is selected from Sinks library and dragged. Bus Creator and Bus Selector blocks are dragged from Commonly Used Blocks library.

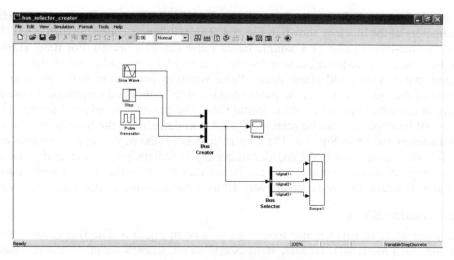

Fig. 3.2   Model of Example 3.1

These blocks are saved in a Simulink file and are connected to each other as shown in Fig. 3.2. The blocks can be connected by left clicking the mouse button on the connection point of the block and then by holding the left button and dragging it to the connection point of the other block. After completing all the connections, double click on the Step block. The parameters of this block will be displayed in a new window as shown in Fig. 3.3. The parameters of the Step block can be entered in this window. The parameters of the Step block taken for the model can be seen in Fig. 3.3.

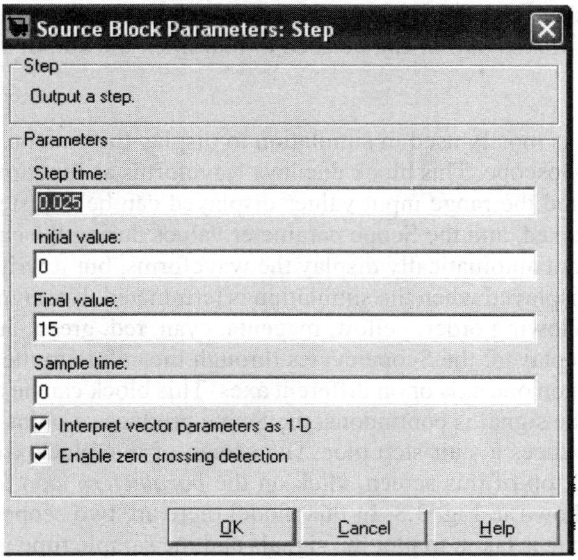

Fig. 3.3   Parameters of a step block appearing in the dialog box

Now double click on the Pulse Generator block. A new window opens showing the parameters of the block as shown in Fig. 3.4. This window is called the dialog box. In the pulse type, a time-based pulse or a sample-based pulse can be selected. For time, simulation time can be opted or, if required, external signal can also be used. Other parameters are amplitude, period, pulse width, and phase delay. Pulse width in percentage is the fraction of the time period of the pulse for which the pulse remains at its mentioned amplitude. Phase delay is the delay in seconds required from the initial time. The parameters selected for this block for the model of Example 3.1 can be seen in Fig. 3.4. Similarly, open the Sine Wave block and set its parameter as shown in Fig. 3.5. The amplitude denotes the peak value of the sine wave, bias is the DC shift, frequency is the angular frequency in radians per second, and phase denotes the phase angle in radians. Sample time can be used if samples of the sine wave are to be taken. For instance, if a sample is required at every 10 μs, a sample time as 10e–6 can be taken.

Bus Creator Block

The parameters of Bus Creator block can be seen in Fig. 3.6. The Bus Creator block assigns a name, such as signal1, signal2, signal3, and so on, to each signal on the bus that it creates. This allows the user to refer to the signals by name while searching for their sources. The block offers two bus signals naming options. We can specify that each signal on the bus inherits the name of the signal connected to the bus by default or that input signal must have a specific name. To specify that bus signals inherit their names from the input ports, we select *Inherit bus signal names from input ports* from the list box on the parameter dialog box of the block. The names of the inherited bus signals appear on the signals in the bus as can be seen in Fig. 3.6.

Bus Selector Block

The parameters of Bus Selector block are shown in Fig. 3.7. This block accepts a bus signals as input which can be created by a Bus Selector, Mux, or Bus Creator block. These are the input objects and are shown in left list box as can be seen in Fig. 3.7. The right list box shows the selections made. *Muxed output* can be used to multiplex the output signals.

Scope Block

The Scope block is mostly used in simulation to display the data on its screen, similar to what is seen in an oscilloscope. This block displays waveforms as functions of simulation time. The amount of time and the range input values displayed can be adjusted, the Scope window can be moved and resized, and the Scope parameter values during the simulation can be modified. This block does not automatically display the waveforms, but it writes data to connect scopes. The signals are displayed when the simulation is terminated. It assigns the colors to each signal element in the following order: yellow, magenta, cyan, red, green, and dark blue. If more than six signals are displayed, the Scope cycles through the colors in the order list. This block can plot multiple data on one axis or on different axes. This block can be opened by double clicking on the block. If the signal is continuous, the Scope produces a point-to-point plot. If the signal is discrete, it produces a stair step plot. The screen of this block will open as can be seen in Fig. 3.10. On the top of this screen, click on the *parameters* icon to open the parameter box of the block as shown in Fig. 3.8. In this model there are two scopes, i.e., Scope and Scope1. For Scope, one axis is taken to plot the signals and the sample time is taken as 10e–6 while for Scope1 three axes are taken. The number of axes can be entered in the parameter block. The output waveforms of both the scopes are shown in Figs 3.9 and 3.10.

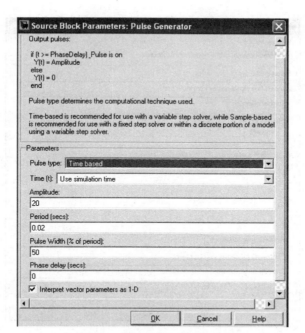

Fig. 3.4 Parameters of pulse generator block

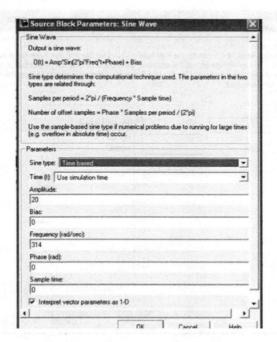

Fig. 3.5 Parameters of sine wave block

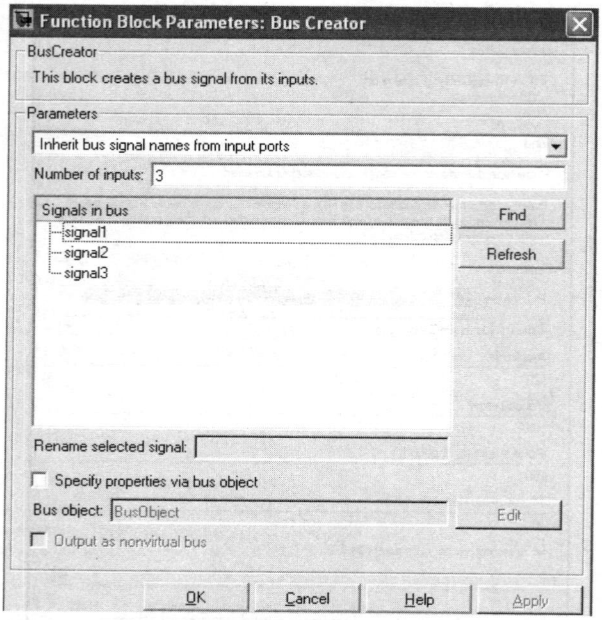

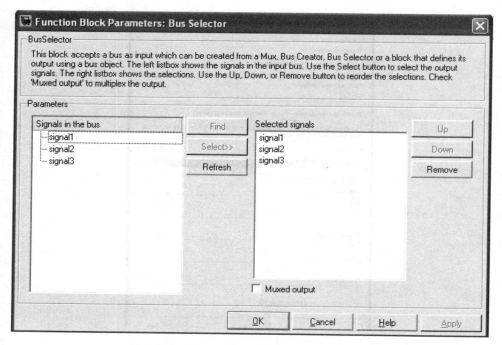

Fig. 3.6   Parameters of bus creator block

Fig. 3.7   Parameters of bus selector block

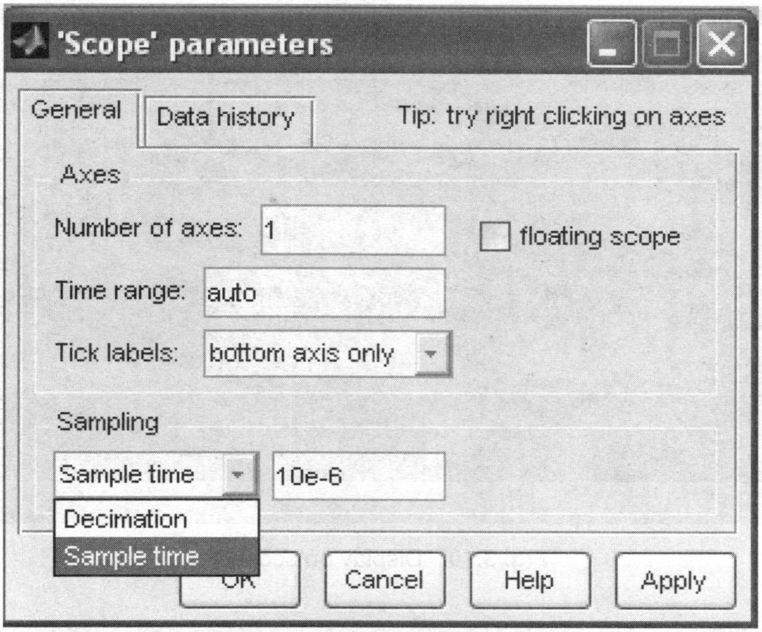

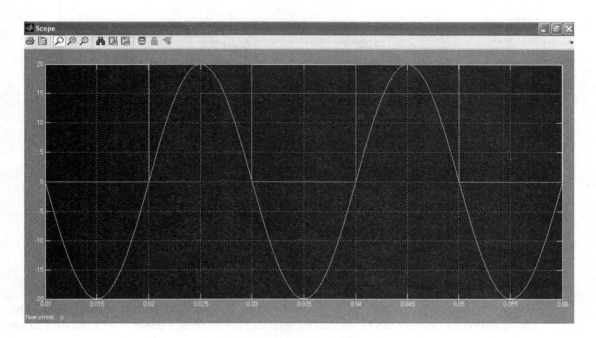

Fig. 3.8 Parameters of scope block

Fig. 3.9 Display on scope

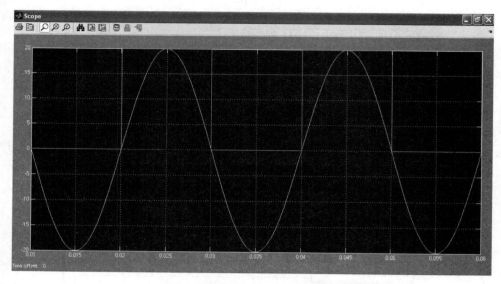

Fig. 3.10 Display on scope 1

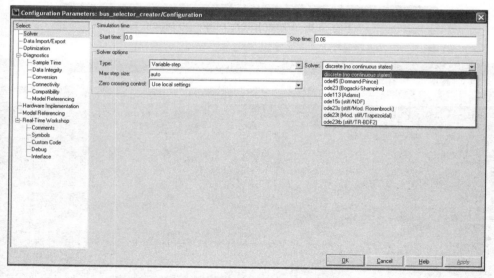

Fig. 3.11 Configuration parameters of the model

Configuration Parameters

The configuration parameters of a model have to be specified before simulating it. The configuration parameters can be opened by selecting configuration parameters from simulation on the Simulink model toolbar. There are two types of solvers in Simulink—fixed step and variable step. Both the types of solvers compute the next simulation time as the sum of the current simulation time and a quantity known as the step size. In a fixed-step size, the step size remains constant throughout the simulation while in the case of a variable-step size, the step size can vary from step to step depending on the dynamics of the model. A variable-step solver reduces the step size when a model's state is changing rapidly to maintain accuracy and increases the step size when the system states are changing slowly in order to avoid taking unnecessary steps. The choice between the two steps of solvers depends upon the model's dynamics.

If a model needs to generate a code that is required to run on real-time computer system, a fixed-step solver should be chosen. This is because the real-time computer systems operate at fixed size signal sample rates. If a code need not be generated from the model, the choice between the available step and a fixed-step solver depends on the dynamics of the model. If the states of the model change rapidly or contain discontinuities, available step solver can shorten the time required to simulate the model significantly. This is because for such models, the available step solver can require fewer time steps than a fixed-step solver to achieve a comparable level of accuracy.

When fixed step is chosen in type control, the solver panel allows choosing a set of fixed-step solver. The fixed-step solvers are of two types—discrete and continuous. The fixed-step discrete solver computes the time of the next time step by adding a fixed-step size to the time of the current time. The accuracy and length of time of the resulting simulation depends on the size of the steps taken by the simulation. For example, the smaller the step size, the more accurate the results but the longer the simulation time required. One can set the step size or let the Simulink set the size itself by default. The fixed-step discrete solver has the fundamental limitations. It cannot be used to simulate models that have continuous states. This is because the fixed-step discrete solver relies on the blocks of the models to compute the values of the states that they desire.

Simulink also provides a set of explicit fixed-step continuous solvers. The solvers differ in the specific integration techniques used to compute the models' state derivatives. The solvers available are ode1 (Euler's method), ode2 (Heuris method), ode3 (Bogacki–Shampine formulae), ode4 (fourth-order Runge–Kutta formulae), and ode5 (Dormand–Prince formulae).

As with the fixed-step discrete solver, the accuracy and length of time of a simulation driver by a fixed-step continuous solver depend on the size of the steps taken by the solver. The smaller the step size, the more accurate the results are, but longer the overall simulation time. Simulink provides one solver in the category as ode14x which uses a combination of Newton's method and extrapolation from the current value of a model at the next time step.

Any of the fixed-step continuous solvers in Simulink can simulate a model to any desired level of accuracy, given enough time and a small enough step size. Unfortunately, it is neither possible nor feasible to decide the solver and step-size combination that will yield acceptable results for a model's continuous states in the shortest time. Thus, finding the best solver for a particular model requires experimentations.

As with the fixed-step solvers, the set of variable-step solvers comprises a discrete solver and a subset of continuous solvers. Both these types of solvers compute time of the next time step by adding a step size to the time of the current time that varies depending on the rate of change of model's state. The continuous solvers, in addition, use numerical integration technique to compute the values of the model's continuous states at the next time step. Both these solvers rely on blocks that define the model's discrete states to compute the values of the discrete states that are defined. The choice between the two types of solvers depends on whether the blocks in the model define states and if so, the kind of states they define. If the model defines no states or only the discrete states, the discrete solver can be selected.

The following variable-step continuous solvers are provided in Simulink:

- Ode45: It is based on an explicit Runge–Kutta formula, the Dormand–Prince pair. It is a one-step solver and is the best known solver to apply as a first try for most models. It is for this reason that it is the default solver used by the Simulink models with continuous states.
- Ode23: It is also based on an explicit Runge–Kutta pair of Bogacki and Shampine. It is also a one-step solver and is more efficient than ode45 at crude tolerances and in the presence of mild stiffness.
- Ode113: It is a variable order Adams–Bashforth–Moulton PECE solver. It is a multistep solver and can be more efficient than ode45 at stringent tolerances.
- Ode15s: It is a variable-order multistep solver based on the numerical differentiation formulas. If it is suspected that a problem is stiff, or ode45 has failed or was inefficient, try ode15s.
- Ode23s: It is based on a modified Rosenbrock formula of order 2. As it is a one-step solver, it can be more efficient than ode15s at crude tolerances.
- Ode23t: It is an implementation of the trapezoidal rule using a free interpolate. It is used if the problem is only moderately stiff and a solution without numerical damping is required.
- Ode23tb: It is an implementation of TR-BDF 2, an implicit Runge–Kutta formula with first stage which is trapezoidal rule step and a second stage that is a backward integration formula of order two. This solver is more efficient than ode15s at crude tolerances.

Variable-Step Solver Error Tolerances   The standard local error control techniques are used to monitor the error at each step. During each time step, the solvers compute the state values at the end of the step and determine the local error, i.e., the estimated error of these state values. They then compare the local error to the acceptable error, which is a function of the relative tolerance (rtol) and the absolute tolerance (atol). If the error is greater than the acceptable limit, the solver reduces the step size and tries again. If we specify auto (default), Simulink sets the absolute tolerance for each state initially to 1e–6. As the simulation progresses, it resets the absolute tolerance for each state to the maximum value that the state has assumed. If the computed setting is not suitable, one can determine the appropriate setting themselves.

For the model of Example 3.1, the following configuration parameters are taken: type—variable step, solver—discrete (no continuous states), maximum step size—auto, and zero crossing control—use local settings. Start and stop times for the simulation are taken as 0.00 and 0.06 s respectively as can be seen in Fig. 3.11.

3.2.2 Data Type Conversion, Constant, and Display Blocks

There are two classes of data types—those defined by MATLAB and Simulink users and those defined strictly by MATLAB only. The latter type is referred to as MATLAB built-in data types. Simulink supports all built-in MATLAB data types except *int64* and *uint64*. Figure 3.12 shows the data types available in Simulink. The data type *single* means single precision floating point,

int8 means signed 8-bit integer, *uint8* means unsigned 8-bit integer, *int16* means signed 16-bit integer, *int32* means signed 32-bit integer, *uint16* means unsigned 16-bit integer, and *uint32* means unsigned 32-bit integer. In addition to these, Boolean data types are also present and these are represented internally by the *uint8* values.

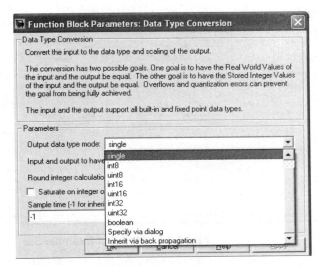

Fig. 3.12 Parameters of Data Type Conversion block

Data Type Conversion Block

The Data Type Conversion block converts an input signal of any data type to the other data type and scaling specified by the block's output data type mode. The input can be any real or complex signal. If the input is real, the output is real; and, if the input is complex, the output is complex. For the input and output to have equal parameters, select the method by which the input is processed. The possible values are Real World Value (RWV) and stored integer (SI). We select RWV to treat the input as $Y = MX + C$, where M is the slope and C is the bias. Y is used to produce $X = (Y - C)/M$, which is stored in the output. This is the default value. Select SI in order to treat the input as a stored integer X. The value of X is directly used to produce the output. In this case, the input and the output are identical except that the input is a raw integer, lacking proper scaling information. Selecting stored integer may be useful if we are generating a code for a fixed-point processor so that the resulting code uses only integers and does not use floating point operations.

Example 3.2

The model shown in Fig. 3.13 uses three Data Type Conversion blocks. In this block, the input is processed as stored integer. In Data Type Conversion1 block, the data type to be displayed is taken as *unit16* and in Data Type Conversion2 block, the data type to be displayed is taken as *int32*. The values displayed at the display blocks after the simulation is executed and can be seen in Fig. 3.13. The simulation is done by fixed-step discrete solver for 0 to 1 s.

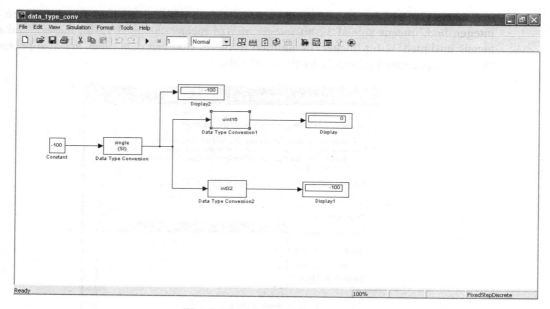

Fig. 3.13   Model of Example 3.2

Constant Block

The Constant block is used to define a real or a complex constant value. This block accepts scalar, vector, or matrix output depending on the dimensions of the constant value parameters that are specified, and the setting of the interpret vector parameters as 1-D parameter. The output of the block has the same dimensions and elements as the constant value parameters. By default, the Constant block outputs a signal whose data type and complexity are same as that of the block's constant value parameter. However, the output can be specified to be any supported data type supported by Simulink including fixed-point data types. In this example, the constant value is taken as –100 and sample time as inf.

Display Block

The Display block shows the value of the input given to it. The display formats are the same as those of MATLAB. They are also specified in the Help menu on this block. The decimation parameter enables us to display data at every nth sample, where n is the decimation factor. Thus, the default value 1 displays the data at every time step. We use the sample time parameter to specify a sampling interval at which the display points are set. This parameter is useful when using a variable-step solver where the interval between time steps might not be the same. The default value of –1 causes the block to ignore the sampling interval when determining the points to be displayed. If the block input is an array, the block must be resized to see more than just the first element. The Display block can be resized vertically or horizontally. The presence of a small black triangle element indicates that the block is not displaying all input array elements. In Example 3.2, the format for display blocks is taken as short, decimation 1, and sample time –1.

3.2.3 Ground, Terminator, In-port, and Out-port Blocks

Ground Block

The Ground block can be used to connect blocks whose input ports are not connected to other blocks. If a simulation is run with blocks having unconnected input ports, Simulink issues warning messages at the command window. The warning messages can be avoided by using Ground blocks. Thus, the Ground block outputs a signal with zero value. The data type of the signal is the same as that of the port to which it is connected.

Terminator Block

The Terminator block can be used to cap blocks whose output ports are not connected to any other blocks. If a model is run with blocks having unconnected or left-out output ports, Simulink issues warning messages. These messages can be avoided by using Terminator blocks.

In-port and Out-port Blocks

The In-port blocks are ports that serve as links from outside a system into the system. The Out-port blocks are output ports for a subsystem. A Subsystem block represents a subsystem of the system that contains it. As the model increases in size and complexity, a subsystem can be created from the system by grouping some of the blocks into a subsystem.

3.2.4 Gain, Product, Sum, and Unit Delay Blocks

These blocks can be used for arithmetic computations. Computation on signals and/or data input can be performed with the help of these blocks.

Gain Block

The Gain block multiplies the input by a constant value. The input and the gain can each be a scalar, vector, or matrix. The value of the gain is specified in the gain parameter. The multiplication parameter specifies element-wise or matrix multiplication. In case of matrix multiplication, this parameter also indicates the order of the multiplicands.

Product Block

The Product block performs multiplication or division of the inputs. This block produces output using either element-wise or matrix multiplication, depending upon the inputs and the multiplication parameter. The operations are specified with the *number* of input parameters. Multiply (*) and divide (/) characters indicates the operation to be performed on the inputs.

Sum Block

The Sum block performs addition or subtraction on its inputs. This block can add or subtract a scalar, vector, or a matrix. From the block parameters dialog box, the shape of its symbols like block, rectangular, or round can be chosen. If there are two or more inputs, then the number of characters must be equal to the number of inputs. For instance, '–++' requires three inputs and configures the block to subtract the first input and add the second and the third inputs.

Unit Delay Block

The function of the Unit Delay block is to delay the input by the sample time specified. In other words, the output signal is the input signal delayed by one sample time. If the model contains multi-rate transitions, we must add more Unit Delay blocks between the slow to fast transitions. In this block, the sampling period and initial conditions are pre-specified in the dialog box of the *function block parameters*. The time between samples is specified with the sample time parameter. A setting of –1 means the sample time is inherited from the simulation. This block also allows discretization of one or more signals in time, or re-sampling the signal at some another rate. The sample rate of this block must be set to that of the slower block.

Example 3.3

In this example two constant blocks of values 10 and 5 are taken in the model. The output of these constant blocks is the input to the Sum and the Product blocks. The output on the display of Sum block is 10 + 5 = 15 and on the Product block is 10 × 5 = 50 as can be seen in Fig. 3.14. A gain of 2 is applied on constant value 5 and the output seen at the display block is 10.

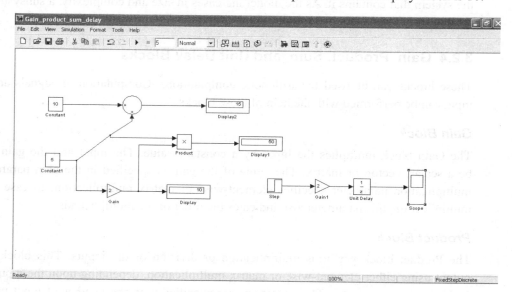

Fig. 3.14 Model of Example 3.3

A unit step source is taken and is fed to a Unit Delay block through a Gain block of Gain2. The output is seen on the Scope block. The output of this scope is shown in Fig. 3.15. It is clearly visible from Fig. 3.15 that the output 2 is delayed by one unit. This simulation is run on fixed-step discrete solver for 0 to 5 s of the simulation time.

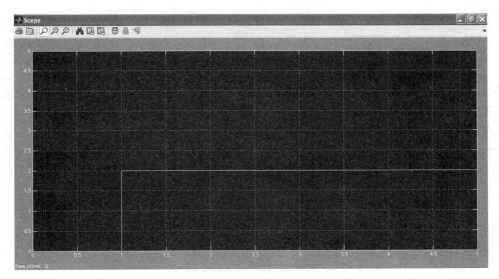

Fig. 3.15 Output wave of the scope of Fig. 3.14

3.2.5 Mux and Demux Blocks

Mux Block

Multiplexing is a method of sending multiple signal streams of information on a career at the same time in the form of single complex signal and then recovering the separate signals at the receiving end. Analog signals are normally multiplexed using *frequency division multiplexing*, in which the carrier bandwidth is divided into sub-channels of different frequency width, each carrying a signal at the same time, in parallel. The digital signals are commonly multiplexed using *time division multiplexing*, in which the multiple signals are carried over the same channel in alternating time slots. If the inputs take turns to use the output channel, then the output bandwidth need to be no greater than the maximum bandwidth of any input signal. A demultiplexer performs the reverse operation of a multiplexer.

The Simulink Mux block combines its inputs into a single output. An input can be a scalar, a vector, or a matrix signal. The Mux block's *number of inputs* parameter allows specifying the input signal names and dimensionality as well as the number of inputs. A value of –1 means that the corresponding port can accept signals of any dimensions. Simulink hides the name of a Mux block when it is dragged from the Simulink block library to a model file. However, the name can be made visible by clicking *show name* on the *format* menu.

Demux Block

The Simulink Demux block extracts the components of an input signal and outputs the components as separate signals. The block accepts either vector signals or bus signals. The number of output parameters allows us to specify the number and the dimensionality of each output port. If the dimensionality of the outputs is not specified, the block determines the dimensionality of the output.

Example 3.4

In this example, three sine wave sources of amplitude 120 and frequency 314 rad/s with a phase delay of 0, $2\pi/3$, and $-2\pi/3$ are taken as shown in Fig. 3.16. These three sine waves are input to a 3×1 Mux block and the output of the Mux block is fed to a 1×3 Demux block. The output of Mux and Demux blocks are given to Scope blocks. The output of the Mux is shown in Fig. 3.17 and that of Demux in Fig. 3.18. This model is run on variable step ode45 solver for 0 to 0.06 s.

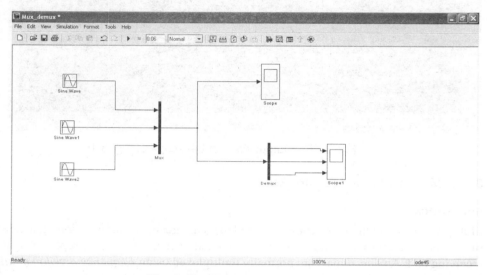

Fig. 3.16   Mux and demux blocks

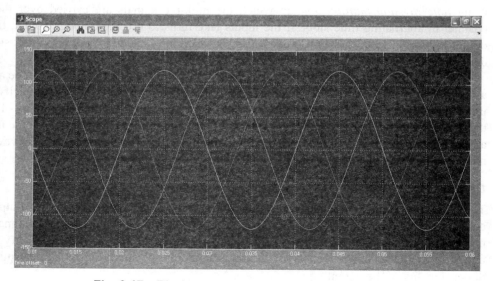

Fig. 3.17   Display as seen on scope block of Fig. 3.16

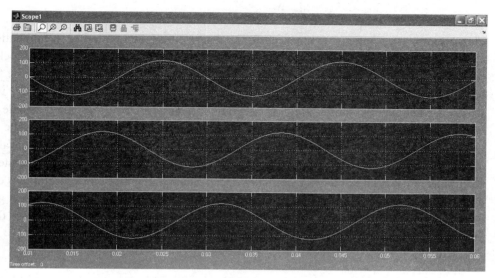

Fig. 3.18 Display as seen on scope 1 block of Fig. 3.16

3.2.6 Integrator and Discrete Time Integrator Blocks

The Integrator and Discrete Time Integrator blocks can be used to find out the integral or summation of a signal or function.

Integrator Block

The Integrator block integrates the input and it is used with continuous time signals. Simulink treats the Integrator block as a dynamic system with one state, its output. The input of the Integrator block is the state's time derivative. The selected solver computes the output of the Integrator block at the current time step, using the current input value and the value of the state at the previous time step. The block also provides the solver with an initial condition for use in computing the block's initial state, at the beginning of the simulation. The default value of the initial condition is 0. The *function block parameter* dialogue box allows specifying another value for the initial condition or creates an initial value input port on the block. It allows setting the upper and lower limits of integration, creating an input that resets the block's output to its initial value, depending on how the input changes, and creating an optional state output that allows using the value of the block's output to trigger a block reset. The Integrator block state port allows to avoid creating algebraic loops when creating an integrator that resets itself based on the value of its output. An algebraic loop generally occurs when an input port with direct feed through is driven by the output of the same block, either directly or by feedback path through the other blocks with direct feed through. The Integrator block's state port makes it possible to avoid creating algebraic loops when creating an integrator that rests itself based on the value of its output.

Discrete Time Integrator Block

The Discrete Time Integrator block performs discrete time integration or accumulation of a signal. This block is used in discrete time systems instead of Continuous Integrator block which is used in continuous time system. It can be used to integrate or accumulate using the forward Euler, backward Euler, or trapezoidal methods. For a given step *n,* Simulink updates *y(n)* and *x (n + 1)*. The block's sample time in integration mode is *T* and in triggered sample time is ΔT. In accumulation mode, *T* = 1 and the block's sample time determines when the block's output is computed, but not the output's value. The constant *K* is the gain value. Values are clipped according to upper or lower limits provided in the block. Purely discrete systems can be simulated using any of the solvers. The Discrete Time Integrator block also allows defining initial conditions on the block dialog box or as input to the block, to define an input gain (*K*) value, output of the block state, upper and lower limits of the integral, and reset the state depending on an additional reset input.

Example 3.5

In this example, a unit step signal is taken as an input and is fed to an Integrator block and a Discrete Time Integrator block through a Unit Delay block as can be seen in Fig. 3.19. A three axes Scope block is taken and the output of the Integrator, Unit Delay, and Discrete Time Integrator is given to it. After executing this simulation, the output is shown in Fig. 3.20. It can be observed that the output of the integrator for Unit Step is a ramp signal, for Unit Delay it is a one-step delayed unit step signal, and for Discrete Time Integrator it is a staircase signal.

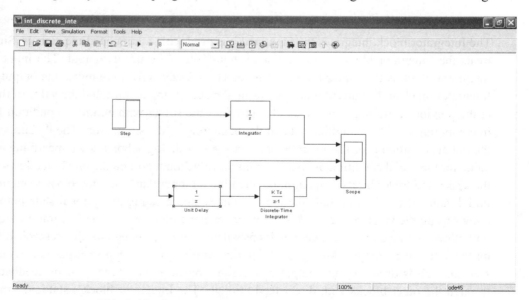

Fig. 3.19 Integrator and discrete time Integrator blocks model

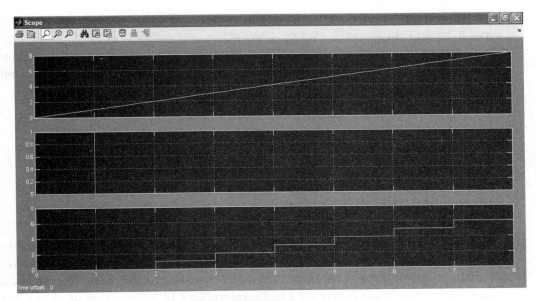

Fig. 3.20 Display on the scope

3.2.7 Logical Operator and Relational Operator Blocks

The Logical Operator and Relational Operator blocks are generally used for logical operations and for comparison.

Logical Operator Block

The Logical Operator block performs the specified logical operation on its inputs. An input value is true, i.e., 1, if it is nonzero and false, i.e., 0, if it is zero. The Boolean operation connecting the inputs is selected with the operator parameter list in the *function block parameter* dialog box. The block updates to display the selected operator. The logical operations which can be performed are as follows:

- AND: True if all inputs are '1'.
- OR: True if at least one input is '1'.
- NAND: True if all the inputs are '0'.
- NOR: True when no inputs are '1'.
- XOR: True if an odd number of inputs are '1'.
- NOT: True if the input is '0' and vice versa.

The numbers of input ports are specified with the *number of input port*s parameters. The output type is specified with the *output data type* mode and/or the *output data type* parameters. The output is true if '1' and false if '0'.

Relational Operator Block

The Relational Operator block performs the specified comparison of its two inputs. We select the relational operator connecting the two inputs with the *relational operator* parameter. The block updates itself to display the selected operator. The supported operations are as follows:

- ==: True if the first input is equal to the second input.
- ~=: True if the first input is not equal to the second input.
- <: True if the first input is less than the second input.
- >: True if the first input is greater than the second input.
- <=: True if the first input is less than or equal to the second input.
- >=: True if the first input is greater than or equal to the second input.

Example 3.6

This example illustrates the functioning of two logical operators AND and OR along with two relational operators equal to (==) and less than equal to (<=). The model constructed is shown in Fig. 3.21. Two input signals are taken, one is a constant value 2 and the other is a *pulse generator* of magnitude 2, period 1 s and pulse width of 50%. In the case of equal to operator, the output is 1 when both the inputs, i.e., constant 2 and pulse magnitude, are equal. While in the case of less than or equal to operator, the output is 1 when the pulse magnitude is less than or equal to the constant value 2. For logical operator AND, the output is 1 when either the inputs are greater than 0 and in case of operator OR, the output is 1 when any one of the input is greater than 0. As constant value 2 is always greater than 1, the output of OR block is always 1. Figure 3.22 shows the output as seen in the Scope after simulating the model.

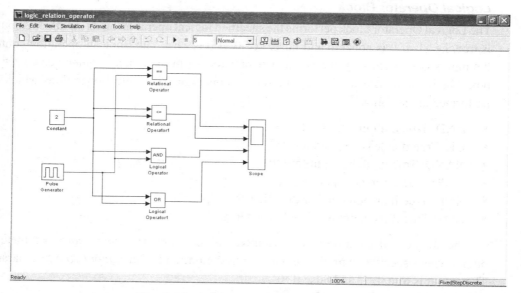

Fig. 3.21   Logical and relational operator blocks

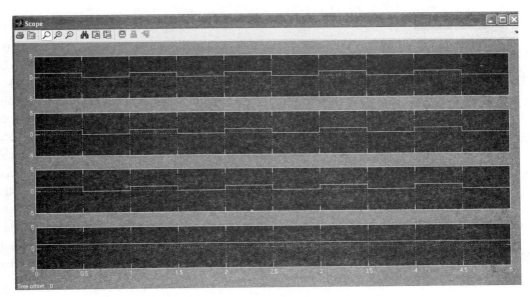

Fig. 3.22 Output of scope of Fig. 3.21

3.2.8 Switch and Saturation Blocks

The Switch block is used to select the input signal and the Saturation block clips the signal if it exceeds a certain specified limit.

Switch Block

The Switch block will output the first input or the third input depending on the value of the second input (control signal). The first and third inputs are called data inputs. The second input is called the control input and it is specified on the function block parameters of the Switch block. In this block the control input has the following options:

u2 >= Threshold

u2 > Threshold

u2 ~= 0

where u2~= indicates a non-zero condition.

Saturation Block

The Saturation block establishes upper and lower bounds for an input signal. When the input signal is within the range specified by the upper and lower limit parameters, the input signal passes through unchanged. When the input signal is outside these bounds, the signal is clipped to the upper and lower bounds. When the lower limit and upper limit parameters are set to the same value, the block outputs that value.

Example 3.7

This example illustrates the functioning of Switch and Saturation blocks. The Step1 block of final value –4, step time 0 is connected to the first input of the Switch block. The Step block is a step function of final value 4 and is connected to the third input of the Switch block, as can be seen in Fig. 3.23. The Sine Wave block of amplitude 5 and frequency 5 rad/s is connected to the second or control input of the Switch block and to the Saturation block. The threshold of the Switch block is set to 0 and the sample time is taken as –1 (default value). For Saturation block, the upper limit is taken as 2 and the lower limit as –2. The sample time for this block is taken as –1 (default value). The condition selected for the control signal of the switch is u2 > 0. Hence, when the sine wave is greater than zero, the Switch passes input one, i.e., –4, otherwise it passes the third input, i.e., 4. The output of the Switch is viewed on the Scope block as shown in Fig. 3.24. In case of Saturation block, the output is a sine wave when its magnitude is less than 2 and greater than –2, otherwise it is a constant, i.e., 2 or –2. So, within the upper and lower limits, the output is equal to the input as shown in Fig. 3.24. This model is executed from 0 to 5 s by using a variable-step discrete solver.

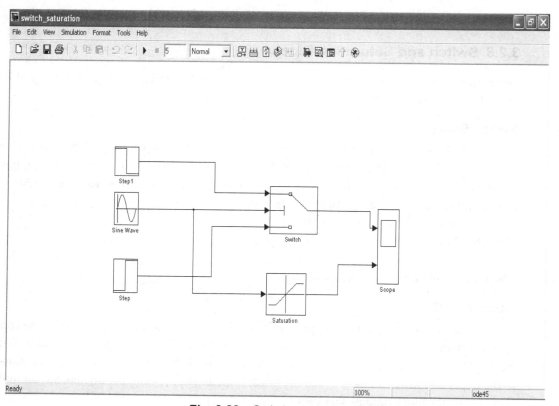

Fig. 3.23 Switch and saturation blocks

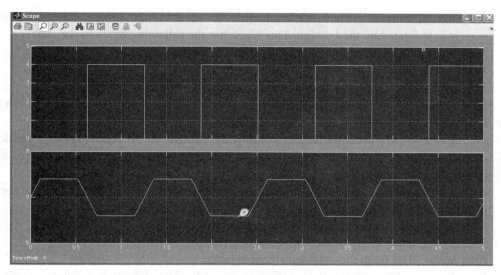

Fig. 3.24   Display on scope of Fig. 3.23

3.3 APPLICATION BLOCK SETS

The application block sets are those blocks which can be used to design a particular real-time system like an electromechanical system.

3.3.1 Power Systems Toolbox

SimPowerSystems toolbox is used to model and design electrical power systems. It is a model design tool that allows researchers and engineers to built models rapidly and easily. It uses the Simulink environment in building a model using simple click and drag procedures. Circuit analyses can be easily done apace and thermal, mechanical, and control systems also can be included in our analyses. As electrical power systems are built up of electrical circuits and electromechanical devices, such as inductors, capacitors, motors, and generators, their analyses is also required so as to constantly improve the overall system. Requirements to rapidly increase the efficiency have forced power system engineers to use power electronic devices and sophisticated control systems in place of conventional ones. Further, the systems become more complicated because of nonlinear devices used. The study of these systems becomes much easier through simulations. The library of SimPowerSystems contains models of typical power equipments such as, thyristor, diode, transformers, machines, and various measurement blocks required to construct an electrical system.

Electrical Sources

It contains various electrical sources like AC current source, AC voltage source, controlled current source, controlled voltage source, DC voltage source, three-phase programmable voltage source, and three-phase source that can be used while constructing electrical circuits.

AC current source—This block implements an ideal current source. The direction of the current is indicated by the '+' sign. This block has the following block parameters:

Peak amplitude—The magnitude of the current in amperes.

Phase—The phase angle of the current in degrees.

Frequency—The frequency current signal in hertz.

Sample time—The sample time period in seconds. The default 0 indicates a continuous source.

Measurements—To measure the current flowing through the AC current source block, select current and place a multimeter block in the model in order to display the selected quantity during the simulation.

AC voltage source—This block implements an ideal voltage source. The polarity of the source is indicated by that sign. This block has the following block parameters:

Peak amplitude—The magnitude of the voltage in volts.

Phase—The phase angle of the voltage in degrees.

Frequency—The frequency voltage signal in hertz.

Sample time—The sample time period in seconds. The default 0 indicates a continuous source.

Measurements—To measure the voltage across the terminals of the AC voltage source block, select voltage and place a multimeter block in the model in order to display the selected quantity during the simulation.

DC voltage source—This block provides an ideal DC voltage source. The positive sign indicates the positive terminal of the source. We can modify the voltage at any time during the simulation. It has the following block parameters:

Amplitude—The amplitude of the source in volts.

Measurements—To measure the voltage across the terminals of the DC voltage source, select voltage, and place a multimeter block in the model. This will display the selected quantity during the simulation.

Three-phase source—This block provides a balanced three-phase voltage source with internal R-L parameters. The three voltage sources are connected in star with neutral wire grounded or made accessible. This block has the following parameters:

Phase-to-phase rms voltage—The phase-to-phase rms voltage in volts.

Phase angle of phase A—The phase angle of the internal voltage generated by phase A, in degrees. The three-phase voltages are generated in positive sequence.

Frequency—The source frequency in hertz.

Internal connection—Any one of the following connections can be selected:

- Y: The three voltage sources are connected in star to an internal floating neutral.
- Y_n: The three voltage sources are connected in star to a neutral connection which is made assessable through a forth terminal.
- Y_g: The three voltage sources are connected in Y to an internally grounded neutral.

Specify impedance using short-circuit level—To specify the internal impedance of the source using the inductive short-circuit level and X/R ratio.

Three-phase short circuit level at base voltage (VA)—The three-phase inductive short-circuit power in volt-ampere (VA), at a specified base voltage, is used to compute the internal inductance L. The internal inductance L in henry is calculated from the inductive three-phase short circuit power P_{sc} in VA, base voltage V_{base} in V_{rms} and source frequency f in hertz as follows:

$$L = (V_{base})^2/P_{sc} \times (1/2\pi f)$$

Base voltage—The phase-to-phase base voltage in volts rms is used to specify the three-phase short-circuit level. The base voltage is usually the normal source voltage. This parameter is only available if *specify impedance using short-circuit level* is selected.

X/R ratio—It is the *X/R* ratio at nominal source frequency of the internal source impedance.

Source resistance—The source internal resistance in ohms.

Source inductance—The internal inductance of the source in henry.

Electrical Elements

Breaker—This block provides a circuit breaker for which the opening and closing times can be controlled either from an external signal (external control mode) or from an internal time (internal control mode), the arc extinction process is simulated by opening the breaker device when the current crosses zero value. When the breaker is closed, it behaves as a resistive circuit. This internal resistance is represented by R_{ON}. A series R-C snubber circuit is also provided in this model. If the Breaker block happens to be in series with the inductive load, an open circuit, or a current source, the snubber circuit must be used.

Breaker resistance R_{ON}—It is the ON time breaker resistance in ohms. Its value cannot be set to 0.

Initial state—It is the initial state of the breaker. If the initial state is set to 1, the breaker is closed and if it is set to 0, the breaker is open.

Snubber resistance R_s—It is the snubber resistance in ohms. To eliminate it, it can be set to inf.

Snubber capacitance C_s—It is the snubber capacitance in farads. To eliminate it, it can be set to 0.

Switching times—Specify the vector of switching times when the circuit breaker is in internal control mode. At such switching time, the breaker opens or closes depending on its initial state. For instance, if the breaker is closed to the first switching time, it opens at the second and continuous to operate like this further.

External control of switching times—If the option is selected, it adds an input port to the block for external control. The switching time is defined by the control signal converted to the port.

Measurements—Select the quantity to be measured for the block.

Ground—It implements a ground connection for power circuits. It is different from the ground described for signals (low power) previously.

Linear transformer—This block provides a two-winding or three-winding single-phase transformer. This model takes into account the winding resistance $R1$, $R2$, and $R3$ and the characteristics of the core, which is the modeled by a linear (Rm, Lm) branch. This block has the following parameters:

Nominal power and frequency—It is the nominal power rating in VA and frequency in Hz of the transformer.

Winding 1 *parameters*—These are the parameters for the first winding. The nominal rms voltage in volts, resistance and leakage inductance in per unit.

Winding 2 *parameters*—It provides winding 2 parameters, viz. nominal rms voltage in volts, resistance and leakage inductance in per unit.

Three windings transformer—It is selected if a three-winding transformer is required.

Winding 3 *parameters*—The nominal rms voltage in volts, resistance and leakage inductance in per unit.

Magnetization resistance and reactance—The value of resistance and inductance in per unit for simulating the core and reactive losses.

Measurements—Select the desired quantity to be measured.

Series RLC branch—This block implements a series RLC branch. It can be a single resistance or inductance or capacitance or any combination of these. To eliminate the resistance set it zero, to eliminate the capacitance set it to inf, and to eliminate the inductance set it zero. Its dialog box has the following parameters:

Resistance—The branch resistance in ohms.

Inductance—The branch inductance in henry.

Capacitance—The branch capacitance in farads.

Measurements—Select the quantity required to be measured through multimeter as accordingly.

Series RLC load—This block provides us a series RLC load. At a particular frequency, the load exhibits a constant impedance. This block has the following parameters:

Nominal voltage V_n—It is the nominal voltage of the load in rms.

Nominal frequency F_n—It is the nominal frequency in hertz.

Active power P—It is the active power of the load in watts.

Inductive reactive power Q_L—It is the inductive reactive power in VARs.

Capacitive power Q_C—It is the capacitive reactive power in VARs.

Measurements—Select the required quantity to be measured.

Three-phase transformer (Three Windings)—This block provides us three-phase transformer of three windings. These three windings can be connected in the following manner according to the requirement:

Y star connected

Y with accessible neutral (for winding 1 and 3)

Grounded Y

Delta D1, delta lagging *Y* by 30 degrees

Delta D11, delta leading *Y* by 30 degrees

The dialog box of this block contains the following parameters:

Nominal power and frequency—Nominal power rating in VA and nominal frequency in hertz of the transformer.

Winding 1(ABC) connection—The winding connections of the first winding.

Winding parameters—Phase rms voltage in volts, resistance and leakage inductance of winding 1 in per unit.

Winding 2(abc2) connection—The connection of second winding.

Winding parameters—Phase rms voltage in volts, resistance and leakage inductance of winding 2 in per unit.

Winding 3(abc3) connection—The connection of third winding.

Winding parameters—Phase rms voltage in volts, resistance and leakage inductance of winding 3 in per unit.

Saturable Core—This block implements a saturable core of the transformer.

Magnetization resistance R_m—Value of R_m in per unit.

Magnetization inductance L_m—Value of L_m in per unit.

Saturation characteristic—It is the saturation characteristics of the core. Specifies a series of current/flux in per unit starting with pair (0, 0).

Simulate hysteresis—To model a saturation characteristics including hysteresis instead of a single-valued saturation curved.

Hysteresis data MAT file—Specifies a MAT file containing the data to be used for the hysteresis model.

Specify initial fluxes—Define initial fluxes by the [phioA phioB phioC] parameter. It specifies the fluxes for each phase of the transformer.

Measurements

Current Measurement—This block is used to measure the instantaneous current flowing through any electrical block or connection line. The output signal parameter of this block is disabled when the block is not used in a phase or simulation.

Impedance Measurement—This block is used to measure the impedance between two modes of a linear circuit as a function of frequency and is displayed by using the impedance versus frequency measurement tool of the Powergui block. This measurement takes into account the initial states of the Breaker and Ideal Switch blocks. The multiplication factor parameter is used to rescale the measured impedance. The default value is 1.

Multimeter—This block is used to measure voltage and currents of the measurements described by the dialog boxes. If a SimPowerSystems block chooses voltage or current measurements through the measurements option available in that block, it is equivalent to connecting an internal voltage or current measurement block inside that block. The measured signals can be seen through the Multimeter block placed in the model. This block has the following parameters:

Available measurements—The available measurements list box shows the measurements in the Multimeter block. The measurements in the list box are identified by the label preceding the block name.

Selected measurements—The selected measurement list box shows the measurements sent to the output of the block. There the measurements can be reordered by using the up, down, and remove buttons.

Plot selected measurements—It is selected to display the plot of selected measurements using a MATLAB figure window.

Output type—Specifies the format of the output signal when the block is used in a phase simulation to measure the complex values. Similarly, one can choose Real-Img, Magnitude-Angle, and Magnitude.

Voltage measurement—The voltage measurement block measures the instantaneous voltage between two electrical nodes. The output of this block is a signal (low voltage) that can be used by other Simulink blocks like Scope. The output signal parameter of this block specifies the format of the output signal if the block is used in a phase or simulation.

Power Electronics

Diode—This block provides us a diode. This block also contains a series $R_s C_s$ snubber circuit that can be connected in parallel with the diode device. This block has the following parameters:

Resistance R_{ON}—It is the diode internal ON time resistance measured in ohms.

Inductance L_{ON}—It is the diode ON time inductance in henry.

Forward voltage V_f—The forward biased voltage of diode in volts.

Initial current I_c—It specifies the initial current through the diode at the start of the simulation.

Snubber resistance R_s—It is the diode snubber circuit resistance in ohms.

Snubber capacitance C_s—It is the diode snubber circuit capacitance in farads.

Show measurement port—If selected, it adds a Simulink output to the block returning the diode current and voltage.

Ideal switch—This block implements an ideal switch which is fully controlled by a gate signal (g). The switch is ON if the gate signal is present otherwise it is OFF. It has the following block parameters:

Internal resistance R_{ON}—It has the internal resistance of the switch in ohms.

Initial state—It is the initial state of the switch.

Snubber resistance R_s—Switch snubber circuit resistance in ohms.

Snubber capacitance C_s—Switch snubber circuit capacitance in farads.

Show measurement port—If selected, adds a Simulink output to the block returning the ideal switch current and voltage.

MOSFET—This block provides us a model of power MOSFET. It also contains a series $R_s C_s$ snubber circuit that can be connected in parallel with the MOSFET. It has the following block parameters available:

Resistance R_{ON}—It is the ON time resistance of the MOSFET.

Inductance L_{ON}—Internal inductance of the MOSFET.

Internal diode resistance R_d—Resistance of the internal diode in ohms.

Internal current I_c—Initial current in the MOSFET at the start of the simulation in amperes.

Snubber resistance R_s—Snubber circuit resistance in ohms.

Snubber capacitance C_s—Snubber circuit capacitance in farads.

Show measurement port—If selected, adds Simulink output to the block returning the MOSFET current and voltage.

Thyristor—This block provides a thyristor or SCR. The Thyristor is a controlled switch which can be controlled by a gate signal (low power). It also contains a snubber circuit which can be connected in parallel with the SCR. The thyristor is turned ON when the anode is more positive than cathode and a positive signal is applied at the gate terminal. It is turned OFF when the current flowing through it becomes zero or negative and negative voltage appears across the anode. The dialog box parameters for this block are as follows:

Resistance R_{ON}—It is the internal ON time resistance of the thyristor in ohms.

Inductance L_{on}—It is the internal inductance of the thyristor in henry.

Forward voltage V_f—It is the forward voltage of the thyristor in volts.

Initial current I_c—It is the initial current in the thyristor and its value corresponding to the initial state of the circuit can be specified.

Snubber resistance R_s—Resistance of the snubber circuit in ohms.

Snubber capacitance C_s—Capacitance of the snubber circuit in farads.

Show measurement port—If selected, it adds a Simulink output to the block returning the thyristor current and voltage.

Latching current 11—Latching current of the detailed thyristor model in amperes.

Turnoff time T_q—The turnoff time of the detailed thyristor model in seconds.

Inputs and Outputs

g—gate signal of the thyristor.

m—it is output containing two signals, i.e., thyristor current (A) and thyristor voltage (V). This port can be connected to a Terminator block if not utilized otherwise the model will show warning message while running the simulation.

Universal bridge—This block provides a universal three-phase power converter, or say a three-phase bridge rectifier, that consists up to six power switches. The type of switch and converter configuration can be selected from the dialog box. The following are the parameters for this block:

Number of bridge arms—For value 1 or 2, it provides a single-phase rectifier/converter using two or four switches. For value 3, it provides a three-phase bridge converter.

Snubber resistance R_s—Resistance of the snubber circuit in ohms. To eliminate it, put its value as inf.

Snubber capacitance C_s—Capacitance of the snubber circuit in farads. To eliminate it, put its value as 0.

Power electronic device—Select the power electronic device for building the converter. The devices available are diodes, thyristors, GTO, MOSFET, IGBT, and ideal switches. Depending on the device selected, fill its parameters.

Measurements—Select the appropriate option for the measurements.

Inputs—The gate control signal is required in case of controlled switches. If diode is selected, this port disappears. A, B, and C are the three phases of the input supply.

Machines

DC machines—This block provides a separately excited DC machine. An access is provided to the field terminals (F+, F−) so that the machine model can be used as shunt or series machines. This block has the following parameters:

Preset model—Select this model only if the electrical and mechanical properties, like ratings of power (HP), DC voltage (V), rated speed (rpm), field voltage (V), of the machine are to be loaded; otherwise select No.

Show detailed parameters—It displays the parameters of the DC machine in detail.

Armature resistance and inductance $[R_a \, L_a]$—Resistance of the armature winding in ohms and its inductance in henry.

Field resistance and inductance $[R_f \, L_f]$—Resistance and inductance of the field winding in ohms and henry respectively.

Field armature mutual inductance L_{af}—Mutual inductance between the armature and field winding in henry.

Total inertia J—Inertia of the DC machine in kg m^2.

Viscous friction coefficient B_m—Friction coefficient of the machine in N m s.

Coulomb friction torque T_f—It is the coulomb friction torque coefficient of the machine in N m.

Initial speed—Initial speed of the machine at the starting of the simulation in rad/s.

Output

m—It is a vector containing four signals, viz. speed in rad/s, armature current in amperes, field current in amperes and electrical torque in N m of the machine. If the output of this port is not desired, close it by using a terminator block.

Asynchronous machine—This block provides a three-phase asynchronous machine, which is also known as induction machine. This model can be operated as a motor or a generator depending on the mechanical torque of the machine. If the torque T_m is positive, it is a motor and if the torque T_m is negative it is a generator. In this model, all variables and parameters are referred to the stator only. This model has the following parameters:

Preset model—The present model, if selected, provides a set of predetermined electrical and mechanical parameters for various asynchronous machines ratings of power (HP), phase-to-phase voltage (V), frequency (Hz), and rated speed (rpm). The show detailed parameters can be selected to view and edit the detailed parameters associated with the model. These parameters are rotor type; reference frame; nominal power; line-to-line voltage (L–L volt); frequency, stator resistance, and inductance (per unit); rotor resistance and inductance (per unit); mutual inductance (per unit); inertia constant; friction factor; and pairs of poles and initial conditions of the machine.

Inputs and Outputs

T_m—It is a Simulink input of the block and is the mechanical torque at the machine's shaft.

m—This is a Simulink output and is a vector containing 21 signals.

Permanent Magnet Synchronous Machine—This block provides a model of a permanent magnet synchronous machine whose operation can be decided by the sign of the mechanical torque. This model assumes that the flux established by the permanent magnets in the stator is sinusoidal, which in turn implies that the emf is also sinusoidal. In this block, the parameters available in the dialog box are *present model* and *show detailed parameters*. The *present model* provides a set of predetermined electrical and mechanical parameters for various permanent magnet synchronous motor ratings of torque (N m), DC bus voltage (V), rated speed (rpm), and continuous stall torque (N m). One of the present models can be selected to load the corresponding electrical and mechanical parameters in the entries of the dialog box. Also select *show detailed parameters* in order to view and edit the detailed parameters associated with the present model. The detailed parameters displayed are stator resistance (Ω), stator inductances (H), flux induced by magnets (Wb), inertia (kg m^2), friction (N m s), and pairs of poles. T_m is the input torque at the machine shaft and m is the Simulink output of the block vector containing 10 signals.

3.3.2 Sim Mechanics

The Sim Mechanics is a block diagram modeling environment for the design and simulation of rigid body machines and their motors by the Newtonian dynamics of forces and torques. The library can be used to model and simulate the mechanical systems with a variety of tools to specify bodies and simulate the mechanical systems with a variety of tools to specify the bodies and their mass properties, their possible motions, kinematics constraints, and coordinate systems, and to initiate and measure body motions. The mechanical system can be represented by a connected block diagram and like other Simulink models, hierarchical subsystems can be incorporated. The following blocks of this library are discussed in this particular section:

Body

The Body block represents a rigid body whose properties can be customized. The parameters to be set broadly includes the mass and moment of inertia tensor, the coordinates for the body's center of gravity and one or more body coordinate systems. A rigid body is defined in space by the position of its center of gravity or centre of mass and its orientation in some coordinate system. Initial conditions of the body can also be set in this block.

Ground

The Ground block represents an immobile ground point at rest in the absolute inertia world reference frame. It prevents one side of the connected side of any joint from moving. It can also be used in a machine environment block. The multiple ground blocks in a model represent different fixed points in the global inertial world.

Body Spring and Damper

The Body Spring and Damper block models a force of damped spring acting on two bodies. The Force Element block can be used to model any linear (Hooke's law) force, with constant coefficient that acts between a pair of bodies. One of the bodies can be ground. The parameters to be set for this block are spring constant (K), damper constant (B), and spring natural length (r_0).

Body Actuator

The Body Actuator block actuates a Body block with a generalized force signal, representing a force or torque applied on the body. The generalized force is a function of time specified by a Simulink input signal. The body actuator applies the actuation signal in the reference coordinate systems specified in the block dialog box. The import is the Simulink input signal. The output is the connected port which is connected to the body block that has to be actuated.

Body Sensor

The Body Sensor block detects the position, velocity, and/or acceleration of a body represented by a Body block. It measures the motion in the reference coordinate system specified in the block dialog box sensed. The output is a set of Simulink signals or one bundled Simulink signal of the position, velocity, and/or acceleration vector(s) and/or rotation matrix of the body.

Variable Mass and Inertia Actuator

The Variable Mass and Inertia Actuator block allows varying the mass and/or inertia tensor of the body to which it is connected. It externally varies the mass or inertia parameter with a Simulink signal.

Mechanical Branching Bar

The Mechanical Branching Bar block bundles multiple actuator and sensor connection lines into one line, allowing to connect the multiple actuators and/or sensor to a single connector port on a joint, constraint, or driver, or to a single body coordinate system port on a body. Any number of sensor/actuator ports can be chosen on the mechanical branching bar. Using this block, a Joint block can be connected to any combination of Joint Sensors, Joint Actuators, Joint Initial Condition Actuators, and Joint Stiction Actuators. The actuators and sensor dialogs displays the joint's primitives as if they were directly connected to the joint.

3.4 USER-DEFINED FUNCTIONS

This section introduces the Users Defined Functions library. It contains Mathematical Functions, MATLAB Function, Embedded MATLAB F-function, S-Function, Level 2 M-File S-Function, S-function Building, and S-Function Examples blocks. We describe the function of each block of this library.

3.4.1 F_{cn} Block

The F_{cn} block applies a specified expression to its input denoted as u. If **u** is a vector, **u**(i) represents the ith element of the vector and V (+) or **u** alone will represent the first element. The expression defined can contain numeric constants, arithmetic operators, relational operators, logical operators, and the math functions.

3.4.2 MATLAB F$_{cn}$ Block

The MATLAB F$_{cn}$ block applies the specified MATLAB function or expression to the input. This block is slower than the F$_{cn}$ block because it calls the MATLAB function during such integration step. As an alternative approach, built-in blocks such as the F$_{cn}$ block or the Math Function block, or writing the function as an M-file S-Function and then accessing it using the S-Function block.

3.4.3 Embedded MATLAB Function Block

The Embedded MATLAB Function block contains a MATLAB language function in a Simulink model. This block accepts multiple input signals and produces multiple output signals.

3.4.4 S-Function Block

The S-Function block provides an access to S-Functions. The S-Function named as the S-Function name parameter can be a level-1 M-File or a level-1 or level-2 C MEX-File S-Function. The M-file S-Function block should be used to include a level-2 m-File S-Function in a block diagram. The S-Function can be written in MATLAB, C, C++, Ada, or FORTRAN. S-Functions are complied as Mex-File using the mex utility. The function uses a special calling syntax that enables interaction with Simulink's equation solvers. The form of S-Function is very general and supplies to continuous, discrete, and hybrid systems. With M-Files S-Function, one can define their own ordinary differential equations and any type of algorithm that can be used with Simulink block diagrams.

3.4.5 Level-2 M-File S-Function Block

The Level-2 M-File S-Function block allows using a level-2 M-File S-Function in a model. This can be done by creating an instance of this block in the model. Then, the name of the level-2 M-File S-Function is entered in the M-File name field of the block's parameter dialog box. For a level-1 M-File S-Function, the S-Function block is used.

3.4.6 S-Function Builder Block

The S-Function Builder block creates C-MEX-File S-Function from specifications and C source code that we provide. As C-MEX Files are not discussed in this book, interested readers can refer to Simulink help for any further information regarding these files.

3.4.7 S-Function Examples Block

The S-Function Examples block displays M-File S-Function, C-File S-Function, C++ S-Function, Ada S-Function, and Fortran S-Function examples.

3.5 SIMULATION PROJECTS

Project 1

To simulate a single-phase universal bridge for different types of switches The Simulink model of single-phase Universal Bridge rectifier feeding a 20 Ω resistive load is shown in Fig. 3.25. The AC voltage source block is picked from the SimPowerSystems library. The peak amplitude (V) of this block is set at 200 and frequency (in Hz) at 50. The phase and sample time are set 0. In measurements voltage is selected. Pulse Generator block is selected from Sources block set in order to supply firing pulses to the controlled power electronic switches of the Universal Bridge. For Pulse Generator and Pulse Generator1 Blocks, the following parameters to be set: Pulse type—time based, time (t)—use simulation time, amplitude—5, period—0.01 s, pulse width—50%, and phase delay (s)—0. For Pulse Generator2 and Pulse Generator3 blocks, phase delay is taken as 0.01 s. These pulses are fed to the Universal Bridge gate terminal (upper most) through a 2 input Mux blocks as shown in Fig. 3.25. The AC Voltage Source block is connected to the A and B terminals of the Universal Bridge block. The following parameters are selected for Universal Bridge block: Number of bridge arms—2 (single-phase full converter), snubber resistance—1e5, snubber capacitance—inf, power electronic device—diodes, thyristors or IGBT/Diodes, R_{ON} = 1e03, L_{ON} = 0, V_f = 0, and measurements—device currents. The output of the Universal bridge is connected to the parallel RLC branch for which the following parameters are selected: resistance—20, inductance—inf, capacitance—0, and measurements—branch voltage and current. Multimeter block is placed in the model for measuring all the selected electrical parameters. The available and selected measurements for this block are as follows: V_b—Parallel RLC Branch, V_{src}—AC Voltage Source, I_b—Parallel RLC Branch, Isw1—Universal Bridge, Isw2—Universal Bridge, Isw3—Universal Bridge, and Isw4—Universal Bridge.

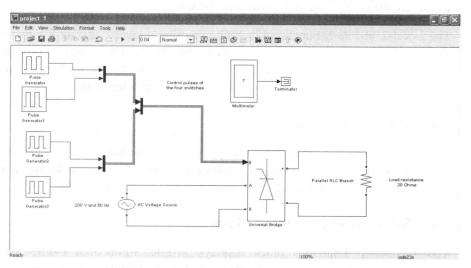

Fig. 3.25 Simulation model for a single-phase universal bridge

Also select Plot selected measurements for plotting these parameters in the figure window as shown in Figs 3.26 through 3.28. In Configuration Parameters start time and stop time is taken from 0 to 0.04 s and the solver selected is variable step ode23s (stiff/Mod. Rosenbrock). This model is simulated for diodes, thyristors, and IGBTs, and the output waveforms obtained are shown in Fig. 3.26, Fig. 3.27, and Fig. 3.28 respectively.

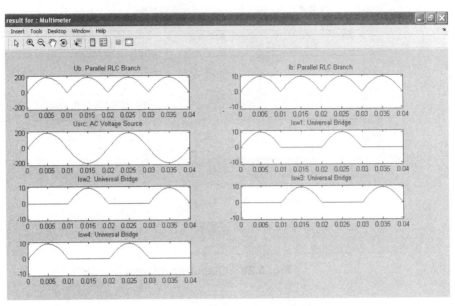

Fig. 3.26 Output waveforms for diodes

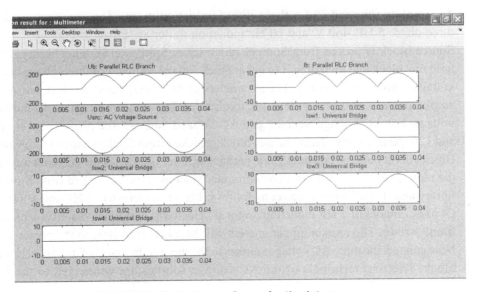

Fig. 3.27 Output waveforms for thyristors

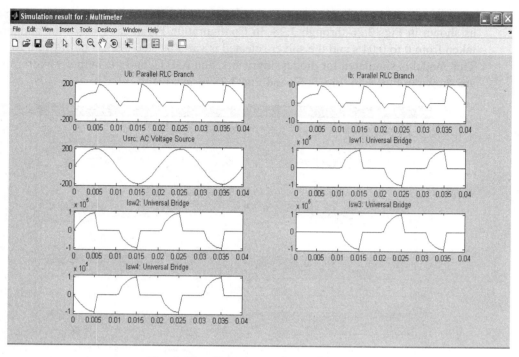

Fig. 3.28   Output waveforms for IGBTs

Project 2

To simulate the model of a three-phase induction motor This Simulink model simulates a three-phase 5 HP, 460 V, and 60 Hz induction motor as shown in Fig. 3.29. The AC Voltage Source block is picked from Sources block set in SimPowerSystems library. Right click and hold this block and then drag it to get another voltage source. For three-phase system, three AC voltage sources are required. The parameters of AC Voltage Source block are as follows: Peak amplitude (V)—460, phase (deg)—0, frequency (Hz)—60, sample time—0, and measurements: voltage. For AC Voltage Source1 and AC Voltage Source2, the phase (deg) is taken as −120° and 120° respectively. The Constant block is picked from Sources block set of Commonly Used Blocks library and its constant value parameter is set 14. This is the constant torque supplied by the motor, i.e., 14 N m. Now, Asynchronous Machine block is picked from SimPowerSystems library and its parameters are set as follows: Preset model—01: 5 HP 460 V 60 Hz 1,750 rpm (A preset model of 5 HP induction motor is selected), rotor type—squirrel-cage, and reference frame—rotor. The remaining parameters will be taken according to the preset model selected. Now connect the three AC voltage sources to the input A, B, and C of the Asynchronous Machine block. The first terminals of the AC voltage sources are connected to the Ground block

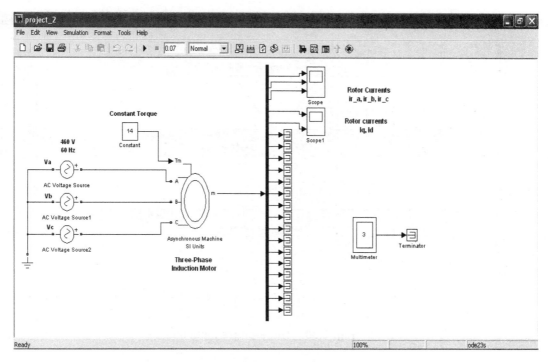

Fig. 3.29 Simulation model of a three-phase induction motor

as shown in Fig. 3.29. The Constant block is connected to the T_m terminal of the machine. The available and selected measurements for Multimeter block are as follows: Vsrc—AC Voltage Source, Vsrc—AC Voltage Source1, and Vsrc—AC Voltage Source2 and select the plot selected measurements. The output port of the Multimeter is connected to the Terminator block taken from Sinks block set. The output signal m of the machine is demultiplexed by using a 21 input Demux block as shown in Fig. 3.29. The 21 output signals in sequence are rotor current ir_a, rotor current ir_b, rotor current ir_c, rotor current iq, rotor current id, rotor flux phir_q, rotor flux phir_d, rotor voltage Vr_q, rotor voltage Vr_d, stator current is_a, stator current is_b, stator current is_c, stator current is_q, stator current is_d, stator flux phis_q, stator flux phis_d, stator voltage Vs_q, stator voltage Vs_d, rotor speed, electromagnetic torque, and rotor angle. From these signals, only first five are connected to the Scope blocks and the rest are connected to the Terminator blocks. To view the other signals, the Scope block can be connected to them. In both the Scope block the sample time is taken as 10e–6 in the parameters section. The output plotted by the Multimeter block is shown in Fig. 3.30. The rotor currents of all the three phases as obtained from Scope block can be seen in Fig. 3.31. The Scope1 block shows the rotor currents i_q and i_d. This model is simulated from 0 to 0.07 s with the help of variable-step ode23s (Stiff/Mod. Rosenbrock) solver. This model can be tried with different preset models available in the Asynchronous Machine blocks and view the results obtained.

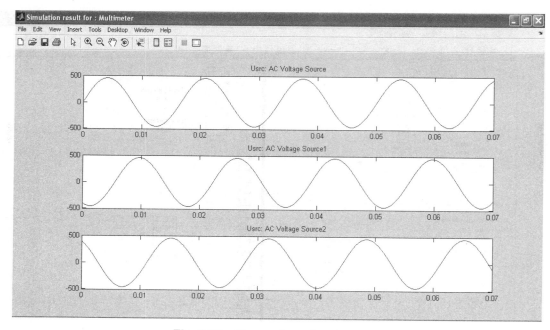

Fig. 3.30 Three-phase input source

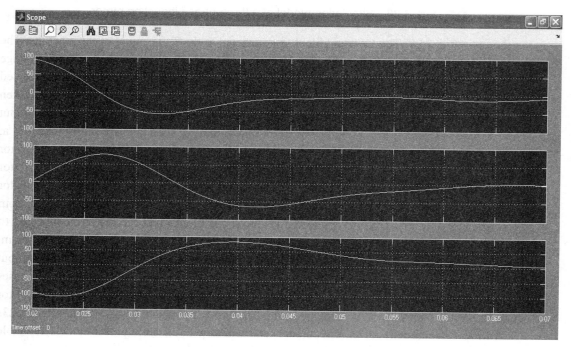

Fig. 3.31 Rotors currents ir_a, ir_b, and ir_c

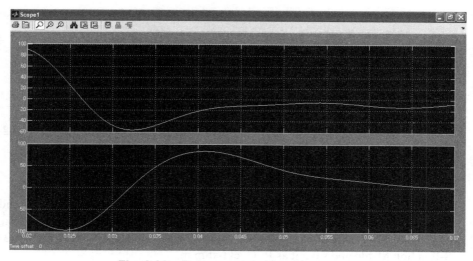

Fig. 3.32   Rotor currents i_q and i_d

Project 3

To plot the input and output states of series R-C circuit by using S-function block
For a simple R-C circuit shown in Fig. 3.33, the state space equations are as follows:

$$dx/dt = -(1/RC)x + V_s$$
$$y = x$$

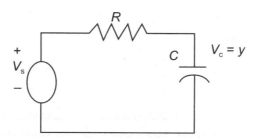

Fig. 3.33   A simple R-C circuit

In this project, we create a S-function block that will implement these relations (state space equations). First, write a M-file as follows and save it as R_C_ckt.m.

```
function dx = R_C_ckt(t,x,Vs)
%%%%%%%%%%%%%%%%%%%%%%%%%%%%%%%%%%%%%%%%%%%%%%%%%%%%%%%%%%%%%%%%%%%%%%%
% Model for RC series circuit, function m-file R_C_ckt.m %
%%%%%%%%%%%%%%%%%%%%%%%%%%%%%%%%%%%%%%%%%%%%%%%%%%%%%%%%%%%%%%%%%%%%%%%

%%%%%%%%%%%%%%%%%%%%%%%%%%%%%%%%%%%%%%%%%%%%%%%%%%%%%%%%%%%%%%%%%%%%%%%
%                   R - C circuit constants                        %
%%%%%%%%%%%%%%%%%%%%%%%%%%%%%%%%%%%%%%%%%%%%%%%%%%%%%%%%%%%%%%%%%%%%%%%
```

```
R = 10^6; % Circuit Resistance in Ohms
C = 10^(-6); % Circuit  Capacitance in Farads
dx = -1/(R*C)*x+Vs; % The arguments x and dx are column vectors
                    % for state and derivative respectively. The
                    % variable t on the first line above specifies
                    % the simulation time. The default is [0 10].
```

On can test the correctness of this function M-file by executing the following command on the command window:

```
>> [t,x,Vs]=ode45(@R_C_ckt, [0 10], 0, [ ], 1)
```

The above command specifies a simulation time interval [0 10], an initial condition value of 0, the null vector [ ] can be used for options, and the input value is set to 1. On execution of this command, MATLAB displays several values for t, x, and Vs. Now we write the following S-function M-file and save it as R_C_ckt-sfun.m.

```
function [sys,x0,str,ts]=...
R_C_ckt_sfcn(t,x,u,flag,xinit)

%%%%%%%%%%%%%%%%%%%%%%%%%%%%%%%%%%%%%%%%%%%%%%%
% This is the m-file S-Function RCckt_sfcn.m     %
%%%%%%%%%%%%%%%%%%%%%%%%%%%%%%%%%%%%%%%%%%%%%%%

xinit = 0;
switch flag
case 0 % Initialize
str = [];
ts = [0 0];
x0 = xinit;


%%%%%%%%%%%%%%%%%%%%%%%%%%%%%%%%%%%%%%%%%%%%%%%%%%%%%%%%%%%%%%%%%
% Alternately, the three lines above can be combined into a single line as
% [sys,x0,str,ts]=mdlInitializeSizes(t,x,u)
%%%%%%%%%%%%%%%%%%%%%%%%%%%%%%%%%%%%%%%%%%%%%%%%%%%%%%%%%%%%%%%%%

sizes = simsizes;
sizes.NumContStates = 1;
sizes.NumDiscStates = 0;
sizes.NumOutputs = 1;
sizes.NumInputs = 1;
sizes.DirFeedthrough = 0;
sizes.NumSampleTimes = 1;
sys =simsizes(sizes);
```

```
case 1 % Derivatives
Vs = u;
sys = R_C_ckt(t,x,Vs);
case 3 % Output
sys = x;
case {2 4 9} % 2:discrete
% 3:calcTimeHit
% 9:termination
sys = [];
otherwise
error(['unhandled flag =',num2str(flag)]);
end
```

Now create a Simulink model for this S-Function file (R_C_ckt.m) as shown in Fig. 3.34. Select 'S-Function' block from the 'User-Defined Functions' blockset and drag it into the model. In the S-Function block parameters dialog box, we assign the S-Function name as R_C_ckt_sfcn. Also add Step, Mux, and Scope blocks to the model as shown in Fig. 3.34. The input and output waveforms of this model as seen on the Scope are shown in Fig. 3.35.

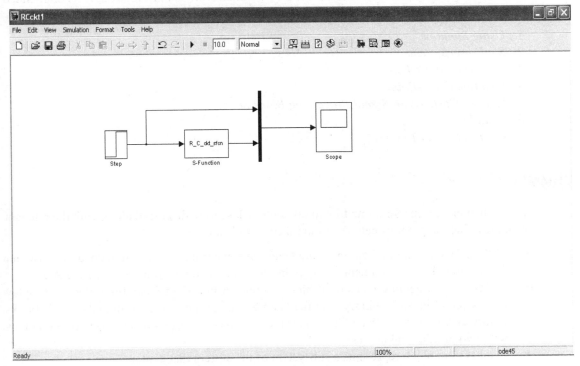

Fig. 3.34 Simulink S-Function model for function R_C_ckt_sfcn

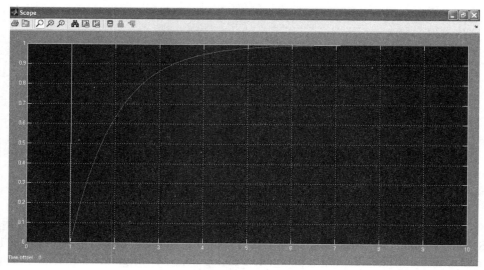

Fig. 3.35 Input and output waveforms of the R-C circuit model

The following publications by the Mathworks (www.mathworks.com) are strongly recommended for further study:

1. *Using Simulink*
2. *SimPowerSystems*
3. *Real-Time Workshop*
4. *Simulink Fixed-Point*
5. *Getting Started with Signal Processing Block Set*
6. *State Flow*
7. *Control System Toolbox*

SUMMARY

In this chapter, various Simulink toolboxes and block sets are discussed along with the relevant examples. The main points can be summarized as follows:

- MATLAB and Simulink are integrated and, thus, we can analyze, simulate, and revise our models in either environment at any point. Simulink is invoked from the MATLAB.
- We can use Algebraic Constraint blocks found in the Math Operations library, Display blocks found in Sinks library, and the Gain blocks found in the Commonly Used Blocks library to draw a model that will produce the simultaneous solution of two or more equations with two or more unknowns.
- A subsystem block represents a subsystem of the system that contains it.
- The Terminator block can be used to cap blocks whose output ports are not connected to the other blocks. Using Terminator block to cap those ports avoids warning messages.

- The Mux block combines its inputs into a single output while a Demux block extracts the components of an input signal and outputs the components as separate signals.
- The Logical Operator block performs the specified logical operation on its inputs. The supported operations are AND, OR, NAND, NOR, XOR, and NOT.
- The Data Type Conversion block converts an input signal of any Simulink data type to the data type and scaling specified by the block's output data type mode, output data type, and/or output scaling parameters.
- An S-Function is a computer language description of a Simulink block. S-Functions can be written in MATLAB, C, C++, Ada, or Fortran. Files in C, C++, Ada, and Fortran S-Functions are compiled as Mex-Files using the mex utility.
- Power electronic switches like thyristor, IGBT, MOSFET, GTO available in SimPower Systems library can be used for building power electronic converters.
- Various types of DC and AC machines available in block sets of machines can be utilized for building models utilizing these machines.
- Voltage, current, impedance, and multimeter can be used for measuring electrical quantities. In case of a multimeter, no connections are required.
- The SimMechanics library is useful for modeling electromechanical systems.

REVIEW QUESTIONS

1. Mention the specialties of Simulink as a programming tool.
2. What is Simulink library?
3. What is the extension of a Simulink file?
4. How is the simulation time of a model specified?
5. Define various configuration parameters of a Simulink model.
6. Mention the various sources available and state their application.
7. Mention the various sinks or display devices available.
8. What are the uses of In-port and Out-port blocks?
9. Define subsystems. Why are subsystems created in a model?
10. Mention the logical and relational operators available in a Simulink library.
11. Mention the applications of Data Type Conversion block.
12. Mention the power electronic switches present in Simulink.
13. What are the advantages of a Multimeter block? What electrical quantities can be measured by a multimeter?
14. Classify different types of DC machines and state their characteristics.
15. Describe the operation of a practical thyristor.
16. Explain the principle of operation of a synchronous and induction machines.
17. Mention the applications of user-defined functions.

SIMULATION EXERCISES

1. A unit step signal of magnitude 10 is applied to a series R-L circuit at $t = 1$ ms. Use Simulink with the help of the Step Function block, the Continuous-Time Transfer Function block to simulate and display the output waveform V_L of the R-L circuit. Take the initial conditions to be $i_L(0) = 0$ ($R = 10\ \Omega$, $L = 10$ mH).

2. A unit step signal of 100 V is applied to a series RLC circuit at $t = 0$ s. The circuit parameters are $R = 5\ \Omega$, $L = 5$ mH, and $C = 2\ \mu F$. Construct a Simulink model of this circuit by using Continuous-Time Transfer Function and the other blocks required and plot the output waveforms V_C, V_L, and i_R. Take the initial conditions to be $i_L = 0$ and $V_C(0) = 0$.

3. A series R-C circuit has a resistance of 50 Ω and a capacitance of 1,000 μF. The initial charge on the capacitor is zero. A sine wave signal of magnitude 100 and frequency 100 Hz is applied to this circuit at $t = 0.01$ rms. Plot the output waveform V_C on the Scope.

4. A constant current applied to a capacitor produces a linear voltage at its terminals, i.e., $V_C = (I/C)t$, where V_C is in volts, I in amperes, C in farads, and t in seconds. Using Variable Transport Delay block create a model to display the output if $I = 2$ mA, $C = 1,000\ \mu F$, and the voltage across the capacitor at some time is 2.5 V.

5. Create a model to simulate the Boolean expression $D' = A(B + C)' + A \cdot B'$. Display the values of the output variables D' and D for all combinations of the variables A, B, and C.

6. Create a model to display a three-phase power system on a single Scope block where the waveforms of the three phases are 120° apart.

7. Create a model to display a triangular wave with unity amplitude clipped at points 0.9 and –0.6.

8. Develop a three-phase full wave bridge rectifier by using thyristors. The output of this rectifier is connected to a 100 Ω resistance. Plot the source voltages, load voltage, and thyristor currents by using a multimeter.

SUGGESTED READING

Baliga, B.J., *Power Semiconductor Devices*, PWS Publishing Company, Boston, 1995.

Karris, Steven T., *Numerical Analysis Using MATLAB and Spreadsheets*, Orchard Publication, California, 2001.

Karris, Steven T., *Circuit Analysis I with MATLAB Applications*, Orchard Publication, California, 2001.

Krein, P.T., *Elements of Power System Electronics*, Oxford University Press, New York, 2009.

Nagsarkar, T.K. and M.S. Sukhija, *Power System Analysis*, Oxford University Press, New Delhi, 2009.

4

BASIC ELECTRICAL ENGINEERING APPLICATIONS

4.1 INTRODUCTION

This chapter discusses and clarifies the fundamental concepts of electrical engineering using the MATLAB approach. In this chapter, a number of analytical problems are illustrated using examples. These problems are solved by the MATLAB programs or models so as to provide the practical insight. After going through the chapter, we will be able to apply MATLAB programming to understand the basics of electrical engineering more efficiently.

4.2 ELEMENTARY DEFINITIONS

There are some elementary terms and definitions which should be known so as to understand the concept of electrical engineering. Table 4.1 illustrates some prefixes for the SI (International Standards) units that are mostly used in electrical engineering.

Table 4.1 Commonly used prefixes for the SI units

Prefix	Symbol	Power of 10	Prefix	Symbol	Power of 10
Exa	E	10^{18}	Deci	d	10^{-1}
Peta	P	10^{15}	Centi	c	10^{-2}
Tera	T	10^{12}	Milli	m	10^{-3}
Giga	G	10^{9}	Micro	μ	10^{-6}
Mega	M	10^{6}	Nano	n	10^{-9}
Kilo	k	10^{3}	Pico	p	10^{-12}
Hecto	h	10^{2}	Femto	f	10^{-15}
Deka	da	10^{1}	Atto	a	10^{-18}

Following are the definitions of the key terms in electrical engineering in the context of the matter presented in this chapter:

Electric charge The electric charge is an indication of the quantity of the electricity. It is denoted by Q or q and its unit is coulombs denoted using C. The charge associated with an electron is -1.6×10^{-19} C.

Electrostatic force It is the force exerted by one charged body on another charged body. This force can be estimated by the following equation:

$$F = (k\, q_1\, q_2/r^2)\, \mathbf{r}_{12}$$

where

F is the force between the two charged bodies in N
r is the distance between the two bodies in m
k is the proportionality constant = 8.99×10^9 N m^2/C^2
$\mathbf{r}_{12}$ is the unit vector pointing from q_1 to q_2.

Electric field strength An electric field is a domain or an area in which an electric charge will experience a force. An electric field strength at a point is defined as the force per unit positive charge. In other words, the electric field strength at any point is the force in magnitude and direction, which would act on a unit positive charge. The unit of electric field is N/C or V/m. It is given by the following expression:

$$E = \frac{F}{q} \quad \text{V/m}$$

Electric current Electric current is defined as the time rate of net motion of charges along a definite path across a cross-sectional boundary. In the case of a conductor, the moving charges are electrons. It is denoted by i or I and its unit is C/s or A.

$$\text{Electric current, } i = \text{Rate of transfer of electric charge}$$
$$= \frac{dq}{dt} \quad \text{C/s or A}$$

Electromotive force and potential difference Electromotive force (EMF) is the force that causes electric current to flow in the circuit. It is the open-circuited voltage of the supply terminals or battery. Potential difference (PD) is defined as the electric pressure available to cause the flow of electrons, when the circuit is closed. In other words, EMF is the voltage across the battery terminals when no current is flowing through the battery and PD is the voltage across the battery when it is supplying the current to a circuit or load. Volt is the unit of electromotive force as well as potential difference. EMF and the potential difference can be related by the following expression:

$$\text{EMF} = \text{PD} + i \times r$$

where
i is the circuit current in A,
r is the internal resistance of the source in Ω

Electric power The rate of doing work is power. Voltage between two points is defined as the work done in moving a unit positive charge from one point to another. So, the energy required to transfer a charge of q coulombs across a potential difference of V volts is given as

$$W = V \times q \text{ watts}$$

Now,

$$\text{Power} = \text{Rate of change of energy}$$
$$= V \times \frac{q}{t} = V \times I \quad \text{watts}$$

The unit of electric power is watt and the symbol of this unit is W.

Electric energy The electric energy is required for the flow of electrons. Electric energy can be expressed as

$$\text{Electric energy} = \text{Electric power} \times \text{Time}$$
$$= V \times I \times t \text{ watt second} = V \times I \times t/(3{,}600) \text{ watt hour (Wh)}$$

The basic unit of energy is joule. The commercial unit of electric energy is kilowatt hour or kWh.

Resistance The property of a material which opposes the flow of current is called resistance and is denoted by R. The resistance of a wire can be estimated by the following expression:

$$R = \frac{\rho l}{A} \quad \text{ohms } (\Omega)$$

where

ρ is the resistivity of the material of the wire in Ω/m

l is the length of the wire in m,

A is the cross-sectional area of the wire in m^2

The unit of resistance is ohm. Each commercial resistor has two main characteristics—its value in ohms and its power dissipating capacity in watts.

Inductance The property of a circuit element or component which opposes the rate of change of current is called inductance and the circuit element that exhibits the property of an inductance is known as an inductor. Its unit is henry. An inductance is capable of storing energy in the magnetic field. The current–voltage relationship of an inductor is given as follows:

$$V = L\frac{di}{dt} \quad \text{V}$$

In other words, an inductor is a circuit component which opposes the change of current flowing through it and induces voltage as the current flowing through it varies in magnitude and/or direction.

Capacitance The property of a circuit element or component which opposes the rate of change of voltage across its terminals is called capacitance and the circuit element that exhibits the property of capacitance is known as capacitor. Its unit is farad. A capacitor is capable of storing energy in the electric field. The current–voltage relationship of a capacitor is given by

$$i = C\frac{dv}{dt} \quad \text{A}$$

In other words, a capacitor is a circuit component which opposes the change of voltage across its terminals and produces a current through it if the applied voltage varies in magnitude and/or direction.

Magnetic flux Magnetic flux is the total number of lines of force comprising the magnetic field. It is denoted by ϕ and is measured in weber.

Magnetic flux density Magnetic flux density is defined as the magnetic flux passing per unit area through any material through a plane at right angles to the direction of flow of the magnetic flux. It is denoted by B and is measured in weber/m^2 or tesla.

$$B = \frac{\Phi}{A} \quad \text{Wb/m}^2$$

Magnetomotive force The magnetic field is produced around a current-carrying coil. If a unit north pole is placed in the magnetic field, it experiences a force similar to an electric charge in an electric field. The product of current and the number of turns of the coil is known as magnetomotive force. It is the force due to which flux flows in the magnetic circuit. It is equivalent to electromotive force (EMF) used in the electric circuits. It is measured in ampere-turns (AT) and is denoted by F.

$$F = I \times N \quad \text{AT}$$

Magnetic field strength The magnetic field strength is the magnetomotive force per unit length of the magnetic flux path. Its symbol is H and is measured in henry/m. It is also called as magnetic field strength or magnetizing force.

$$H = \frac{F}{L} = \frac{IN}{L} \quad \text{H/m}$$

Table 4.2 shows seven fundamental SI units with their symbols and units and Table 4.3 shows some important SI units along with their symbols and units.

Table 4.2 Seven basic SI units

Measured quantity	Unit	Symbol
Length	Meter	m
Mass	Kilogram	kg
Time	Second	s
Electric current	Ampere	A
Thermodynamic temperature	Kelvin	K
Amount of a substance	Mole	mol
Luminous intensity	Candela	cd

Table 4.3 Some important SI units along with their symbols

Measured quantity	Unit	Symbol	Measured quantity	Unit	Symbol
Angle	Radian	rad	Electric potential	Volt	V
Frequency	Hertz	Hz	Resistance	Ohm	Ω
Force	Newton	N	Conductance	Siemens	S
Pressure	Pascal	Pa	Inductance	Henry	H
Energy/Work/Heat	Joule	J	Capacitance	Farad	F
Power	Watt	W	Magnetic flux	Weber	Wb
Electric charge	Coulomb	C	Magnetic flux density	Tesla	T

4.3 BASIC WAVEFORMS

The electric engineers often encounter some waveforms like sine, cosine, square, and sawtooth. Plotting and analyzing these waveforms in MATLAB is an easy task. This is illustrated by Example 4.1 and Simulink model.

Example 4.1

This program illustrates the steps for plotting sine and cosine waveforms.

In the first step, the time array is defined. Here, the time is taken from 0 to 0.04 s. Then the peak amplitude (340 V), frequency (50 Hz), and the time period (0.02 s) of the wave are defined. After defining these parameters, the sine and cosine waves are defined using the functions `sin` and `cos` and these waves are plotted. These waves will be plotted from time 0 to 0.04 s. The plots of these waves are shown in Fig. 4.1.

```
% Program to plot sine and cosine waves
clc;
clear all;
t = 0:10e-6:0.04;
T = 0.02;
Vm = 340;
f = 50;
V_sin = Vm * sin(2*pi*f*t);
V_cos = Vm * sin(pi/2-2*pi*f*t);
plot(t,V_sin,t,V_cos)
```

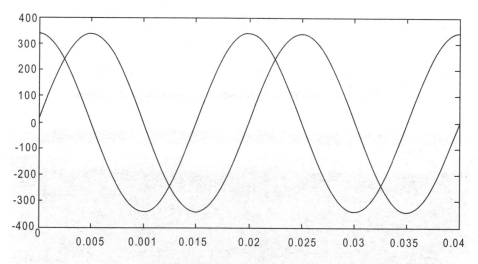

Fig. 4.1   Plot of sine and cosine waves

Example 4.2

A Simulink model is constructed in order to obtain sine, square, and sawtooth waves (see Fig. 4.2) of desired amplitude and frequency.

For obtaining a sine wave, Sine Wave block is selected. The amplitude of the sine wave is taken as 20 and frequency as 314 rad/s. Similarly, for obtaining a square wave, Signal Generator block is selected. This block can generate square, sine, sawtooth, and random signals. Out of

(*Continued*)

Example 4.2 (*Continued*)

these signals, the square signal is selected and its frequency and amplitude are taken as 50 Hz and 20 V respectively. For sawtooth signal, Repeating Sequence block is selected. The time values for this block are taken as 0, 0.02 and amplitude values as 0, 20. These values generate a sawtooth signal of 50 Hz and amplitude 20 V. All the three blocks are then connected to a three input Mux block. The output of the Mux is given to the Scope. This model is executed for 0.04 s, i.e., for two cycles, for each wave. The signals obtained are seen on the Scope as shown in Fig. 4.3.

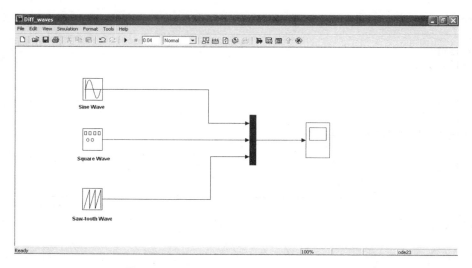

Fig. 4.2 Model for sine, square, and sawtooth waves

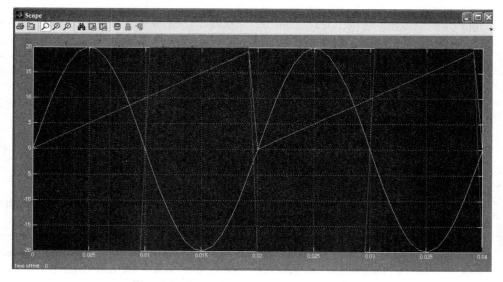

Fig. 4.3 Plot of sine, square, and sawtooth waves

4.4 AVERAGE, RMS, AND PEAK VALUE

In the DC system, the value of the circuit voltage and current are constant, so there is no problem in specifying their magnitude. In the case of AC system, the circuit current and the voltage varies in magnitude with time. Therefore, AC voltages and currents are expressed in terms of peak value, average value, or root mean square (rms) value.

The peak or maximum value of an AC voltage or current is the value of the maximum magnitude attained during each cycle. This value is attained twice in each cycle. If a voltage is defined as $V = 200 \sin \omega \times t$, then the peak value of the voltage is 200. The average or mean value of an AC current is equal to that value of a DC current that transfers same amount of charge in a circuit as transferred by the AC signal. Mathematically, the average value of a signal $V(t)$ is given as follows:

$$V_{avg} = \frac{1}{T} \int_0^T V(t)dt$$

where T is the time period of the AC signal. If $V(t) = 200 \sin \omega \times t$ and $T = 2\pi$, V_{avg} will be

$$V_{avg} = \frac{1}{T} \int_0^T 200 \sin \omega t \, d(\omega t)$$
$$= 500 \times \{200[-\cos \omega t]_0^{2\pi}\}$$
$$= 50 \times 200 \, (1-1) = 0 \text{ V}$$

Hence, the average value of the signal is zero. The average value of a symmetrical signal is always zero for a complete cycle as the positive half-values are equal to the negative half-values.

The rms value of an AC current is given by that steady current which when flows through a given resistance produces the same amount of heat as produced by the steady current for a given period of time. Mathematically, rms value of signal $V(t)$ is given as

$$V_{rms} = \left\{ \frac{1}{T} \int_0^T V^2(t)dt \right\}^{0.5}$$

where T is the time period of the AC signal. If $V(t) = 200 \sin w \times t$ and $T = 2\pi$, V_{rms} will be

$$V_{rms} = \left\{ \frac{1}{T} \int_0^T 200^2 \sin^2(\omega t) \, d(\omega t) \right\}^{0.5}$$
$$= \left\{ \frac{1}{T} \int_0^T 200^2 (1 - \cos 2\omega t) \, d(\omega t) \right\}^{0.5}$$
$$= \left\{ \frac{1}{2\pi} 200^2 \left(\omega t - \frac{1}{2\pi} \sin 2\omega t \right)_0^{2\pi} d(\omega t) \right\}^{0.5}$$
$$= \frac{200}{\sqrt{2}} = 141.421 \text{ V}$$

Example 4.3

Figure 4.4 shows the model of a circuit containing an AC voltage of 200 V peak value and a frequency of 50 Hz in series with two resistances R_1 and R_2 of 10 Ω. The rms and average values of the current and voltage will be 7.07 (i.e., 100/10√2), 0 and 70.7 (i.e., 100/√2) and 0 respectively. The values obtained after simulating the model can be seen in the display block of Fig. 4.4. The plot of the current and voltage waveform can be seen in Fig. 4.5.

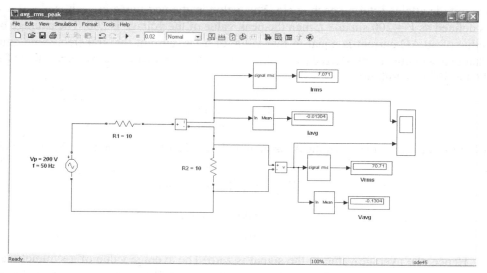

Fig. 4.4   Circuit model for average and rms values

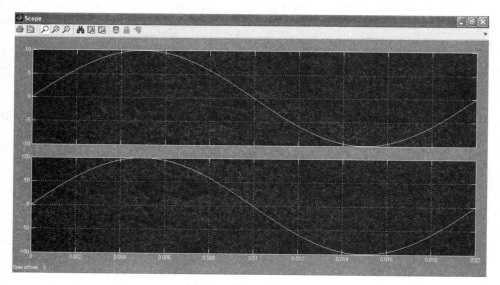

Fig. 4.5   Plot showing the peak values of the current and voltage

4.5 OHM'S LAW

Georg Simon Ohm formulated the relation between the voltage across a conductor and current through a conductor known as Ohm's law. It may be defined as,

'The current flowing through a conductor is directly proportional to the potential difference across its ends if the physical states, i.e., temperature, pressure, etc., remain constant.'

That is,

$$I = \frac{V}{R} \quad \text{or} \quad V = I R$$

where R is a constant known as the resistance of the conductor.

Example 4.4

Figure 4.6 shows a simple circuit with a DC voltage source of 100 V in series with a resistance 100 Ω. The current in the circuit is 100 V/100 Ω = 1 A. This model is simulated and the current and voltage across the resistance was 1 A and 100 V respectively. Hence, the Ohm's law is verified by this simulation model.

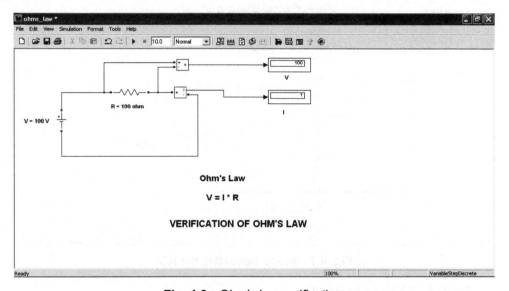

Fig. 4.6 Ohm's law verification

4.6 KIRCHHOFF'S LAWS

Gustov Robert Kirchhoff defined two basic laws governing the networks, the first law is known as Kirchhoff's current law or KCL and the second law is known as Kirchhoff's voltage law or KVL.

The KCL can be stated as,

'The algebraic sum of all the currents at any node is zero at any instant time.'

Mathematically,

$$\text{For a node } n, \Sigma\, i(t) = 0$$

While applying the KCL, the convention that the currents approaching the node are assigned positive sign is adopted.

Example 4.5

A circuit containing two DC voltage sources and three resistances is shown in Fig. 4.7. Consider a node in this circuit. At this node, currents I_1 and I_2 are approaching the node while the current I_3 is leaving the node. So by applying the KCL at this node, we get

$$I_1 + I_2 - I_3 = 0$$

From Fig. 4.7, we can get the values of these currents as
$I_1 = 3.75$, $I_2 = 2.5$, and $I_3 = 6.25$.
It can be seen that $3.75 + 2.5 - 6.25 = 0$.

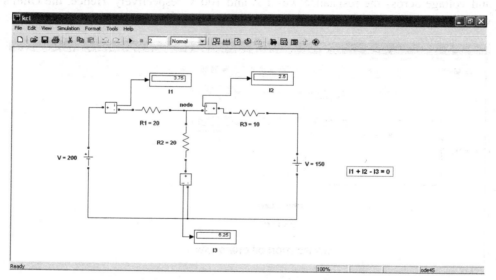

Fig. 4.7   Model illustrating the KCL

Similarly, KVL can be stated as,

'Algebraic sum of all the voltages across a closed loop is zero at any instant of time.'
Mathematically,

$$\text{For a closed loop } \Sigma\, V(t) = 0$$

The loop can be traced either clockwise or anticlockwise, but usually it is traced clockwise. While applying the KVL we shall adopt the convention that while traveling clockwise in the loop, if the voltage across an element is – to + (i.e., there is a rise in voltage along the path we trace) then it is taken as positive, otherwise negative. The sign can be arbitrarily assigned to the voltage across an element and apply KVL, if the voltage comes out to be negative it means that the sign assigned was opposite.

Example 4.6

The circuit model containing two DC voltage sources and three resistances is shown in Fig. 4.8. There can be three closed loops in this circuit, viz. ABEF, BCDE, and ABCDEF. By applying KVL in these three loops, we get

$$V - V_1 - V_2 = 0$$
$$V_2 + V_3 - V' = 0$$
$$V - V_1 - V_2 - V' = 0$$

The values of these voltages can be seen in Fig. 4.8. By substituting these values in the above equations, the equations are satisfied. Thus, KVL is verified by this circuit model.

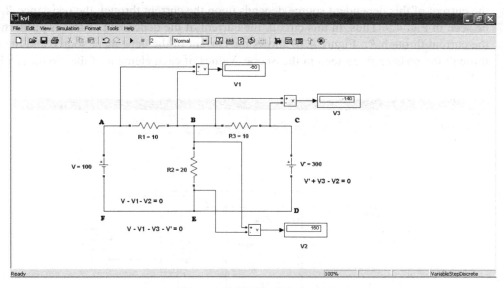

Fig. 4.8   Model illustrating KVL

4.7 INDEPENDENT AND DEPENDENT DC SOURCES

An independent source is that source whose magnitude does not depend on any of the parameters of the circuit. Independent sources can be of two types—independent current source and independent voltage source. A dependent source is that source whose magnitude depends on some parameters of the circuit or some currents or voltages in the circuit. There are four types of dependent sources:

1. **Current-controlled current source (CCCS)** A current-controlled current source is a dependent current source whose value depends on the current through any circuit element.

2. **Current-controlled voltage source (CCVS)** A current-controlled voltage source is a dependent current source whose value depends on the voltage across any element of the circuit.

3. **Voltage-controlled current source (VCCS)** A voltage-controlled current source is a dependent voltage source whose value depends on the current through any element of the circuit.

4. **Voltage-controlled voltage source (VCVS)** A voltage-controlled voltage source is a dependent voltage source whose value depends on the voltage across any element of the circuit.

Example 4.7

A circuit model of a current-controlled current source is shown in Fig. 4.9. This circuit contains an independent voltage source of peak amplitude 600 V, 50 Hz, and a CCCS. The amplitude of the current of this dependent source depends upon the current through the resistance R_1 as can be seen in Fig. 4.9. Thus, in this circuit, the current of the dependent source is proportional to the current through R_1. Figure 4.10 shows the current of the dependent source and the current through the resistor R_1 as seen in the Scope. Values of each element of the circuit can be seen in Fig. 4.9.

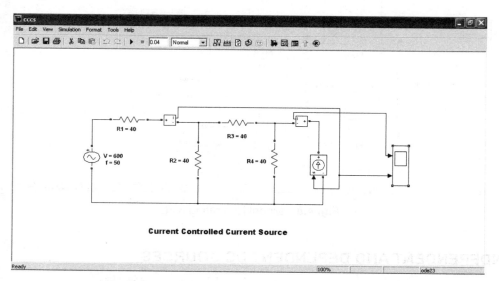

Fig. 4.9 Model of current-controlled current source

Example 4.8

A circuit model of a current-controlled voltage source is shown in Fig. 4.11. This circuit contains an independent voltage source of peak amplitude 200 V, 50 Hz, and a CCVS. The amplitude of the voltage of this dependent source depends upon the current through the resistance R_1 as can be seen in Fig. 4.11. Thus, in this circuit, the voltage of the dependent source is proportional to the current through R_1. This circuit is simulated for $R_1 = 20 \ \Omega$, $R_2 = 20 \ \Omega$, and $R_3 = 20 \ \Omega$. Figure 4.12 shows the waveform of current through resistance R_1, voltage across the dependent source, and voltage across resistance R_3.

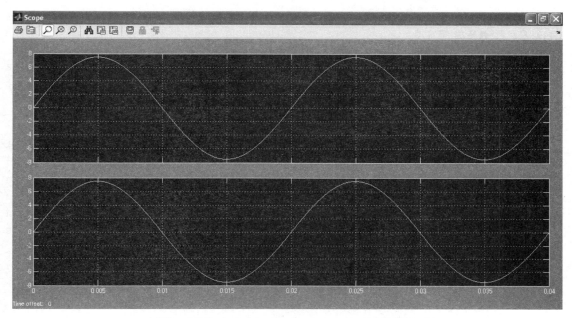

Fig. 4.10   Input and output currents of CCCS

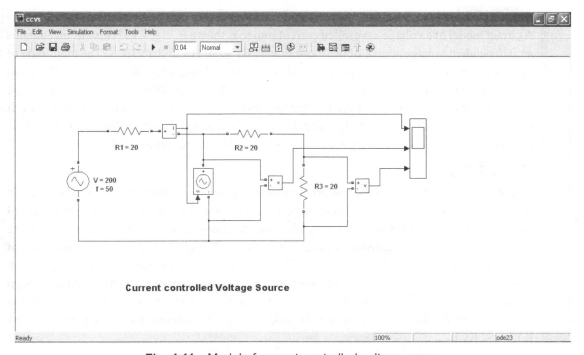

Fig. 4.11   Model of current-controlled voltage source

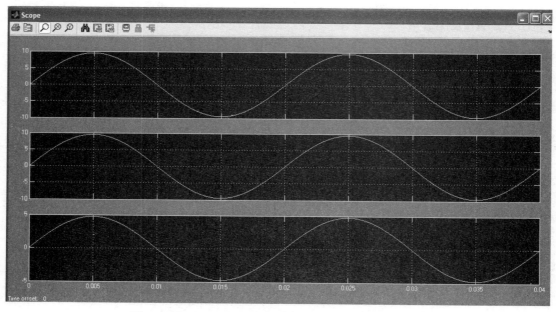

Fig. 4.12 Input voltage and output current of CCVS

Example 4.9

A circuit model of a voltage-controlled current source is shown in Fig. 4.13. This circuit contains an independent voltage source of peak amplitude 200 V, 50 Hz, and a VCCS. The amplitude of the current of this dependent source depends upon the voltage across the resistance R_2 as can be seen in Fig. 4.14. This circuit is simulated for $R_1 = 20\ \Omega$, $R_2 = 20\ \Omega$, and $R_3 = 20\ \Omega$. Figure 4.14 shows the waveform of current of the dependent source and voltage across resistance R_2. It can be seen from this figure that the current of the VCCS is equal to the voltage across R_2. This current can also be calculated by mathematically solving the circuit by KVL and KCL.

Example 4.10

Figure 4.15 contains a circuit model containing a voltage-controlled voltage source, an independent voltage source, four resistances, and other measurement devices. The magnitude of the voltage of this dependent source depends upon the voltage across the resistance R_3 as can be seen in Fig. 4.14. The independent voltage source has peak amplitude of 100 V and frequency of 50 Hz. This circuit is simulated for $R_1 = 20\ \Omega$, $R_2 = 30\ \Omega$, $R_3 = 30\ \Omega$, and $R_L = 40\ \Omega$. Figure 4.16 shows the waveform of voltage across resistance R_3, voltage across VCVS, and current through the resistance R_L. It can be seen from this figure that the voltage of the VCVS is equal to the voltage across R_3. This current can also be calculated by mathematically solving the circuit by KVL and KCL.

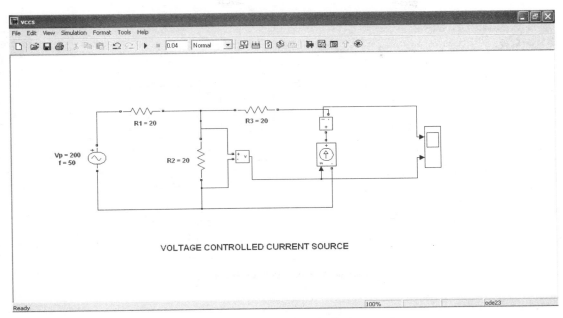

Fig. 4.13 Model of voltage-controlled current source

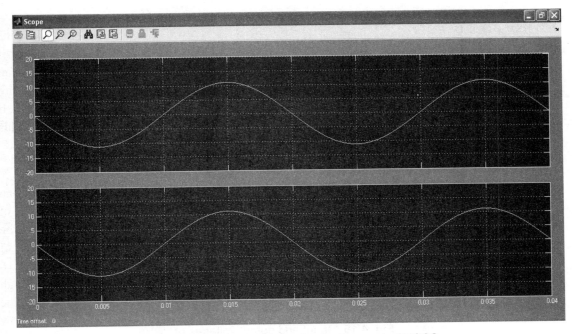

Fig. 4.14 Input voltage and output current of VCCS

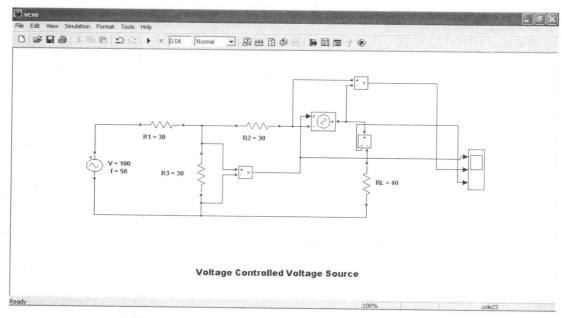

Fig. 4.15 Model of voltage-controlled voltage source

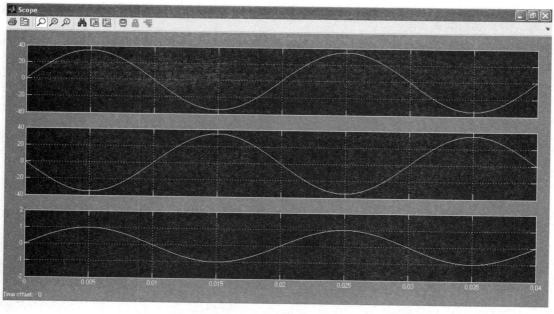

Fig. 4.16 Input and output voltages of VCVS

4.8 SERIES AND PARALLEL CIRCUITS

There are three basic elements—inductor, capacitor, and resistor—which are well known. In this section we analyze the series and parallel circuits constructed by these elements. We should be familiar with some terms, which are given in the following, when dealing with the steady-state analysis of AC circuits, also known as AC analysis. We define the terms of reactance and impedance so that the KVL and KCL can be institutively applied to AC analysis.

Inductive reactance Inductive reactance of an inductor is given by the equation

$$X_L = 2\pi f L$$

where f is the supply frequency in hertz and L is the value of the inductor in henry. The inductive reactance is measured in ohms. The equivalent inductive reactance of the two or more inductors connected in series is equal to the sum of their individual inductive reactance similar to the case of resistance.

Capacitive reactance Capacitive reactance of a capacitor is given by the equation

$$X_C = \frac{1}{2\pi f C}$$

where f is the supply frequency in hertz and C is the value of the capacitor in farads. The capacitive reactance is measured in ohms. The equivalent capacitive reactance of the two or more capacitors connected in series is equal to the sum of their individual capacitive reactance.

Impedance The impedance of a circuit is given by

$$Z = R + j(X_L - X_C); \text{ where } \|Z\| = \sqrt{\{R^2 + (X_L - X_C)^2\}} \text{ and } Z = \tan^{-1}\{(X_L - X_C)/R\}$$

where R, X_L, and X_C are total equivalent resistance, total equivalent inductive reactance, and total equivalent capacitive reactance of the circuit respectively in ohms. The impedance is measured in ohms. The equivalent impedance of the two or more branches connected in series is equal to the sum of their individual impedance.

Conductance Conductance is the reciprocal of resistance, i.e., $G = 1/R$, and is measured in mho (Ω^{-1}). The equivalent conductance of two or more resistances connected in parallel is the sum of their individual conductances.

Inductive susceptance Inductive susceptance is the reciprocal of inductive reactance, i.e., $Y_L = 1/X_L$, and is measured in siemens. The equivalent inductive susceptance of the two or more inductors connected in parallel is equal to the sum of their individual inductive susceptance.

Capacitive susceptance Capacitive susceptance is the reciprocal of capacitive reactance, i.e., $Y_C = 1/X_C$, and is measured in siemens. The equivalent capacitive susceptance of the two or more capacitors connected in parallel is equal to the sum of their individual capacitive susceptance.

Admittance Admittance is the reciprocal of impedance, i.e., $Y = 1/Z$, and is measured in siemens. The equivalent admittance of the two or more branches connected in parallel is equal to the sum of their individual admittance.

Power factor Power factor is defined as the cosine of the angle between voltage and current. Its value lies between 0 and 1; negative sign indicates that the power factor is leading and positive sign indicates that the power factor is lagging. The power factor of a circuit can be calculated mathematically as

$$\phi = \tan^{-1}\{(X_L - X_C)/R\}$$

and

$$\text{Power factor} = \cos\phi$$

4.8.1 R-L Series Circuit

Consider a circuit consisting of resistance R in series with an inductance L, connected across an AC supply of rms value of V volts and f hertz. Let V_R be the voltage across the resistance R, V_L be the voltage across the inductance L, and I be the rms value of the current flowing through the circuit. The impedance of this circuit will be

$$Z = R + j\,\omega L$$
$$= \sqrt{\{R^2 + \omega^2 L^2\}}/\tan^{-1}(\omega L/R)$$

Similarly, the circuit current will be

$$I = V/Z$$
$$= V/(R + j\omega L)$$

The phase difference between the voltage and current will be $\phi = \tan^{-1}(\omega L/R)$ and the power factor of the circuit will be $\cos\phi$ or $\cos\{\tan^{-1}(\omega L/R)\}$.

Example 4.11

A series R-L circuit is shown in Fig. 4.17. In this circuit, $R = 10\ \Omega$ and $L = 10$ mH. The single phase AC voltage source has a peak value of 130 V and frequency of 50 Hz. Now for this model,

$$V_{(rms)} = 130/\sqrt{2} = 91.92\ \text{V}$$
$$Z = 10 + j2 * \pi * 50 * 0.01$$
$$= 10 + j\,3.14 = 10.481, \phi = 17.43°$$
$$I = V/Z$$
$$= 91.92/(10 + j3.14)$$
$$= 8.77, \phi = -17.43°$$

This circuit model is simulated from time 0 to 0.04 s and the rms value of the current and voltage are found to be 8.77 A and 91.92 V as can be seen in the model. Figure 4.18 shows the voltage and current waveforms for the circuit. It can be observed from this figure that the current is lagging the voltage by 17.43°. The peak value of the current is $8.77\sqrt{2} = 12.40$ A.

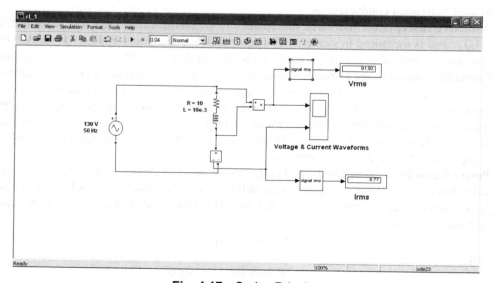

Fig. 4.17 Series R-L circuit

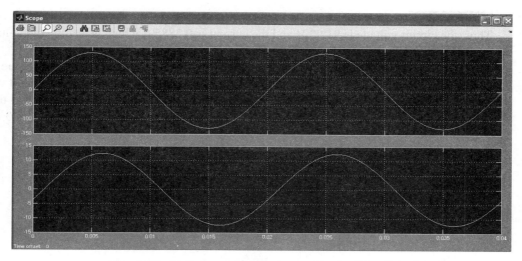

Fig. 4.18 Voltage and current waveforms of the R-L circuit

4.8.2 Series R-C Circuit

Consider a circuit containing a capacitor C and a resistor R connected across an AC supply of rms value of V volts and a frequency of f hertz. Consider the voltage across the capacitor to be V_C and across the resistor to be V_R and let I be the circuit current. For this circuit, the values of the circuit current, voltage, and power factor can be calculated as follows:

$$Z = R - j/(\omega C)$$
$$= \sqrt{\{R^2 + (1/\omega C)^2\}}, \phi = \tan^{-1} - (1/\omega RC)$$
$$I = V/Z$$
$$= V/\{R - j/\omega C\}$$
$$= V/\sqrt{\{R^2 + (1/\omega C)^2\}}, \phi = \tan^{-1}(1/\omega RC)$$

The power factor of this circuit is $\cos\phi$, where $\phi = -\tan^{-1}(1/\omega RC)$.

Example 4.12

Consider a circuit shown in Fig. 4.19 containing a resistance in series with a capacitor. They are connected to an AC voltage source of peak voltage 130 V and frequency of 50 Hz. In this circuit $R = 10\ \Omega$ and $C = 10\ \mu F$. The value of the circuit parameters can be calculated as follows:

$$\text{Impedance of the circuit } Z = 10 - j\ 10^5/2\pi \times 50\ \Omega$$
$$= 318.63, \phi = -72.57°$$
$$\text{RMS value of the supply voltage} = 130/\sqrt{2} = 91.92\ V$$
$$\text{Circuit current } I = V/Z$$
$$= 91.92/(10 - j\ 10^5/2\pi \times 50)$$
$$= 0.2885\ A, \phi = 72.57°$$

(*Continued*)

Example 4.12 (*Continued*)

Thus, the circuit voltage is 91.92 V and the current is 0.2885 A which is leading the voltage by an angle of 72.57°.

After simulating the circuit model from time 0 to 0.04 s, the value of the rms value of the current measured was 0.2885 A and voltage was 91.92 V. Figure 4.20 shows the circuit voltage and current waveform obtained after simulating the model. From Fig. 4.20 it is clear that the current is leading the voltage by 72.57°. The peak value of the current is $0.2885\sqrt{2} = 0.408$ A.

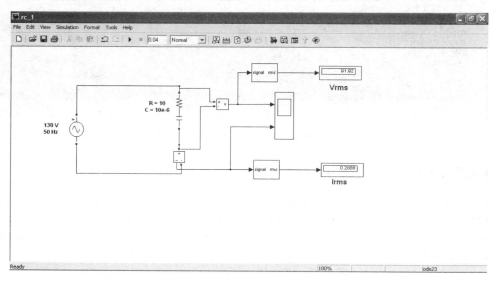

Fig. 4.19 Model of series R-C circuit

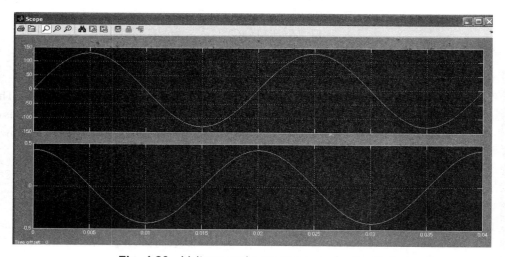

Fig. 4.20 Voltage and current waves for the R-C circuit

4.8.3 Series–Parallel Circuit

Example 4.13

Consider a circuit containing two parallel branches. The first parallel branch contains a resistor of 10 Ω and an inductor of 30 mH. The second parallel branch contains a resistor of 50 Ω and an inductor of 20 mH. These branches are supplied by a voltage source of peak value of 430 V and frequency 50 Hz (see Fig. 4.21).

For this circuit, the branch currents I_1 and I_2 can be calculated as follows:

$$\text{Impedance of the circuit } Z = Z_1 \parallel Z_2$$

where Z_1 is the impedance of the first branch and Z_2 is the impedance of the second branch.

$$Z_1 = 10 + j\,2\pi \times 50 \times 30 \times 10^{-3}$$
$$= 10 + j\,9.42 = 13.74, \phi = 43.29°$$
$$Z_2 = 50 + j\,2\pi \times 50 \times 20 \times 10^{-3}$$
$$= 50 + j\,6.28 = 50.393, \phi = 7.159°$$

Now, the current in first branch will be

$$I_1 = V/Z_1$$
$$V_{\text{rms}} = 430/\sqrt{2} = 304.06$$
$$I_1 = 304.06/(10 + j\,9.42)$$
$$= 22.13 \text{ A}, \phi = -43.29°$$

Similarly,

$$I_2 = 304.06/(50 + j\,6.28)$$
$$= 6.034 \text{ A}, \phi = -7.159°$$

Refer Fig. 4.22 for the waveforms of output voltage and branch currents.

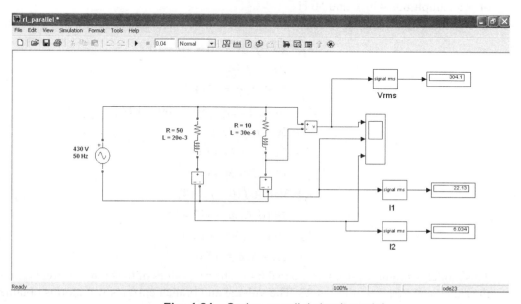

Fig. 4.21 Series–parallel circuit model

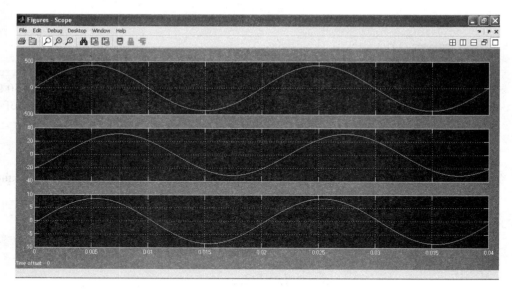

Fig. 4.22 Waveforms of output voltage and branch currents

Example 4.14

Consider another series–parallel circuit as shown in Fig. 4.23. This circuit contains two **parallel** branches, the first branch containing a resistance of 50 Ω and a capacitance of 200 μF, the second branch containing a resistance of 10 Ω and an inductance of 10 mH and a voltage source of peak amplitude 430 V and 50 Hz.

For this circuit, branch currents and voltage can be calculated as follows:

$$Z_1 = 10 + j\, 2\pi \times 50 \times 10 \times 10^{-3}$$
$$= 10 + j\, 3.14$$
$$= 10.481, \phi = 17.432°$$
$$Z_2 = 50 - j\, \{1/2\pi \times 50 \times 200 \times 10^{-6}\}$$
$$= 50 - j\, 15.92$$
$$= 52.47, \phi = -17.66°$$
$$V_{rms} = 430/\sqrt{2} = 304.06 \text{ V}$$
$$I_1 = 304.06/(10 + j\, 3.14)$$
$$= 29.01 \text{ A}, \phi = -17.432°$$
$$I_2 = 304.06/(50 - j\, 15.92)$$
$$= 5.79 \text{ A}, \phi = 17.66°$$

This model is simulated from time 0 to 0.04 s. The rms values of the branch currents and voltage measured can be seen in Fig. 4.23. The voltage and branch currents waveforms obtained after simulation are shown in Fig. 4.24.

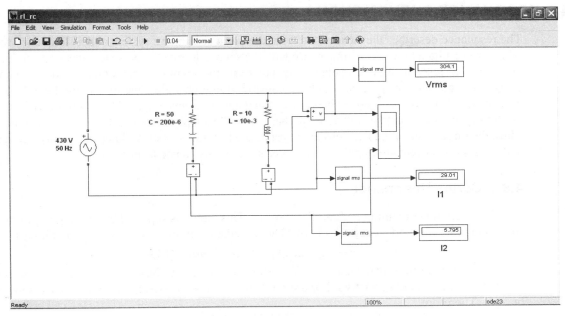

Fig. 4.23   Series parallel RC–RL circuit

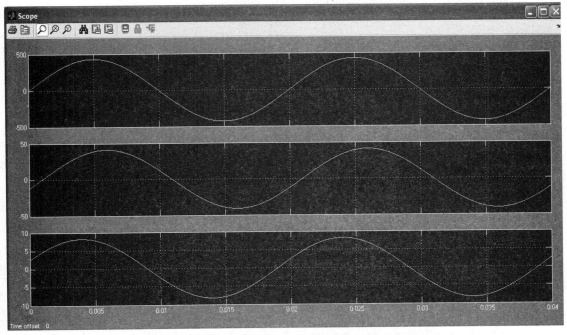

Fig. 4.24   Supply voltage and branch currents waveforms

4.9 RESONANCE PHENOMENON

The resonance describes the steady-state operation of a circuit or a system at that frequency for which the resultant output is in time phase with the source despite the presence of energy storage elements. The resonance can take place only if there are two complementary types of independent energy storage elements present which are capable of interchanging energy among them. In the case of electrical circuits, these elements are inductance and capacitance. The condition in a RLC circuit in which the supply voltage is in phase with the supply current is known as resonance. Thus, the impedance of the RLC circuit at resonance is purely resistive. Since the supply voltage and the supply current are in phase, the power factor at resonance is unity.

4.9.1 Series Resonance

Consider a circuit containing R, L, and C in series. This circuit is supplied by a voltage source of rms voltage V volts and frequency f hertz. The impedance of this circuit is given as follows:

$$Z = \sqrt{\{R^2 + (X_L - X_C)^2\}} = \sqrt{\{R^2 + (\omega L - (1/\omega C))\}^2}$$

From the definition of resonance, the applied voltage is in phase with the supply current and the circuit is purely resistive. For a circuit to be purely resistive, its net reactance should be zero. Therefore,

$$X_L - X_C = 0$$
$$X_L = X_C$$

In this case, $Z = R$ and supply current $I = V/Z$. Now, since

$$X_L = X_C$$
$$\omega L = 1/\omega C$$
$$\omega^2 = 1/LC$$
$$\omega = 1/\sqrt{(LC)}$$
$$2\pi f = 1/\sqrt{(LC)}$$
$$f = 1/\{2\pi\sqrt{(LC)}\} \text{ Hz}$$

where f is the resonant frequency of the circuit.

The properties of series resonance can be summarized as follows:
1. The power factor of the circuit is unity, i.e., $\phi = 0°$ and $\cos\phi = 1$.
2. Impedance of the circuit is minimum and is resistive, i.e., $Z = R$.
3. The supply current is maximum and is in phase with the supply voltage, i.e., $I = V/R$.
4. The voltage drop across the inductor is equal to the voltage drop across the capacitor and it is maximum and it could be more than V.
5. The net reactance of the circuit is zero, i.e., $X_L - X_C = 0$.
6. The power absorbed by the circuit is maximum and is V^2/R W.

4.9.2 Parallel Resonance

Consider a circuit containing an inductance L and a resistance R in the first branch, a capacitance C in the second branch, and a resistance R_1 in the third branch. All the three branches are in parallel. The supply voltage applied to this circuit is of V volts rms and frequency f hertz.

Consider the current in the first branch to be I_1 and that in the second branch to be I_2. The total admittance of this circuit will be

$$Y = j\omega C + 1/R_1 + 1/(R + j\omega L)$$
$$= j\omega C + 1/R_1 + (R - j\omega L)/(R^2 + \omega^2 L^2)$$
$$= \{1/R_1 + R/(R^2 + \omega^2 L^2)\} + j\omega(C - L/R^2 + \omega^2 L^2)\}$$

Now, in case of resonance, the supply current and voltage should be in phase. So the admittance of the circuit should be real which means that the magnitude of the imaginary term should be zero. Thus,

$$j\omega(C - L/R^2 + \omega^2 L^2) = 0$$
$$C - L/R^2 + \omega^2 L^2 = 0, \text{ as } \omega \text{ is not zero}$$
$$R^2 + \omega^2 L^2 = L/C$$
$$\omega^2 L^2 = L/C - R^2$$
$$\omega = \sqrt{\{1/LC - R^2/L^2\}}$$
$$f = (1/2\pi) \sqrt{\{1/LC - R^2/L^2\}}$$

where f is the resonance frequency of the circuit in case of parallel resonance. It should be remembered that R is the resistance in series with the inductor L. We can summarize the properties of parallel resonance as follows:

1. The power factor of the circuit is unity, i.e., $\phi = 0°$ and $\cos \phi = 1$.
2. The admittance of the circuit is minimum and is $R_1 + R/(R^2 + \omega^2 L^2)$.
3. The supply current is minimum and is in phase with the supply voltage.
4. The net susceptance of the circuit is zero.

Example 4.15

Figure 4.25 shows the circuit model of a series circuit containing a resistance, inductance, and capacitance in series with an AC voltage source of peak amplitude of 440 V and frequency 53.08. The values of the circuit elements are $R = 100 \ \Omega$, $L = 30$ mH, and $C = 300 \ \mu F$. The resonance frequency for this circuit will be as follows:

$$f = 1/2\pi \ \{1/LC\}^{0.5}$$
$$= 1/2 \times 3.14\{1/(30 \times 10^{-3} \times 300 \times 10^{-6})\}$$
$$= 50.08 \text{ Hz}$$

This is the resonance frequency of the series circuit. Now the circuit current at this frequency will be as follows:

$$I_{rms} = V_{rms}/R$$
$$V_{rms} = V_{peak}/\sqrt{2}$$
$$= 440/\sqrt{2}$$
$$= 311.08$$
$$I_{rms} = 311.08/100$$
$$= 3.1108 \text{ A}$$

This supply current should be in phase with the supply voltage at resonance. The series resonant circuit shown in Fig. 4.25 is simulated from time 0 to 0.04 s. The value of the supply current measured was 3.11 A as shown in Fig. 4.25. The waveforms of the supply voltage and supply current are shown in Fig. 4.26. From this figure it can be seen that the supply current is in phase with the supply voltage.

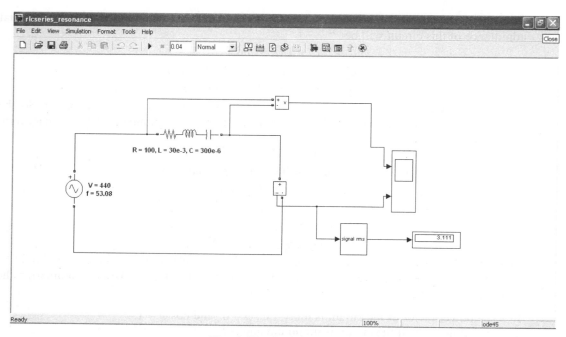

Fig. 4.25 Series resonance circuit

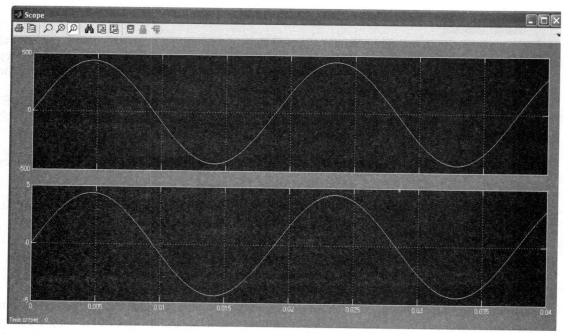

Fig. 4.26 Supply voltage and current waveforms for series resonance

Example 4.16

Figure 4.27 depicts the circuit model of a parallel resonance circuit containing resistance R' in series with two parallel branches containing a resistance R_1 in series with capacitor and the second branch containing a resistance R in series with an inductor. The values of these parameters are $R' = 100\ \Omega$, $R_1 = 0.001\ \Omega$, $C = 30\ \mu F$, $R = 0.01\ \Omega$, and $L = 30$ mH. The AC voltage source has the peak amplitude of 340 V and frequency of 53.08 Hz. In case of parallel resonance, the resonance frequency is given by

$$f = (1/2\pi)\sqrt{\{1/LC - R^2/L^2\}}$$
$$= 1/2\pi\{(1/30 \times 10^{-3} \times 30 \times 10^{-6}) - (0.01^2/(30 \times 10^{-3})^2)\}$$
$$= 53.08\ Hz$$

This is the resonance frequency of the parallel circuit. Now the circuit current at this frequency will be

$$I_{rms} = V_{rms} \times Y$$
$$= (340/\sqrt{2}) \times \{1/R_1 + R/(R^2 + \omega^2 L^2)\}$$
$$= 2.389\ A$$

This supply current should be in phase with the supply voltage at resonance. The parallel resonant circuit shown in Fig. 4.27 is simulated form time 0 to 0.04 s. The value of the supply current measures was 2.389 A as shown in Fig. 4.28. The waveforms of the supply voltage and supply current are shown in Fig. 4.29. From this figure it can be seen that the supply current is in phase with the supply voltage.

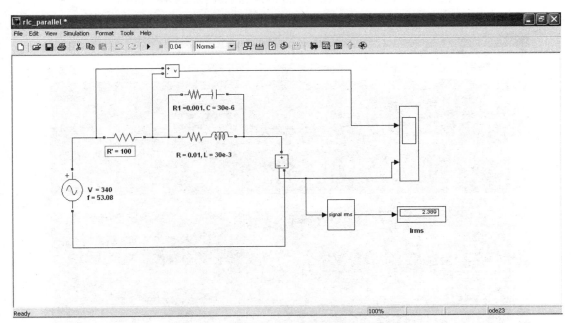

Fig. 4.27 Parallel resonance

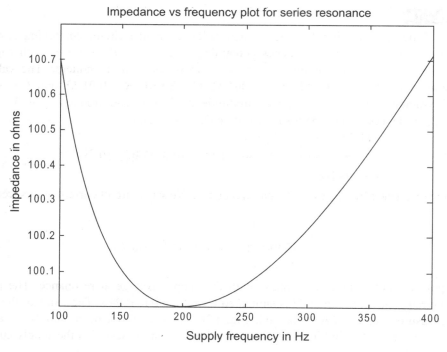

Fig. 4.28 Frequency vs impedance plot for series resonance

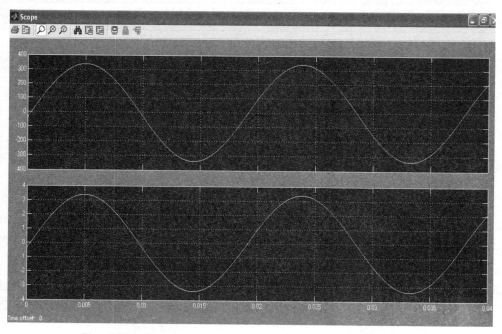

Fig. 4.29 Supply voltage and current for parallel resonance

Example 4.17

In case of series resonance, the impedance of the circuit is minimum and is equal to the resistance of the circuit. The following program plots a graph between impedance and the supply frequency for a series circuit containing a resistance of 100 Ω, capacitance of 100 μF, and an inductance of 6.34 mH. The frequency of the supply is varied from 100 Hz to 400 Hz in steps of 1 Hz. The plot for this circuit is shown in Fig. 4.29. It can be seen from the figure that at resonance, the impedance of the circuit is minimum and is 100 Ω.

```
% This program plots a graph of frequency vs impedance in case of series
% resonance
clc;
clear all;
f = 100:1:400;
L = 6.34e-3;
C = 100e-6;
R = 100;
XL = 2*pi*f*L;
XC = 1./(2*pi*f*C);
Z = (R^2 + (XL - XC).^2).^0.5;
plot(f,Z)
```

Example 4.18

In case of parallel resonance, the admittance of the circuit is minimum at the resonant frequency and is equal to the conductance of the parallel circuit. This program models a parallel resonance circuit in which the capacitance and the inductance are in parallel. This parallel combination is in series with a resistance. The series resistance of the inductor and capacitor is assumed to be zero. The supply frequency is varied from 100 Hz to 400 Hz in steps of 1 Hz. Figure 4.30 plots the graph of frequency versus admittance for this circuit. From this plot, it is clear that the admittance of the circuit is minimum at the resonance frequency and is equal to the conductance, i.e., 0.01, of the circuit.

```
% This program plots a graph of frequency vs admittance in case of parallel
% resonance
clc;
clear all;
clf;
f = 100:1:400;
L = 6.34e-3;
C = 100e-6;
```

(Continued)

Example 4.18 (*Continued*)

```
R = 100;
YR = 1/100;
YL = 1./(2*pi*f*L);
YC = 2*pi*f*C;
Y = (YR^2 + (YL - YC).^2).^0.5;
plot (f,Y)
title('\bf frequency vs admittance plot for parallel resonance');
xlabel('\bf Frequency in Hz');
ylabel('\bf Admittance in mhos');
```

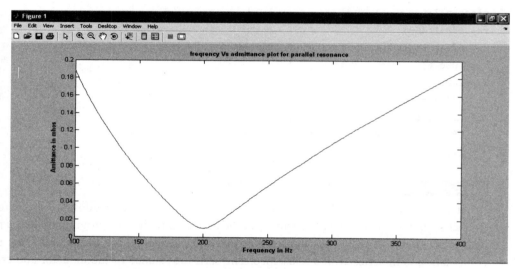

Fig. 4.30 Frequency vs admittance plot for parallel resonance

4.10 NETWORK THEOREMS

The electric networks may be solved with the help of basic laws of networks, i.e., Ohm's law, KCL, and KVL. However, in the case of large and complex circuits, these methods consume more time and are laborious. In these cases, we can apply network theorems as they can be applied in general conditions and their conclusions are relatively simple. Thus, one can use these theorems as short-cut methods for solving complex networks.

4.10.1 Superposition Theorem

The superposition theorem follows the principle of linearity of the circuit elements. This theorem is very useful when the network is excited by more than one source. The superposition principle states that in a linear time invariant circuit containing many independent energy sources, the overall response of the system can be calculated by considering the individual responses of these energy sources acting one at a time and then combining them later on suitably to get the overall response.

For electrical networks, the superposition theorem can be stated as follows:

'In a linear electrical network containing two or more independent electrical sources, the overall response (voltage and current) of any branch of the network is the algebraic sum of the responses produced by individual sources acting one at a time when all other independent sources are removed from the network.'

For removing an ideal current source, we can replace it by an open circuit and in case of an ideal voltage source, we can replace it by a short circuit. In case of non-ideal sources, replace voltage and current sources by their internal resistances. This theorem can be applied for all linear time invariant networks.

Example 4.19

Consider a circuit containing two DC voltage sources of 125 and 250 V and three resistances of 45, 105, and 35 Ω as shown in Fig. 4.31. Let the currents through resistances R_1, R_2, and R_3 be I_1, I_2, and I_3, respectively. These currents can be obtained by applying the superposition theorem as follows.

Step 1: Take one source at a time and remove the other sources.

Consider 125 V voltage source first and short circuit the 250 V source as shown in Fig. 4.32. For this circuit, let the current produced in the branches be I_1', I_2', and I_3'. Now,

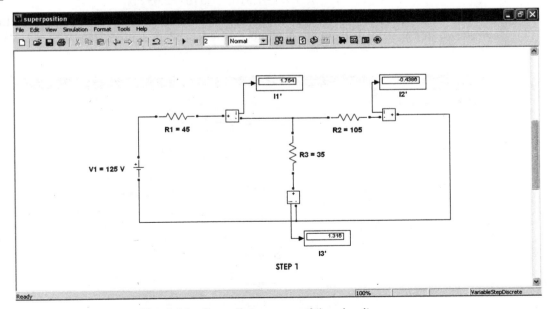

Fig. 4.31 Overall response of the circuit

(Continued)

Example 4.19 (*Continued*)

$$R_{eq} = 45 + 105/35 = 71.25 \ \Omega$$
$$I_1' = 125/71.25 = 1.754 \ A$$
$$I_2' = -1.754 \times 35/(105 + 35) = -0.4386 \ A$$
$$I_3' = 1.754 \times 105/(35 + 105) = 1.316 \ A$$

So, these were the currents due to 125 V source.

Step 2: Take the second source and remove the other sources.

Consider a 250 V voltage source and short circuit the 125 V source as shown in Fig. 4.33. For this circuit, let the current produced in the branches be I_1'', I_2'', and I_3''. Now,

$$R_{eq} = 105 + 45//35 = 124.687 \ \Omega$$
$$I_2'' = 250/124.687 = 2.005 \ A$$
$$I_1'' = -2.005 \times 35/(35 + 45) = -0.8772 \ A$$
$$I_3'' = 2.005 \times 45//(45 + 35) = 1.316 \ A$$

So, these were the currents due to 250 V source.

Step 3: Combine the currents obtained by individual sources in order to get the overall response of the circuit.

To obtain the currents I_1, I_2, and I_3 when both the sources are connected, add the individual responses of the independent sources obtained algebraically. Thus,

$$I_1 = I_1' + I_1'' = 1.754 - 0.8772 = 0.877 \ A$$
$$I_2 = I_2' + I_2'' = -0.4386 + 2.005 = 1.566 \ A$$
$$I_3 = I_3' + I_3'' = 1.316 + 1.316 = 2.632 \ A$$

The values of I_1, I_2, and I_3 are obtained for the circuit with the help of superposition theorem as shown in Fig. 4.31. If there are N number of independent sources, repeat Step 1 N number of times.

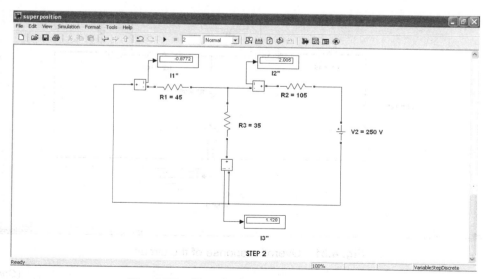

Fig. 4.32 Step 1 of superposition theorem

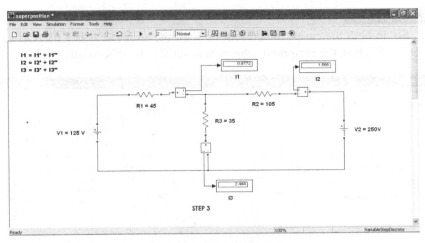

Fig. 4.33 Step 2 of superposition theorem

4.10.2 Reciprocity Theorem

In an electric circuit, it is sometimes required to interchange the source and the load. In this condition, the reciprocity theorem can be used to find the load current. The reciprocity theorem can be defined as follows:

'In a linear network containing an independent energy source and a zero resistance ammeter, if the position of the source and ammeter are interchanged, the magnitude of the current through ammeter will remain same no matter how complex the network is. The networks which follow this theorem are called as reciprocal networks.'

Example 4.20

Figure 4.34 shows two circuits which are reciprocal of each other. Consider the left circuit first. In this circuit, there is an AC voltage source of 240 V peak value and an ammeter connected for measuring current through the resistance R_4. For this circuit, the current in resistance R_4 can be calculated as

$$R_{eq} = 20 + 25//(15 + 30) = 36.07 \ \Omega$$
$$V_{rms} = 240/\sqrt{2} = 169.68 \ \text{V}$$

Current through resistance $\quad R_1 = 169.68/36.07 = 4.7 \ \text{A}$

Current through $\quad R_4 = 4.7 \times 25/(25 + 45) = 1.68 \ \text{A}$

In case of the reciprocal circuit, i.e., when the voltage source and ammeter are interchanged, the current through R_1 can be calculated as,

$$R_{eq} = 15 + 30 + (20//25) = 56.12 \ \Omega$$

Current through $\quad R_4 = 169.68/56.12 = 3.024 \ \text{A}$

Current through $\quad R_1 = 3.024 * 25/(25 + 20) = 1.68 \ \text{A}$

(Continued)

Example 4.20 (*Continued*)

So, in both the cases, the current through the ammeter was 1.68 A as can be viewed from Fig. 4.35. Thus, the reciprocity theorem is verified by this circuit model.

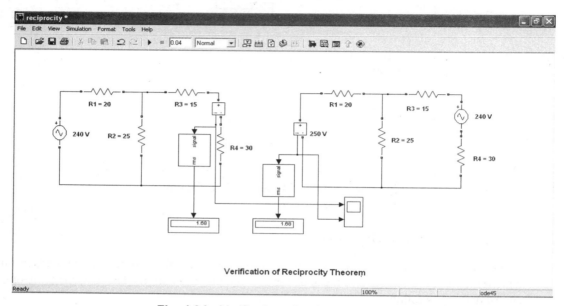

Fig. 4.34 Verification of reciprocity theorem

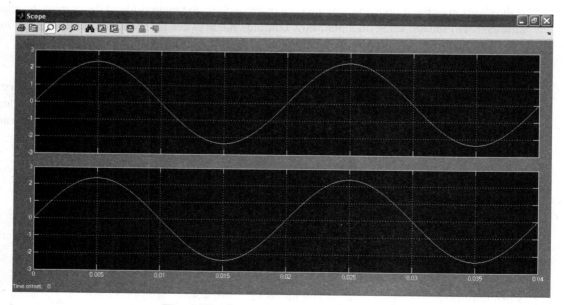

Fig. 4.35 Currents through the ammeters

4.10.3 Thevenin's Theorem

According to Thevenin's theorem, any two terminal networks can be replaced by a voltage source in series with a resistance. This theorem simplifies a complex circuit into a simple circuit containing a voltage source in series with a resistance and thus simplifies the calculations involved. Thevenin's theorem states that, 'Any electrical linear network with two output terminals and containing voltage and/or current sources can be converted into an Thevenin's equivalent circuit containing a voltage source (V_{TH}) in series with a resistance (R_{TH}). This voltage source is known as Thevenin's equivalent voltage and is denoted by V_{TH}, and the series resistance is known as Thevenin's equivalent resistance and is denoted by R_{TH}. R_{TH} is the equivalent resistance looking back from the output terminals and V_{TH} is the open circuit voltage at the output terminals.'

This theorem is quite useful to compute the load current and power. If V_{TH} and R_{TH} are known, the load current I_L can be computed as follows:

$$I_L = V_{TH}/(R_{TH} + R_L)$$

Example 4.21

Figure 4.36 shows the circuit diagram containing a DC voltage source along with three resistances (R_1, R_2, and R_3) and a load resistance R_L. Thevenin's equivalent network and load current for this network can be estimated as follows:

Thevenin's equivalent resistance, $R_{TH} = 10 + 20//20 = 20\ \Omega$

Thevenin's equivalent voltage, $V_{TH} = 200 \times 20/40 = 100\ V$

Load current $I_L = V_{TH}/(R_{TH} + R_L) = 100/120 = 0.833\ A$

Thevenin's equivalent voltage is measured by removing the load resistance as shown in Fig. 4.37 and is 100 V. Figure 4.38 shows the Thevenin's equivalent network and the value of the load current which is 0.833 A.

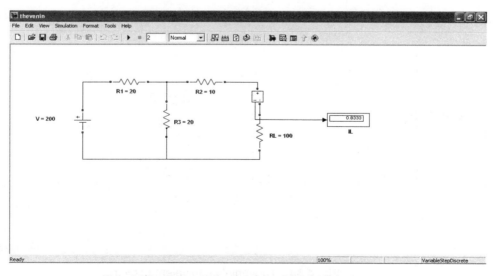

Fig. 4.36   Network for applying Thevenin's theorem

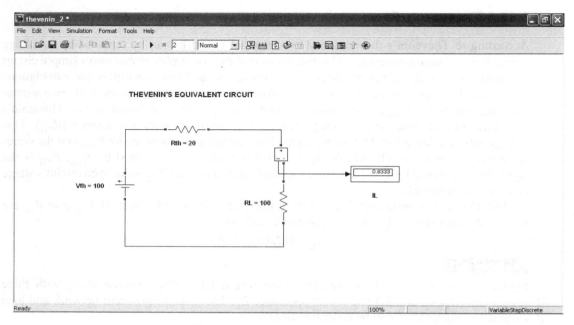

Fig. 4.37 Circuit to measure V_{TH}

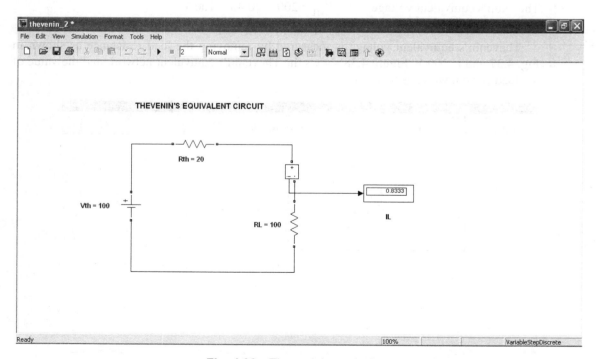

Fig. 4.38 Thevenin's equivalent network

4.10.4 Norton's Theorem

Norton's theorem is similar to that of Thevenin's theorem expect that it converts the given network into a current source in parallel with a resistance. Norton's theorem states that, 'Any electrical linear network with two output terminals containing voltage and/or current sources can be converted into a Norton's equivalent circuit containing a current source (I_N) in parallel with a resistance (R_N). This current source is known as Norton's equivalent current source and is denoted by I_N, and the parallel resistance is known as Norton's equivalent resistance and is denoted by R_N. R_N is the equivalent resistance of the circuit looking back from the output terminals and I_N is the short circuit current at the output terminals.'

After estimating I_N and R_N for the network, the load current through the load resistance R_L can be calculated as follows:

$$I_L = I_N \times R_N/(R_L + R_N)$$

Example 4.22

Figure 4.39 shows the circuit containing a current source of peak value 10 A, three resistances (R_1, R_2, and R_3) and a load resistance R_L. Figure 4.40 shows the load current of the circuit. The Norton's equivalent parameters and the load current for this circuit can be calculated as follows:

Norton's equivalent resistance of the circuit, $R_N = 10 + 20 = 30 \ \Omega$
Norton's equivalent current of the circuit, $I_N = I_{rms} \times 20/30 = 4.714$ A

Norton's equivalent current measurement, shown in Fig. 4.41, is 4.714 A. Figure 4.42 depicts the plot of I_N. Norton's equivalent circuit is shown in Fig. 4.43. The load current can be calculated from this circuit (see Fig. 4.44) as

$$I_L = I_N \times R_N/(R_N + R_L) = 4.714 \times 30/(30 + 50) = 1.766 \text{ A}$$

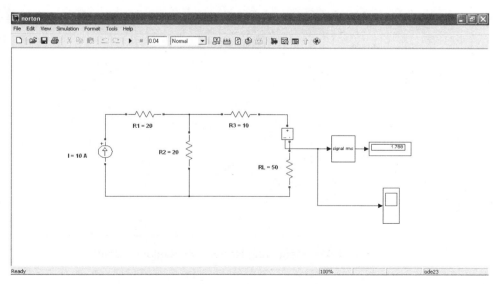

Fig. 4.39   Network for applying Norton's theorem

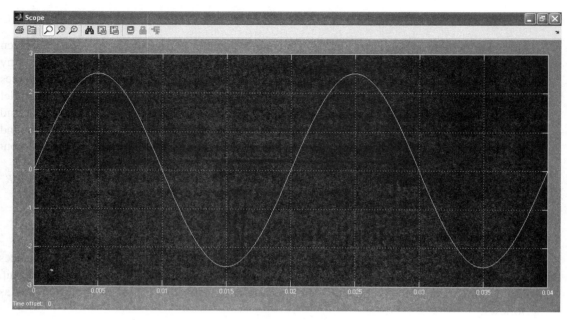

Fig. 4.40 Load current

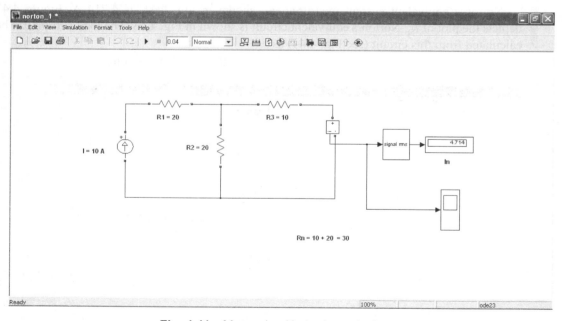

Fig. 4.41 Measuring Norton's equivalent current I_N

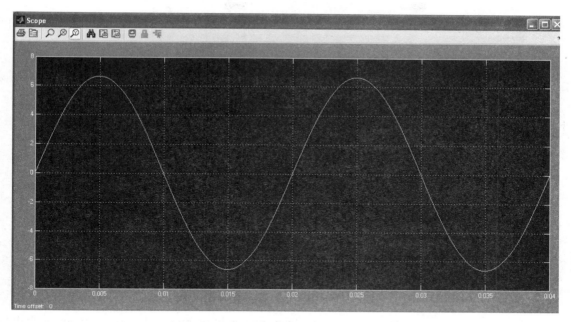

Fig. 4.42   Norton's equivalent current I_N

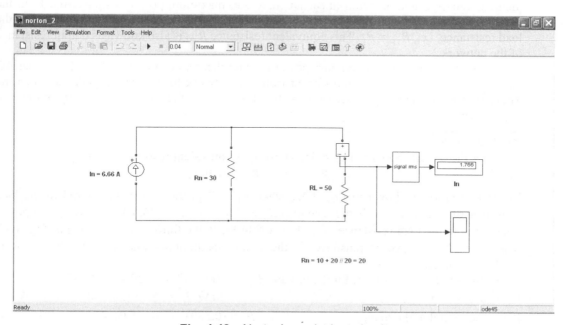

Fig. 4.43   Norton's equivalent circuit

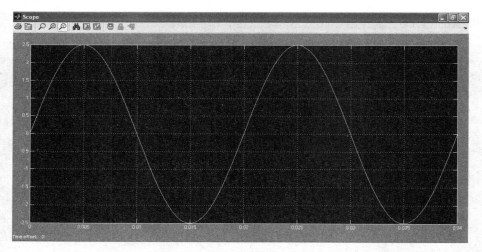

Fig. 4.44   Load current I_L obtained from the equivalent circuit

4.10.5 Maximum Power Transfer Theorem

The maximum power transfer theorem states the condition by which maximum power shall be transferred from the network to the load. In case of a DC network, maximum power transfer theorem can be stated as, 'In a linear DC network, maximum power is transferred from the network to the load when the Thevenin's equivalent resistance of the network is equal to the load resistance, i.e., $R_{TH} = R_L$. This equivalent resistance R_{TH} is also known as output resistance of the network.'

In case of AC network, maximum power transfer theorem can be stated as, 'In a linear AC network, maximum power is transferred from the network to the load impedance when the Thevenin's equivalent impedance of the network is equal to the complex conjugate of the load impedance, i.e., $Z_{TH} = Z_L^{*}$.'

Example 4.23

Consider the circuit shown in Fig. 4.45. Thevenin's equivalent resistance for this circuit is

$$R_{TH} = 30 + 20//20 = 40 \ \Omega$$

In the first case, as shown in Fig. 4.45, when $R_L > R_{TH}$, the power transferred to the load resistance of 60 Ω is 7.5 W. In the second case, as shown in Fig. 4.46, when $R_L < R_{TH}$, the power transferred to the load resistance of 20 Ω is 6.944 W. In the third case, as shown in Fig. 4.47, when $R_L = R_{TH}$, the power transferred to the load is maximum and is 7.813 W. These powers can be calculated as

Thevenin's equivalent voltage for the circuit, $V_{TH} = 100 \times 0.707 \times 20/40 = 35.35 \ \Omega$

First case, when $R_L = 20 \ \Omega, I_L = 35.35/(20 + 40) = 0.589 \ A$

Power delivered to the load, $P_L = (0.589)^2 \times 20 = 6.94 \ W$

Second case, when $R_L = 40 \ \Omega, I_L = 35.35/(40 + 40) = 0.442 \ A$

Power delivered to the load, $P_L = (0.442)^2 \times 40 = 7.81 \ W$

(Continued)

Example 4.23 (*Continued*)

Third case, when $\qquad R_L = 60 \ \Omega, \ I_L = 35.35/(40 + 60) = 0.354 \ \text{A}$

Power delivered to the load, $\qquad P_L = (0.354)^2 \times 60 = 7.5 \ \text{W}$

It is clear from these calculations that the power delivered to the load is maximum when the load resistance is equal to the Thevenin's equivalent resistance of the network. If the load is a complex quantity, i.e., $Z_L = R_L + j \, X_L$, then the maximum power will be transferred to this load when Thevenin's equivalent impedance of the network is $Z_{TH} = Z_L^{'} = R_L - j \, X_L$.

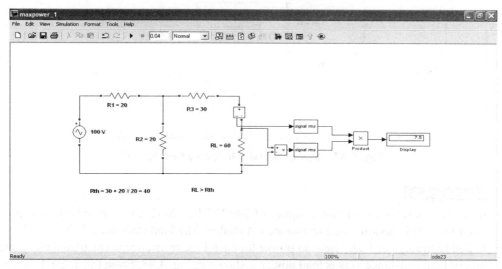

Fig. 4.45 Power transfer to load when $R_L > R_{TH}$

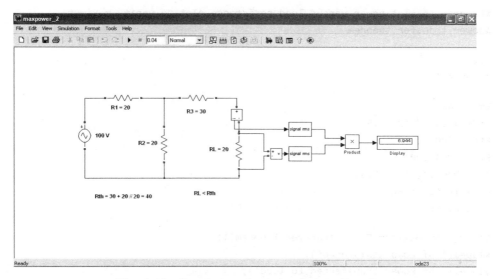

Fig. 4.46 Power transfer to load when $R_L < R_{TH}$

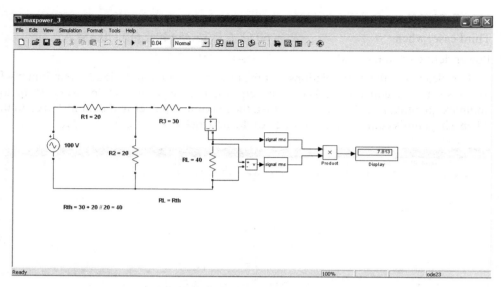

Fig. 4.47 Power transfer to load when $R_L = R_{TH}$

Example 4.24

Consider a circuit containing a supply of 340 V. The Thevenin's equivalent resistance of this circuit is 100 Ω and the load resistance is variable. The load resistance R_L is varied from 50 Ω to 200 Ω in steps of 1 Ω. The load power for each load resistance is calculated and plotted. The plot of load resistance versus load power is shown in Fig. 4.48. From this it can be seen that the power delivered to the load is maximum when $R_{TH} = R_L$.

```
%Program to plot power versus load resistance plot to verify maximum
%power transfer theorem

clc;
clear all;

Vm = 340;
Vrms = 340 / (2)^0.5;
Rth = 100;
RL = 50:1:200;
IL = Vrms./(Rth + RL);
PL = IL.^2 .* RL;
plot(RL,PL)

title('\bf Maximum Power Transfer Theorem');
xlabel('\bf Load Resistance');
ylabel('\bf Power transferred to load');
```

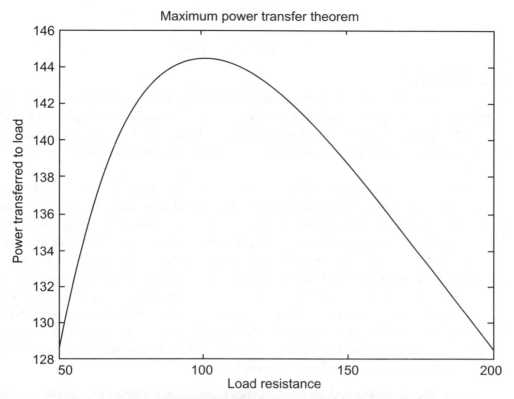

Fig. 4.48 Plot of load resistance versus load power

4.11 APPARENT, ACTIVE, AND REACTIVE POWERS

There are three types of electrical power defined for enabling the calculations in the case of AC. These are apparent power, true power, and reactive power. They can be defined as follows:

Apparent power Apparent power is the product of root mean square values of the voltage and current. It is measured in volt-ampere or VA. Thus, apparent power $= V \times I$. This is similar to the power calculation in DC networks, but unlike in DC, the apparent power does not represent true power. Instead, it represents the thermal rating of the AC machines.

True power True power is the product of rms value of the voltage and current and the power factor. It is the real power consumed by the load. It is measured in watts or W. Thus, true power $= V \times I \times \cos \phi$. The symbol is W. In DC, the real power is $V \times I$.

Reactive power Reactive power is the product of rms value of the voltage and current and the sine of the angle between the voltage and current. It is also called as wattless power. It is measured in volt-ampere-reactive or VAR. Thus, reactive power $= V \times I \times \sin \phi$. The relationship between apparent power (AP), true power (TP), and reactive power (RP) is as follows:

$$AP^2 = TP^2 + RP^2$$

or

$$VA^2 = W^2 + VAR^2$$

Example 4.25

Figure 4.49 shows the block diagram containing an AC voltage source of 500 V, 50 Hz in series with a resistance of 10 Ω and an inductance of 10 mH in one branch. This branch is in series with two parallel branches; one branch with a resistance of 35 Ω and capacitance of 40 μF, and another having a resistance of 20 Ω and an inductance of 10 mH and a load resistance of 35 Ω.

Active and reactive powers delivered to the load resistance of 35 Ω are to be estimated. For this, the Active and Reactive Power block is available in Simulink. This block calculates the active and reactive powers delivered by averaging the product of voltage and current with averaging it over one complete cycle of the fundamental frequency. This block provides active and reactive powers at the output. Thus, in order to estimate the active and reactive powers consumed by the load of 35 Ω, the rms value of the current through the load and voltage across it are measured and are given to the Active and Reactive Power block as shown in Fig. 4.49. The output of this block is given to a 1 × 2 Mux in order to separate the active and reactive powers. Active and reactive powers delivered to the load are shown in Fig. 4.50. The rms values of the load voltage and current are also shown in Fig. 4.49. Instantaneous values of the apparent power can be obtained by multiplying these values.

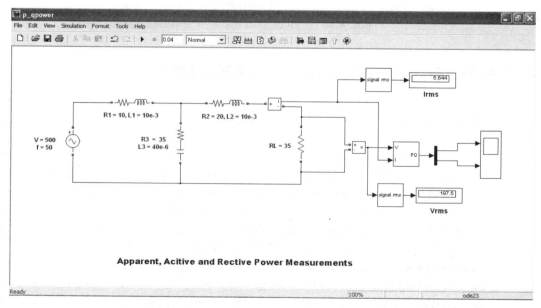

Fig. 4.49 Measurement of active and reactive powers

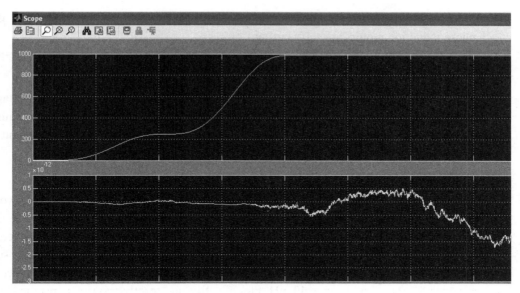

Fig. 4.50 Waveforms of active and reactive powers

4.12 THREE-PHASE SOURCE AND LOAD SIMULATION

A three-phase symmetrical source consists of three sources of same magnitude and frequency displaced by 120° from each other. These sources can be star or delta connected. A load is said to be balanced if the impedances connected in all the three phases are equal in magnitude and phase. The following are the merits of a three-phase system over a single-phase system:

1. The three-phase system is mostly suitable for power loads. As in a single-phase system, the power delivered to the load is pulsating. If the load current and voltage are in phase, the power is zero twice in one cycle and if the current leads or lags the voltage, the power is negative twice and zero four times in a cycle. For lightning, heating, and small motors, it is not problematic but in case of large machines it causes vibrations. In three-phase power system, if the load is balanced the power delivered to the load is constant and the torque will be smooth.

2. The size of a three-phase motor is less than the single-phase motor of similar rating. Normally, the output of a three-phase motor is 1.5 times the output of a single-phase motor of same rating.

3. The weight of the copper required for a three-phase transmission system is three-fourth of the weight of the copper required to transmit same amount of power at given distance and voltage.

4. The single-phase induction motors require some auxiliary means for starting while three-phase induction motors are self-starting.

5. Parallel operation of three-phase alternators is simple in comparison to single-phase alternators.

In India, the generation of three-phase electric power is done at a maximum voltage of 11 kV. As overhead transmission lines are less expensive, they are mostly used for transmission and distribution of three-phase power in India but there are also underground transmission lines that are used in metro cities.

Example 4.26

Figure 4.51 shows a model of a symmetrical three-phase power system. The peak voltage amplitude of each phase is 315 V and frequency 50 Hz. The resistance and inductance of the transmission lines are assumed to be zero. All the three voltage sources are star connected, i.e., line currents and phase currents equal. The phase voltages of this three-phase voltage source are shown in Fig. 4.53. The transmission line feeds a three-phase star connected balanced load of resistance 50 Ω and inductance 40 mH. Plot of active and reactive powers consumed by this three-phase load are shown in Fig. 4.52. From Fig. 4.52 it is clear that the power consumed by the three-phase load is constant.

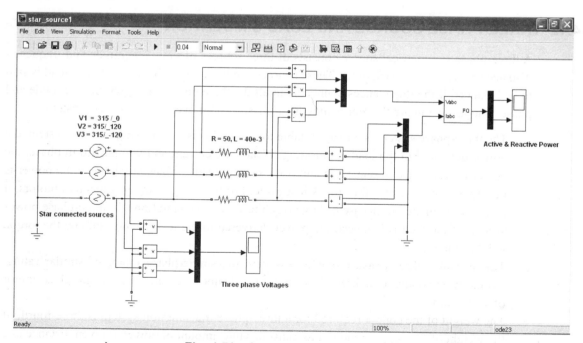

Fig. 4.51 Star-connected three-phase source

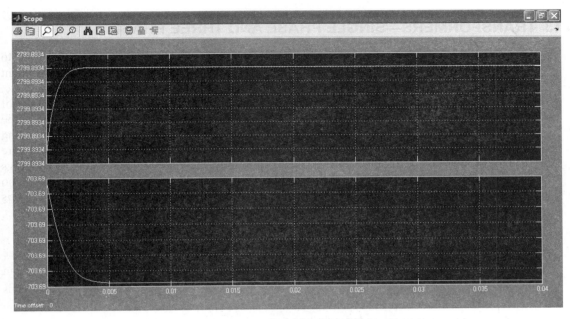

Fig. 4.52   Three-phase active and reactive powers

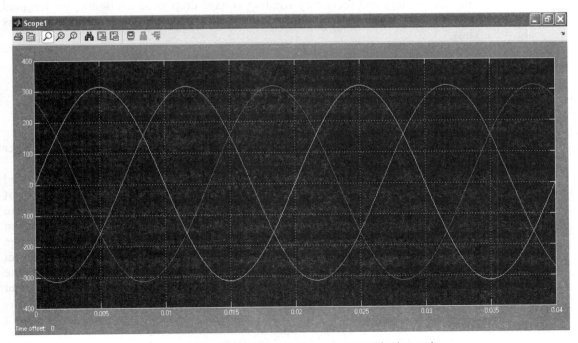

Fig. 4.53   Three-phase symmetrical supply

4.13 TRANSFORMERS—SINGLE PHASE AND THREE PHASE

A transformer is a static passive device which transfers electrical energy from one line to another without any change in the supply frequency. These are extensively used as instrument transformers for metering, to electrically isolate circuits, to step-up or step-down voltage, etc. It works on the principle of electromagnetic induction. The simplest transformer consists of two windings that are electrically isolated. When a time-varying signal is applied to one of the windings, it sets up a magnetic flux in the core. The same flux is linked with the other winding and thus generates an EMF in it. Although in the physical construction the two windings are usually wound on one another, for the sake of simplicity, the theory figures show the windings on the opposite sides of the core. Ideally, a transformer should have no winding resistance, no magnetic leakage, no iron loss, and the magnetizing current should be zero but practically, it is not possible, so we call them 'non-ideal transformers'.

The EMF equation of transformer is as follows:

RMS value of EMF induced in the primary winding, $E_1 = 4.44\,f N_1\,\phi_m$ V
RMS value of EMF induced in the secondary winding, $E_2 = 4.44\,f N_2\,\phi_m$ V
where

f is the supply frequency in Hz
N_1 is the number of primary winding turns
N_2 is the number of secondary winding turns
ϕ_m is the maximum flux in Wb

Assuming the primary and secondary winding voltage drop to be negligible, the terminal voltage across primary winding is $V_1 = E_1$ and terminal voltage across secondary winding is $V_2 = E_2$. Thus,

$$V_2/V_1 = E_2/E_1 = (4.44\,f N_2\,\phi_m)/(4.44\,f N_1\,\phi_m) = N_2/N_1$$

Now, in case of an ideal transformer, the loss of power is negligible. So,

$$V_1 I_1 = V_2 I_2$$
$$V_2/V_1 = I_1/I_2$$
$$I_1/I_2 = V_2/V_1 = E_2/E_1 = N_2/N_1$$

N_2/N_1 is known as voltage transformation ratio. If voltage transformation ratio is greater than one, it is a step-up transformer and if it is less than one it is a step-down transformer.

Transformers are specified for rated output power, voltages, frequency, etc. The rated output of the transformer is expressed in kVA rather than in kW, because the rated output of the transformer is limited by the losses in the core and windings. In transformers, there are two types of losses—constant losses also known as iron losses and variable losses also known as ohmic losses. Constant losses are independent of load current but variable losses change depending on the load current. As these losses depend upon the transformer voltage and current and not on the load power factor, the transformers are rated in kVA. Since the efficiency of the transformer is very high (approximately 95%), these losses might be ignored. Also, the input and output power factors can be assumed to be equal, i.e., $\cos\phi_1 = \cos\phi_2$.

Example 4.27

A circuit model containing a single-phase two winding transformer is shown in Fig. 4.54. The primary winding parameters of the transformer are $V_{1(rms)} = 110$ kV, $R_{1(pu)} = 0.02\ \Omega$, and

(Continued)

Example 4.27 (*Continued*)

$L_{1(pu)} = 0.08$ H and secondary winding parameters are $V_{2(rms)} = 220$ kV, $R_{2(pu)} = 0.02$ Ω, and $L_{2(pu)} = 0.08$ H. The primary winding is supplied with a voltage source of 200 V, 50 Hz. A resistance of 10 Ω is connected in series with the primary winding and of 120 Ω in series with the secondary winding. The primary and secondary winding voltages and currents are shown in Fig. 4.55 and Fig. 4.56 respectively. It can be seen from these figures that primary winding voltage is 150 V and current is 5 A while the secondary winding voltage is 300 V and current is 2.5 A.

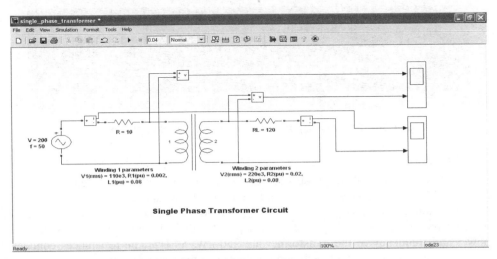

Fig. 4.54   Single-phase transformer

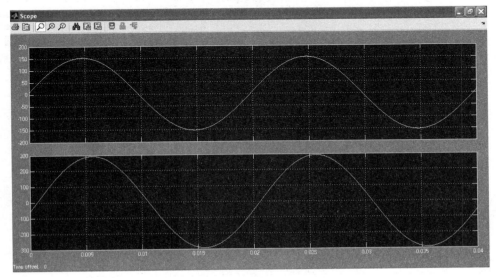

Fig. 4.55   Primary (V_1) and secondary (V_2) winding voltages

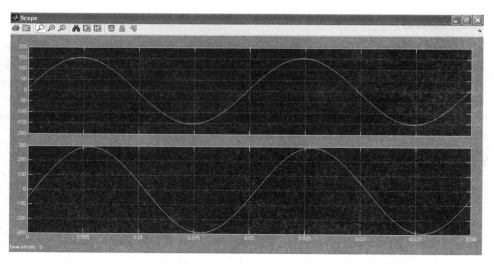

Fig. 4.56   Primary (I_1) and secondary (I_2) winding currents

Example 4.28

A circuit containing a three-phase transformer having two star-connected windings is shown in Fig. 4.57. The primary winding is fed by a three-phase supply of 300 V, 50 Hz. The primary winding parameters are $V_{1(rms)}$ = 300 V, $R_{1(pu)}$ = 0.002 Ω, and $L_{1(pu)}$ = 0.08 H and the secondary winding parameters are $V_{2(rms)}$ = 600 KV, $R_{1(pu)}$ = 0.002 Ω, and $L_{1(pu)}$ = 0.08 H. The nominal power and frequency of the transformer are 250 kW, 50 Hz. The resistances connected to the primary windings are of 5 Ω each and connected to the secondary windings are of 10 Ω each. The primary and secondary winding voltages and currents are shown in Fig. 4.58 and Fig. 4.59 respectively. The peak amplitude of the three-phase primary winding voltage is 100 V and that of current is 40 A. For secondary winding, the three-phase voltage and current are 200 V and 20 A.

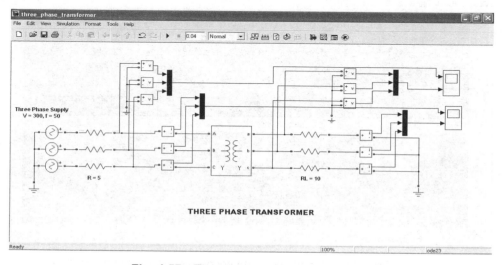

Fig. 4.57   Three-phase transformer

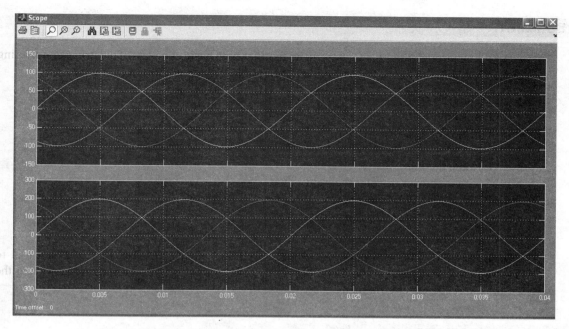

Fig. 4.58   Three-phase primary and secondary winding voltages

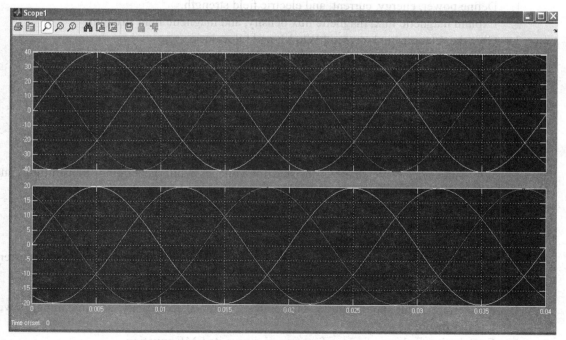

Fig. 4.59   Three-phase primary and secondary winding currents

SUMMARY

After going through this chapter we will be familiar with basic concepts of electrical engineering elaborated with the help of MATLAB and Simulink. These include the following:

- Kirchhoff's laws
- Ohm's law
- Superposition theorem
- Thevenin's theorem
- Norton's theorem
- Maximum power transfer theorem
- Series and parallel resonance
- Active, reactive, and apparent powers
- Transformers

These principles can be applied for solving circuit problems. MATLAB is used as a tool to analyze these problems in the graphical form. This tool helps in designing and analyzing the circuits in real-time environment.

REVIEW QUESTIONS

1. Define power, energy, current, and electric field strength.
2. Explain magnetic flux, magnetic flux density, and magnetic field strength.
3. Give the relationship between kWh, kJ, and kilocalorie.
4. On what factors do the resistance of a conductor depends?
5. What do you mean by conductivity and how is it related with resistivity?
6. What are the active and passive components of an electric current?
7. Differentiate between independent and controlled sources.
8. Define an ideal voltage and a current source.
9. State Kirchhoff's first and second laws with suitable examples.
10. What are the limitations of Ohm's law?
11. Why sinusoidal wave shape is insisted for voltages while generating, transmitting, and utilizing AC electric power?
12. Why is the rms value of an alternating current or voltage used to denote its magnitude?
13. What do you mean by power factor?
14. What do you mean by resonant frequency?
15. Why are series resonance and parallel resonance called voltage resonance and current resonance respectively?
16. Define quality factor of resonant circuit.
17. State and explain superposition, Thevenin's, Norton's, reciprocity, and maximum power transfer theorems.
18. Explain the working of a transformer and derive its EMF equation.

19. Mention the various losses that take place in a transformer and the factors on which they depend.
20. What is the condition for maximum efficiency of a transformer at a given power factor?
21. Define voltage regulation of a transformer.
22. What will happen when a transformer is connected to a DC supply of the same voltage rating? Explain.
23. Why are the core losses in a transformer substantially independent of the load current?

SIMULATION EXERCISES

1. In a series resonance circuit, the bandwidth and quality factor of the circuit are given by
$$BW = R/2\pi L, \quad Q = \omega_r L/R$$
where R and L are the resistance and inductance of the series circuit and ω_r is the angular resonance frequency of the circuit. Plot a graph between bandwidth and Q factor of this circuit for different resonance frequencies obtained by varying inductance L of the circuit. Take L from 10 mH to 70 mH and assume suitable values for R and C. Also conclude the effect of Q factor on bandwidth by observing this plot.

2. A coil of resistance 15 Ω and inductance 50 mH is connected in series with a capacitor of 5 μF. This circuit is supplied with an AC voltage source of 110 V rms and variable frequency. Plot the graph between supply frequency versus X_L and frequency versus X_C. Take supply frequency in range of 100 to 300 Hz.

3. Develop a MATLAB program to plot a graph between the Q factor and the supply frequency for the circuit mentioned in Q. No. 2.

4. An electrical circuit consists of two parallel branches. The first branch contains a resistance of 50 Ω and an inductance of 30 mH. The second branch contains a resistance of 10 Ω and an inductance of 20 mH. It is supplied by 440 V, 50 Hz voltage source. Plot a graph of the supply voltage and the branch currents.

5. A copper wire has a diameter of 0.05 m. Develop a program to analyze the effect of length on the resistance of the wire with the help of a graph. The resistivity of copper is 0.0178 μΩ m.

6. There are two wires, one of copper and the other of aluminum. The length of both the wires is 100 m. Write a program to compare the resistance of both the wires for various diameters with the help of a suitable graph. The resistivity of copper is 0.0178 μΩ m and that of aluminum is 0.0248 μΩ m.

7. A Wheatstone bridge consists of AB = 400 Ω, BC = 220 Ω, CD = 225 Ω, and DA = 100 Ω. A 20 V cell is connected between A and C and a galvanometer of 200 Ω resistance between B and D. Write a program to estimate the current through the galvanometer by applying Kirchhoff's laws.

8. A single-phase transformer has 400 primary and 1,000 secondary turns. The net cross-sectional area of the core is 60 cm². If the primary winding be connected to f Hz supply at 520 V, develop a program to plot the graph between maximum flux density in the core versus supply frequency and the voltage induced in the secondary and the supply frequency.

9. A circuit contains three parallel branches B_1, B_2, and B_3. The first branch contains a resistance of 20 Ω in series with an inductance of 33 mH, the second branch contains a resistance of 50 Ω and a capacitance of 33 μF, and the third branch contains a resistance of 7 Ω in series with an inductance and capacitance of 11 mH and 24 μF. Develop a model to estimate the active and reactive powers consumed in the each branch of the circuit and the total active power consumed.

10. A coil is in series with a 20 µF capacitor across a 240 V, 50 Hz supply. At resonance, the current in the circuit is 8 A. Plot a graph of inductance versus power factor for this circuit.

11. A three-phase star-connected symmetrical power supply of 314 V, 50 Hz is connected to a balanced three-phase load containing a resistance of 40 Ω and a capacitance of 54 µF. Estimate the active and reactive powers consumed by the three-phase load by building a model of this circuit.

12. A battery having an EMF of 150 V and an internal resistance of 1 Ω is connected in parallel with another battery of EMF 110 V and internal resistance of 0.5 Ω to supply a load of 25 Ω. Develop a suitable model of this circuit and estimate the current in the batteries and the load by using KVL and KCL.

13. A circuit contains two batteries in parallel. The first battery has an EMF of 2.05 V and an internal resistance of 0.05 Ω and the second battery has an EMF of 2.15 V and an internal resistance of 0.04 Ω. This combination supplies a load of 1 Ω. Calculate the load current and power consumed by the load by applying superposition principle, Norton's theorem, and Thevenin's theorem. Develop suitable models for each of the cases.

14. Three resistances of 30 Ω are connected in parallel. This parallel combination is supplied by a battery of 115 V in series with a resistance of 20 Ω. A variable load resistance is in between the parallel combination and the battery. Construct a model of this circuit and find out the value of the load resistance for maximum power transfer.

15. Twelve resistances of 10 Ω each are connected so as to form a cube. This cube is supplied by a 40 V supply across the diagonals of the cube. Develop a model of this cube and find out the resistance between the diagonals of the cube and the current in the each branch.

SIMULATION
OF RECTIFIERS

5.1 INTRODUCTION

This chapter intends to provide an idea about the design and analysis of different types of rectifiers, and evaluates their performance parameters. This chapter also covers briefly the various power electronic switches. Hence, one should be familiar with different types of power electronic switches and their functioning before going through this chapter. A basic understanding of Fourier series and Laplace transform is also required. The models of various single-phase and three-phase controlled and uncontrolled rectifiers and their output results under different loading conditions are discussed in this chapter. One can further analyze these output waveforms by experimenting the models provided in the CD along with this book. This chapter gives a practical insight into the design and working of the rectifiers. To provide better comprehension to the readers, four simulation projects are also discussed at the end of the chapter.

5.2 PERFORMANCE PARAMETERS OF RECTIFIERS

Rectifier is an AC to DC converter that provides DC output voltage or current and takes AC voltage or current at the input. Ideally, the rectifier should generate almost nil harmonics and should take sinusoidal input current from the supply at unity power factor. However, as it requires nonlinear devices like diode, thyristors, and IGBT for its construction hence, it injects harmonics in the system. Mathematically, the harmonics injected in the system can be calculated by Fourier series. There are various types of rectifier circuits and their performances are evaluated based on certain parameters such as average and rms values of the output voltage (V_{avg} and V_{rms}), average and rms values of the output current (I_{avg} and I_{rms}), output DC and AC power (P_{dc} and P_{ac}), rectifier efficiency η, form factor (FF), ripple factor (RF), transformer utilization factor (TUF), displacement factor (DF), harmonic factor (HF), power factor (PF), and crest factor (CF). These performance indicators can be defined as follows:

1. V_{avg} It is defined as the average output voltage of the rectifier. Mathematically, it can be calculated as,

$$V_{avg} = \frac{1}{T} \int_0^T V(t)dt$$

where T is the time period of the output waveform and $V(t)$ is the expression for the rectifier output voltage.

2. V_{rms} It is the root mean square value of the rectifier output voltage. Mathematically, it can be calculated as,

$$V_{rms} = \left\{ \frac{1}{T} \int_0^T V^2(t)dt \right\}^{0.5}$$

3. I_{avg} This is the average value of the rectifier output current flowing through the load. Mathematically, it can be calculated as,

$$I_{avg} = \frac{1}{T} \int_0^T I(t)dt$$

where T is the time period and $I(t)$ is the equation of the load current.

4. I_{rms} It is the root mean square value of the rectifier output current flowing through the load. Mathematically, it can be calculated as,

$$I_{rms} = \left\{ \frac{1}{T} \int_0^T I^2(t)dt \right\}^{0.5}$$

5. P_{dc} It is the total average or DC power delivered to the load. It is given by

$$P_{dc} = V_{avg} \times I_{avg}$$

where V_{avg} is the average output voltage and I_{avg} is the average output current.

6. P_{ac} It is total AC power delivered to the load. It is given by

$$P_{ac} = V_{rms} \times I_{rms}$$

7. η It is the efficiency of the rectifier and is given by

η = Output DC power of the rectifier/Input AC power to the rectifier

8. **FF** This factor measures the form of the output voltage and is given by

$$FF = V_{rms}/V_{avg}$$

9. **RF** It measures the ripple content of the output voltage waveform and is given by

$$RF = V_{ac}/V_{dc}$$

where

$$V_{ac} = \sqrt{\{V^2_{rms} - V^2_{dc}\}}. \text{ So, } RF = \sqrt{\{FF^2 - 1\}}$$

10. **TUF** Transformer utilization factor measures the transformer utilization in the rectification circuit. It is given as

$$TUF = P_{dc}/V_s I_s$$

where V_s and I_s are the rms values of transformer secondary voltage and current respectively.

11. **DF** The displacement factor is given by

$$DF = \cos \phi$$

where ϕ is the angle between the fundamental components of the input current and voltage.

12. **HF** The harmonic factor is a measure of the distortion of the waveform and is also called as total harmonic distortion (THD). HF for the input current of the rectifier is given by

$$HF = [(I_s/I_{s1})^2 - 1]^{0.5}$$

where I_s is the rms value of the input current and I_{s1} is the rms value of the fundamental component of the input current. Less the THD, the better will be the rectifier.

13. **PF** Power factor of the input side is given by

$$PF = I_{s1}/I_s \cos \phi$$

14. **CF** Crest factor is used to specify the peak current ratings of the components and devices and is given by

$$CF = I_{s(peak)}/I_s$$

where $I_{s(peak)}$ is the peak input current and I_s is the rms value of the input current.

An ideal rectifier should have $\eta = 100\%$, $V_{ac} = 0$, FF = 1, RF = 0, TUF = 1, HF = 0, PF = 1, and DF = 1.

5.3 POWER ELECTRONIC SWITCHES

There are various types of switching devices available for switching power electronic signals. These switches can be broadly classified into the following two categories:

1. Uncontrolled and controlled
2. Unidirectional and bidirectional

An uncontrolled switch is one which does not require a control signal to operate like a diode. For instance, a p-n junction diode is ON when it is forward biased (anode is more positive than cathode) by the applied waveform and is OFF when it is reversed biased (anode is less positive than cathode) by the applied waveform. So, a diode does not require an external control pulse to switch its states.

A controlled switch, on the other hand, requires a control signal to operate. The common devices are power transistor, insulated gate bipolar transistor (IGBT), silicon-controlled rectifier (SCR), triode for alternating current (TRIAC), and the gate turn-off thyristor (GTO). We know that a SCR is ON when the anode is more positive than the cathode and a control signal or a firing pulse is also present at the gate.

A switch that will allow the flow of current in only one direction (i.e., from anode to cathode) is known as a unidirectional switch. In the case of a diode, the current can flow only from anode to cathode. In the SCR also, the current can flow only from anode to cathode, so these are unidirectional switches.

In the bidirectional switches, the current can flow in both the directions. These switches can be formed by using the unidirectional switches theoretically. A TRIAC is formed by combining two SCRs and is a bidirectional switch. Current can flow in both the directions provided that the control signal is present.

The circuits that enable switching are to be specially designed and are called the control circuits.

5.4 UNCONTROLLED RECTIFIERS

The rectifiers constructed by uncontrolled switches are known as uncontrolled rectifiers as there is no control on the output power supply. The uncontrolled rectifiers normally use diodes as a switch and are also known as diode rectifiers. Depending upon the type of supply and rectifier operation cycle, they are further classified as single-phase or three-phase rectifier and half-wave or full-wave rectifier respectively. These designs are popular in industrial applications as they are cost effective and easy to build. The AC voltage of the supply can be directly rectified without using bulky transformers. The transformers are usually employed in rectifier circuits for step up or step down of the input voltage. These rectifiers are used to feed the DC bus for distribution. In some applications, they supply power to motors (resistive and inductive loads) and to power supplies (resistive and capacitive loads). It is desired that the output voltage or current is ripple free. In practice, the peak-to-peak output ripple is designed to be as small as possible and the ripple frequency to be as large as possible. As these converters use switches which have nonlinear characteristics they give rise to poor power factor and harmonics in the supply. These rectifiers find application as battery chargers (low power) and high voltage direct current (HVDC) transmission converters (high power). All these types of uncontrolled rectifiers are discussed in the following section.

5.4.1 Single-phase Half-wave Rectifier

A single-phase half-wave uncontrolled rectifier model is shown in Fig. 5.1. For simulation purposes we assume that it consists of a single-phase input supply of 314 V, 50 Hz, a diode, and a load resistance of 25 Ω. When the input is positive, the diode is forward biased and is turned ON and conducts. In the case of negative input, the diode becomes reverse biased and does not conduct. So, for the positive half, the diode conducts and for the negative half it does not conduct. When the diode conducts, the current flows from the supply to the load. The output voltage and current waveforms of the rectifier are shown in Fig. 5.2. It can be seen from the figure that the output is present for the positive half cycle and is zero for the negative half cycle. The average and rms values of the output voltage and current are also measured and are found to be as follows:

$$V_{avg} = 99.54 \text{ V}$$
$$V_{rms} = 156.5 \text{ V}$$
$$I_{avg} = 3.982 \text{ A}$$
$$I_{rms} = 6.259 \text{ A}$$

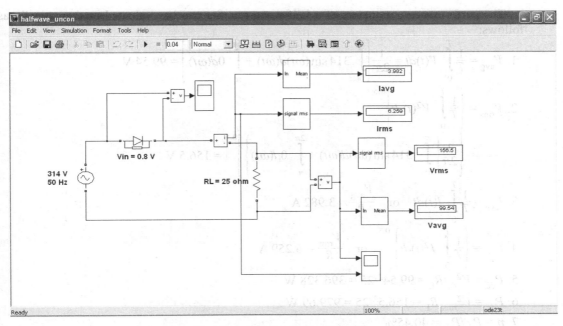

Fig. 5.1 Single-phase half-wave rectifier

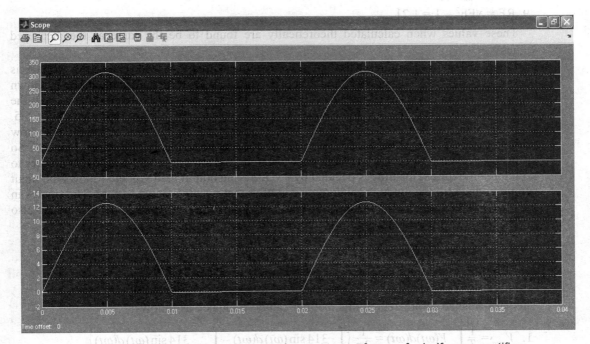

Fig. 5.2 Output voltage and current waveforms of a half-wave rectifier

The performance parameters of a half-wave rectifier can be evaluated mathematically as follows:

1. $V_{avg} = \frac{1}{T} \int_0^T V(t)dt = \frac{1}{2\pi} \left[\int_0^\pi 314 \sin(\omega t)d(\omega t) + \int_\pi^{2\pi} 0\, d(\omega t) \right] = 99.54$ V

2. $V_{rms} = \left\{ \frac{1}{T} \int_0^T V^2(t)dt \right\}^{0.5}$

$$= \left[\frac{1}{2\pi} \left(\int_0^\pi 314 \sin(\omega t)d(\omega t) + \int_\pi^{2\pi} 0\, d(\omega t) \right) \right]^{0.5} = 156.5$$ V

3. $I_{avg} = \frac{1}{T} \int_0^T I(t)dt$ or $\frac{V_{avg}}{R} = 3.982$ A

4. $I_{rms} = \left\{ \frac{1}{T} \int_0^T I^2(t)dt \right\}^{0.5}$ or $\frac{V_{rms}}{R} = 6.259$ A

5. $P_{dc} = V^2_{dc}/R_L = 99.54^2/25 = 396.328$ W

6. $P_{ac} = V^2_{rms}/R_L = 156.5^2/25 = 979.69$ W

7. $\eta = P_{dc}/P_{ac} = 40.45\%$

8. FF $= V_{rms}/V_{dc} = 1.57$

9. RF $= \sqrt{FF^2 - 1} = 1.21$

These values when calculated theoretically are found to be the same as the measured values.

Figure 5.3 shows a single-phase half-wave rectifier with a R-L load. The supply voltage is of 314 V, 50 Hz and load of 25 Ω and 200 mH. Voltage waveform across the diode is shown in Fig. 5.4. It can be observed from this figure that there are negative voltage spikes across the diode which are due to the load inductance. The load current and voltage are shown in Fig. 5.5. It is clearly visible that even if the supply voltage goes negative, the current continues to flow through the load. This is because the current cannot change suddenly through an inductor. So in this case, even if the supply voltage goes negative, the current in the R-L load continues to flow for some time. This time is decided by $\tau = L/R$. The output load voltage follows the input supply voltage till the diode is conducting and after that it settles down to zero. The ripples in the load voltages are due to the inductive load. The following performance parameters are also measured for this circuit and are measured as follows:

$$V_{avg} = 71.08 \text{ V}$$
$$V_{rms} = 118.4 \text{ V}$$
$$I_{avg} = 1.653 \text{ A}$$
$$I_{rms} = 2.674 \text{ A}$$

The performance parameters of this rectifier can be evaluated as follows:

1. $V_{avg} = \frac{1}{T} \int_0^T V(\omega t)d(\omega t) = \frac{1}{2\pi} \left(\int_0^\pi 314 \sin(\omega t)\, d(\omega t) - \int_\pi^{\pi+\beta} 314 \sin(\omega t)\, d(\omega t) \right)$

$= 71.08$ V

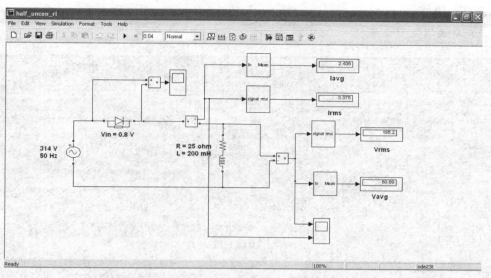

Fig. 5.3   Single-phase half-wave rectifier with R-L load

2. $V_{rms} = \left\{ \dfrac{1}{T} \displaystyle\int_0^T V^2(\omega t)d(\omega t) \right\}^{0.5}$

$\qquad = \left[\dfrac{1}{2\pi} \left\{ \displaystyle\int_0^\pi (314)^2 d(\omega t) - \int_\pi^{\pi+\beta} (314\sin(\omega t))^2\, d(\omega t) \right\} \right]^{0.5}$

$\qquad = 118.4\ V$

3. $I_{avg} = \dfrac{1}{T} \displaystyle\int_0^T I(\omega t)d(\omega t) = \dfrac{1}{2\pi} \left\{ \displaystyle\int_0^{\pi+\beta} 314\sin(\omega t)d(\omega t) \right\}$

$\qquad = 1.653\ A$

4. $I_{rms} = \left\{ \dfrac{1}{T} \displaystyle\int_0^T I^2(\omega t)d(\omega t) \right\}^{0.5}$

$\qquad = \left[\dfrac{1}{2\pi} \left\{ \displaystyle\int_0^{\pi+\beta} 314\sin(\omega t)d(\omega t) \right\} \right]^{0.5}$

$\qquad = 2.674\ A$

5. FF $= V_{rms}/V_{dc} = 1.67$

6. RF $= \sqrt{FF^2 - 1} = 1.33$

where the time for which the current flows after the supply voltage crosses zero level is given by $\tau = L/R = 200 \times 10^{-3}/25 = 0.008$ s and delay angle $\beta = \tau \times (0.01/\pi)$. It is evident that the ripple factor and the form factor of the output waveform have increased.

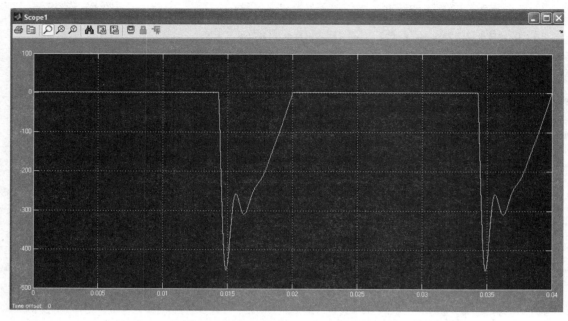

Fig. 5.4 Voltage waveform across the diode

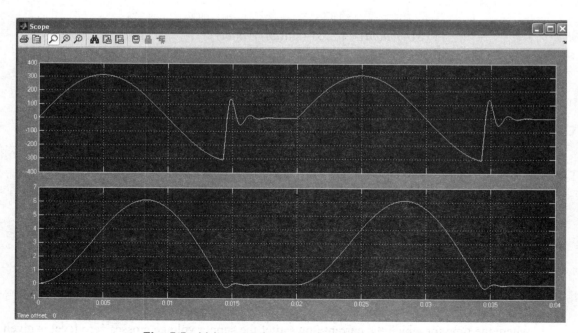

Fig. 5.5 Voltage and current waveforms across the R-L load

Figure 5.6 shows a single-phase half-wave rectifier with R-C load. The supply voltage is 314 V, 50 Hz and load of 25 Ω and 200 µF. The voltage waveform across the diode is shown in Fig. 5.7. It can be observed from this figure that the voltage waveform across the diode is negative. Also, it can be observed that the voltage across the diode goes negative even if the supply voltage is positive. This is due to the charging of the load capacitor. The load voltage and current waveforms are shown in Fig. 5.8. It is clearly visible from the figure that the load voltage goes to 314 V and then falls to 254 V and then almost remains constant there. As voltage cannot change suddenly across the terminals of a capacitor, so in this case, even if the supply voltage goes negative the load voltage remains positive. The output load current is required only to charge the load capacitor up to 314 V, so it remains there only for a small duration of the complete cycle. The following performance parameters are also measured for this circuit and are given as follows:

$$V_{avg} = 284.2 \text{ V}$$
$$V_{rms} = 284.6 \text{ V}$$
$$I_{avg} = 0.1651 \text{ A}$$
$$I_{rms} = 0.4522 \text{ A}$$

The performance parameters of this rectifier can be evaluated as follows:

1. $V_{avg} = 284.2$ V
2. $V_{rms} = 284.6$ V
3. $I_{avg} = 0.1651$ A
4. $I_{rms} = 0.4522$ A
5. FF $= V_{rms}/V_{dc} = 1.001$
6. RF $= \sqrt{\text{FF}^2 - 1} = 0.053$

It is evident that the ripple factor and the form factor of the output waveform have decreased further.

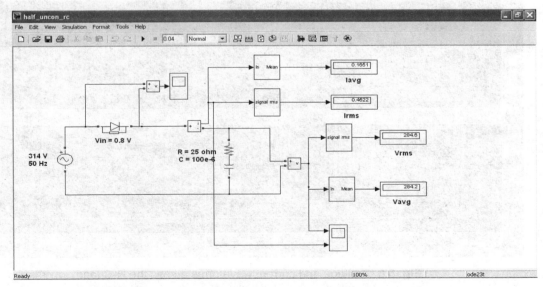

Fig. 5.6 Single-phase half-wave rectifier with R-C load

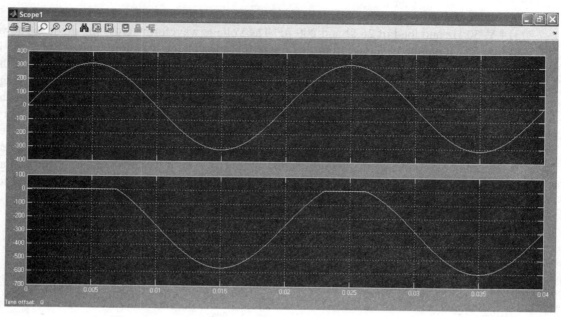

Fig. 5.7 Voltage waveform across the diode

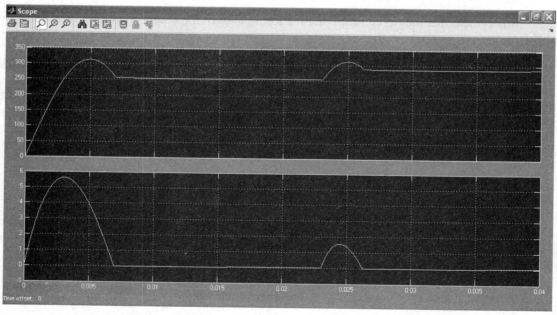

Fig. 5.8 Voltage and current waveforms across the R-C load

5.4.2 Single-phase Full-wave Rectifier

An uncontrolled single-phase full-wave rectifier with purely resistive load is shown in Fig. 5.9. This rectifier consists of a 314 V, 50 Hz AC supply, four diodes (D1, D2, D3, and D4) and a load resistance of 31.4 Ω. When the supply voltage is positive, diodes D1 and D2 are ON, so the load current flows from A to B. When the supply voltage is negative, diodes D3 and D4 are turned ON and again the load current flows from A to B. The load current and voltage waveforms can be seen in Fig. 5.10. The load current is in phase with the load voltage, and the frequency of the output waveform is 100 Hz.

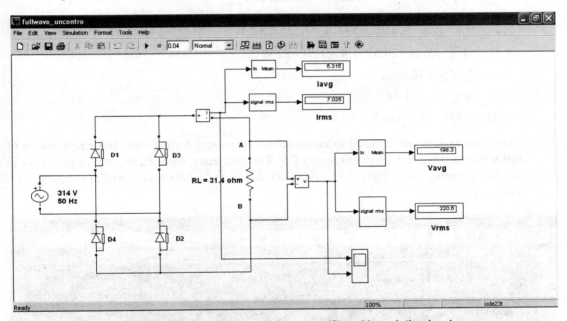

Fig. 5.9 Single-phase full-wave rectifier with resistive load

The following measurements are obtained by simulating this model:
$$V_{avg} = 198.3 \text{ V}$$
$$V_{rms} = 220.6 \text{ V}$$
$$I_{avg} = 6.315 \text{ A}$$
$$I_{rms} = 7.025 \text{ A}$$

The performance parameters of full-wave rectifier can be evaluated as follows:

1. $V_{avg} = \dfrac{1}{T}\displaystyle\int_0^T V(\omega t)\,d(\omega t) = \dfrac{1}{2\pi}\left\{\displaystyle\int_0^\pi 314\sin(\omega t)\,d(\omega t) - \displaystyle\int_\pi^{2\pi} 314\sin(\omega t)\,d(\omega t)\right\}$

   $= 198.3 \text{ V}$

2. $V_{rms} = \left\{\dfrac{1}{T}\displaystyle\int_0^T V^2(t)\,d(t)\right\}^{0.5}$

$$= \left[\frac{1}{2\pi} \left\{ \int_0^{\pi} (314\sin(\omega t))^2 \, d(\omega t) + \int_{\pi}^{2\pi} (314\sin(\omega t))^2 \, d(\omega t) \right\} \right]^{0.5}$$

$$= 220.6 \text{ V}$$

3. $I_{\text{avg}} = \frac{1}{T} \int_0^T I(t) dt = 6.315$ A (Calculate as V_{avg})

4. $I_{\text{rms}} = \left\{ \frac{1}{T} \int_0^T I^2(t) dt \right\}^{0.5} = 7.025$ A (Calculate as V_{rms})

5. $P_{\text{dc}} = V^2_{\text{dc}}/R_L = 198.3^2/31.4 = 1,252.32$ W

6. $P_{\text{ac}} = V^2_{\text{rms}}/R_L = 220.6^2/31.4 = 1,549.82$ W

7. $\eta = P_{\text{dc}}/P_{\text{ac}} = 80.80\%$

8. FF $= V_{\text{rms}}/V_{\text{dc}} = 1.11$

9. RF $= \sqrt{\text{FF}^2 - 1} = 0.481$

These parameters can be used to compare the efficiency, form factor, and ripple factor of a full-wave rectifier and a half-wave rectifier. The efficiency, form factor, and ripple factor of a full-wave rectifier is 80.80%, 1.11, and 0.481 and that of a half-wave rectifier is 40.45%, 1.57, and 1.21.

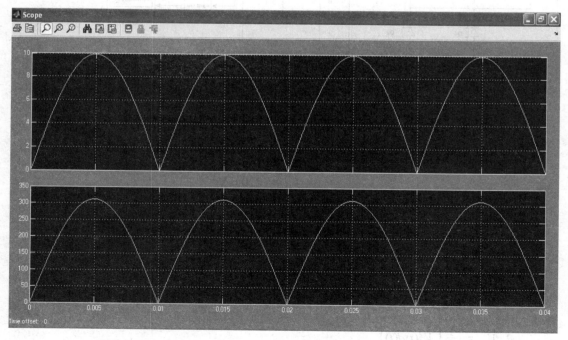

Fig. 5.10 Load current and voltage waveforms for a resistive load

Figure 5.11 shows a single-phase full-wave rectifier with R-L load. For this model, the supply voltage is 314 V, 50 Hz and load consists of 31.4 Ω and 30 mH. The load voltage and current waveforms are shown in Fig. 5.12. It is clear from the figure that the load voltage is

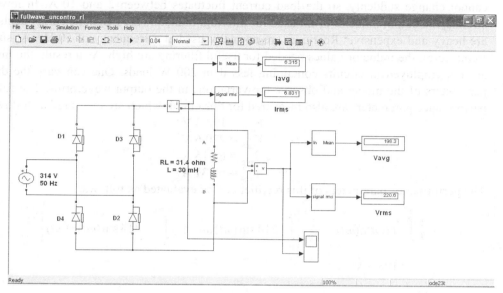

Fig. 5.11 Single-phase full-wave rectifier with R-L load

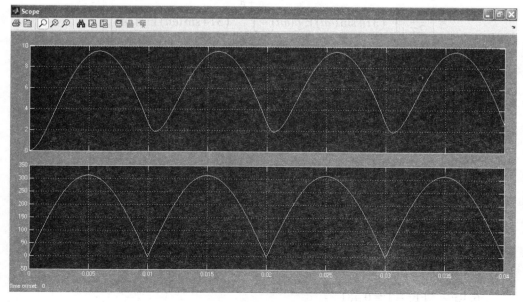

Fig. 5.12 Load current and voltage waveforms for R-L load

positive irrespective of the sign of the input supply voltage. When the supply voltage is positive, diodes D1 and D2 conduct and when the supply voltage is negative, diodes D3 and D4 conduct so the voltage at the load terminals remains positive. As the current flowing through an inductor cannot change suddenly, so the load current fluctuates between 2 and 10 A. In low-power circuits, the output inductor is not used for filtering. Inductors are magnetic components which are heavy and expensive. Rectifiers are normally used at a frequency of 50 Hz. At such low frequencies, the inductor values needed for useful filtering are high. As a result, the inductors are not employed in circuits containing less than 200 W loads. One can vary the different parameters of the model and observe the variations in the output waveforms. The following performance parameters are also measured for this circuit which are measured as follows:

$$V_{avg} = 198.3 \text{ V}$$
$$V_{rms} = 220.6 \text{ V}$$
$$I_{avg} = 6.315 \text{ A}$$
$$I_{rms} = 6.831 \text{ A}$$

The performance parameters of this rectifier can be evaluated as follows:

1. $$V_{avg} = \frac{1}{T}\int_0^T V(\omega t)d(\omega t) = \frac{1}{2\pi}\left(\int_0^\pi 314\sin(\omega t)\,d(\omega t) - \int_\pi^{\pi+\beta} 314\sin(\omega t)\,d(\omega t)\right)$$
$$= 198.3 \text{ V}$$

2. $$V_{rms} = \left(\frac{1}{T}\int_0^T V^2(\omega t)d(\omega t)\right)^{0.5}$$
$$= \left[\frac{1}{2\pi}\left(\int_0^\pi (314\sin(\omega t))^2\,d(\omega t) - \int_\pi^{\pi+\beta} (314\sin(\omega t))^2\,d(\omega t)\right)\right]^{0.5}$$
$$= 220.6 \text{ V}$$

3. $$I_{avg} = \frac{1}{T}\int_0^T I(\omega t)d(\omega t)$$
$$= \frac{1}{2\pi}\left[\int_0^{\pi+\beta} 314\sin(\omega t)d(\omega t)\right] = 6.315 \text{ A}$$

4. $$I_{rms} = \left(\frac{1}{T}\int_0^T I^2(\omega t)d(\omega t)\right)^{0.5}$$
$$= \left[\frac{1}{2\pi}\left(\int_0^{\pi+\beta} 314\sin(\omega t)d(\omega t)\right)\right]^{0.5} = 6.831 \text{ A}$$

5. FF $= V_{rms}/V_{dc} = 1.11$

6. RF $= \sqrt{FF^2 - 1} = 0.481$

where $\tau = L/R = 30 \times 10^{-3}/31.4 = 0.955$ ms and $\beta = (\pi/0.01) \times \tau$ rad.

5.4.3 Three-phase Half-wave Rectifier

Figure 5.13 shows a three-phase half-wave rectifier model containing a three-phase voltage source of amplitude 315 V, 50 Hz, three diodes D1, D2, and D3, and a resistive load of 105 Ω. The waveforms of the three phases are defined as follows:

$$V_1 = V_m \times \sin \omega t$$
$$V_2 = V_m \times \sin (\omega t - 120°)$$
$$V_3 = V_m \times \sin (\omega t + 120°)$$

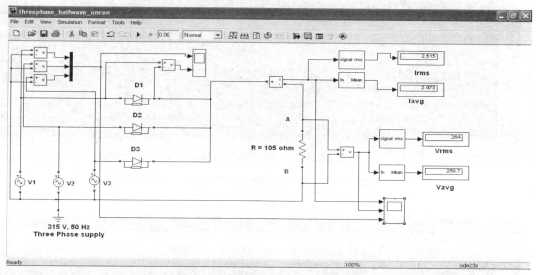

Fig. 5.13   Three-phase half-wave rectifier

Figure 5.14 shows three-phase input supply voltage waveform and voltage across the diode D1. Diode D1 will be ON only when voltage V1 is most positive, i.e., more positive than V2 and V3, otherwise it will remain OFF. Load current and voltage waveforms are shown in Fig. 5.15. The frequency of the load voltage is 150 Hz and each diode conducts for $\pi/3$ time duration in one input supply cycle.

The following performance parameters are measured for this rectifier:

$$V_{avg} = \frac{3}{\pi} \int\limits_{0}^{\pi/3} 315 \sin(\omega t) d(\omega t) = 259.7 \text{ V}$$

$$V_{rms} = \left[\frac{3}{\pi} \int\limits_{0}^{\pi/3} \{315 \sin(\omega t)\}^2 d(\omega t) \right]^{0.5} = 264 \text{ V}$$

Similarly, I_{avg} and I_{rms} can be evaluated as

$$I_{avg} = 2.47 \text{ A}$$
$$I_{rms} = 2.51 \text{ A}$$

So, for this rectifier FF = 1.016 and RF = 0.1827.

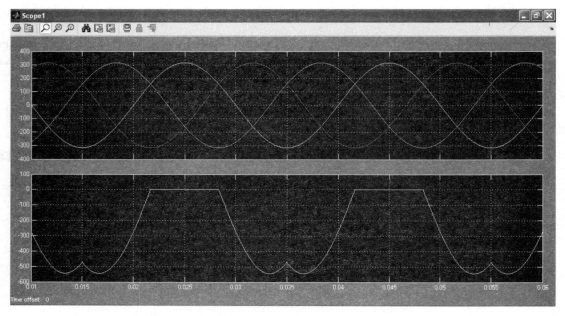

Fig. 5.14 Voltage waveform across diode D1

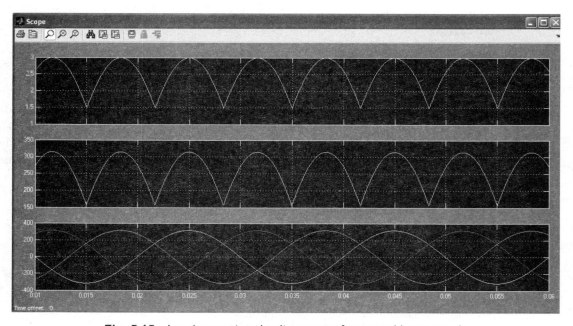

Fig. 5.15 Load current and voltage waveforms and input supply

5.4.4 Three-phase Full-wave Rectifier

The functions of a three-phase full-wave rectifier are similar to that of a single-phase full-wave rectifier except that the input supply is of three phase and six diodes are used for rectification. The three-phase rectifier can be used for converting a three-phase AC supply into a DC or high voltage DC supply. A bridge-type three-phase full-wave rectifier model is shown in Fig. 5.16. It contains six diodes D1, D2, D3, D4, D5, and D6. Diodes D1, D3, and D5 are ON when phase voltages V1, V2, and V3 are positive. Similarly, diodes D2, D6, and D4 are ON when phase voltages V1, V2, and V3 are negative.

For the continuous supply of current to the load, it is necessary that any one pair of diodes is ON at any instant of time during each input cycle. For instance if phase V1 is most positive and V3 is most negative, then diodes D1 and D2 are ON and the load current flows from A to B. Similarly, when phase V2 is most positive and V3 is most negative, diodes D3 and D2 are ON. In the model shown in Fig. 5.16, the supply voltage is 315 V, 50 Hz, three-phase and the load resistance is of 105 Ω. The voltage across diodes D1 and D2 is shown in Fig. 5.17. Load

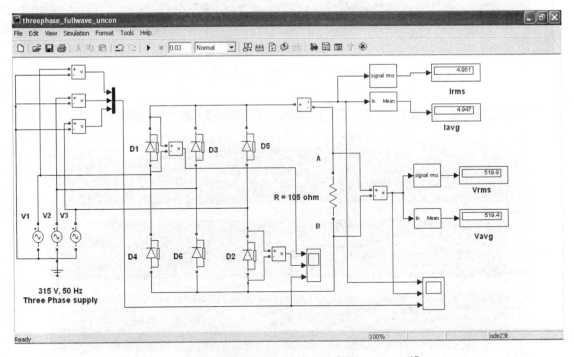

Fig. 5.16   Three-phase full-wave rectifier

current and voltage waveforms are shown in Fig. 5.18. It can be observed that the load voltage is in phase with the load current and the output ripple frequency is 300 Hz. So, each diode conducts for a period of $2\pi/6$ in one output cycle. The following measurements are taken for this model:

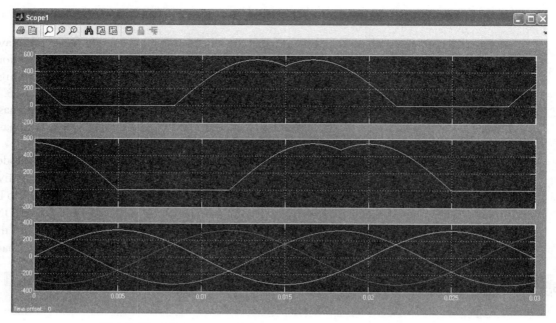

Fig. 5.17 Voltage waveform across diodes D1 and D2 and input supply

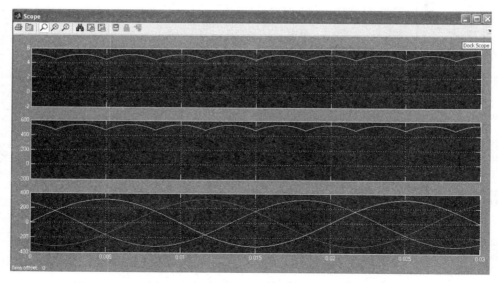

Fig. 5.18 Current and voltage waveforms for the resistive load

1. $V_{avg} = 519.4$ V
2. $V_{rms} = 519.9$ V
3. $I_{avg} = 4.947$ A
4. $I_{rms} = 4.951$ A

V_{avg} and V_{rms} can be evaluated as follows:

$$V_{avg} = \frac{6}{\pi} \int\limits_{0}^{\pi/6} 315 \sin(\omega t) d(\omega t)$$

$$= 519.4 \text{ V}$$

$$V_{rms} = \left[\frac{6}{\pi} \int\limits_{0}^{\pi/6} \{315 \sin(\omega t)\}^2 d(\omega t) \right]^{0.5}$$

$$= 519.9 \text{ V}$$

From these measured parameters, we can evaluate FF = 1.0009 and RF = 0.04. So, we can say that for a three-phase rectifier, the form factor is almost 1 and the ripple factor is quite low.

5.5 CONTROLLED RECTIFIERS

The controlled rectifiers are built using the controlled switches. In these types of rectifiers, the output power delivered to the load can be controlled by controlling the switches. Normally used control switches are thyristors, IGBTs, MOSFETs, TRIAC, and MCTs. In these types of rectifiers, the output voltage and current can be controlled by varying the firing angle to the gate pulse applied to the switches. These converters are very popular in battery charger systems, electromagnet power supplies, and variable speed DC drives ranging from fractional horsepower to several thousands of horsepower. Low power converters are normally of a single-phase type while medium to high power converters are of three phase. These types of rectifiers can be further divided as single-phase or three-phase rectifiers based on the type of input supply. These types of rectifiers are further discussed in the following subsection.

5.5.1 Single-phase Half-wave Rectifier

A single-phase half-wave converter model is shown in Fig. 5.19. Input can be a sinusoidal voltage or a current source. The output is desired to be a DC voltage or current at the load terminals. The control signal can be an angle, pulse width, a voltage, or a current signal depending on the control switch used. Mostly, the sources in power electronics are voltage sources and the converters are used to supply controllable DC current or voltage or a combination of both. These converters are built up of controlled switches and reactive components like L, C, R, and transformers (if required).

In the model shown in Fig. 5.19, the single phase supply is of magnitude 314 V, 50 Hz. A thyristor is used as a controlled switch and the load resistance is of 54 Ω. Now when the supply

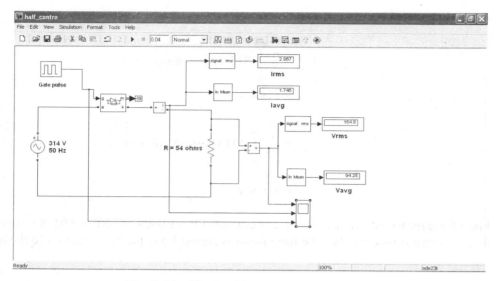

Fig. 5.19 Single-phase half-wave controlled rectifier

voltage is positive, the thyristor is forward biased but conducts only when a control pulse is present at its gate terminal. When the supply voltage is negative, the thyristor becomes reverse biased and does not conduct even if the control pulse is applied. Thyristor firing pulse, load voltage, and current are shown in Fig. 5.20. It can be observed that load voltage follows the supply voltage after the control pulse is applied. Until the control pulse is applied, the load

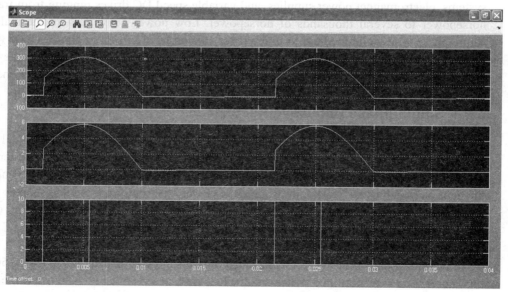

Fig. 5.20 Voltage and current waveforms of load and gate pulse of the thyristor

voltage is zero. Also, the load voltage and current are in the same phase as the load is purely resistive. In this simulation, the firing pulse is applied after 0.0015 s delay or in other words, the firing angle is $\pi \times 0.15$ rad (as the time period is 0.02 s = 2π). After simulating this model, the following measurements are taken:

1. I_{rms} = 2.867 A 2. I_{avg} = 1.745 A

3. V_{rms} = 154.8 V 4. V_{avg} = 94.25 V

These values can also be obtained mathematically as follows:

1. $V_{avg} = \dfrac{1}{T}\displaystyle\int_0^T V(\omega t)d(\omega t) = \dfrac{1}{2\pi}\displaystyle\int_\alpha^\pi 314\sin(\omega t)d(\omega t) = 94.25$ V

2. $V_{rms} = \left\{\dfrac{1}{T}\displaystyle\int_0^T V^2(\omega t)d(\omega t)\right\}^{0.5} = \left\{\dfrac{1}{2\pi}\displaystyle\int_\alpha^\pi (314\sin(\omega t))^2\,d(\omega t)\right\} = 154.8$ V

3. $I_{avg} = V_{avg}/R = 94.25/54 = 1.745$ A

4. $I_{rms} = V_{rms}/R = 154.8/54 = 2.867$ A

So, for this rectifier FF = 1.64 and RF = 1.3. As the form factor and the ripple factor also depend on the firing angle α, they vary with the values of α.

Figure 5.21 illustrates a model of a single-phase half-wave controlled rectifier with R-L load. This model is same in principle as the model shown in Fig. 5.19 expect that the load is inductive in this case. The supply voltage is of magnitude 240 V, 50 Hz frequency. The load applied to

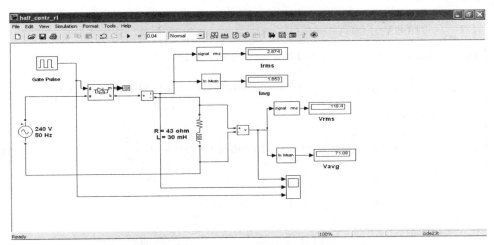

Fig. 5.21 Single-phase half-wave controlled rectifier with R-L load

this rectifier is 43 Ω and 30 mH. The thyristor is ON only when the supply voltage is positive and the control signal is present at the gate terminal of the thyristor. Figure 5.22 shows load voltage and current along with the control signal. As the current cannot change instantaneously through an inductor, so even if the supply voltage goes negative, the load current continues to flow for a short duration of time (τ). This short duration of time for which current flows after

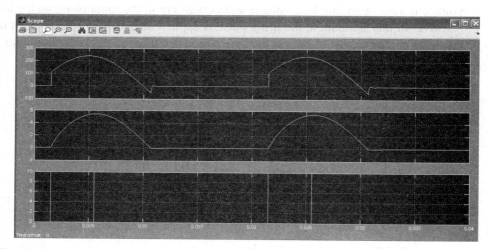

Fig. 5.22 Voltage and current waveforms of load and gate pulse of the thyristor

the supply voltage is zero is decided by $\tau = L/R = 30 \times 10^{-3}/43 = 0.00069$ s. The supply voltage becomes zero at 0.01 s in the first cycle and the load current becomes zero at 0.01069 s. The larger the load resistance, the lesser the time (τ) required to bring the load current to its steady-state value. After simulating this model, the following parameters are measured:

1. $I_{rms} = 2.674$ A
2. $I_{avg} = 1.653$ A
3. $V_{rms} = 118.4$ V
4. $V_{avg} = 71.08$ V

These values can also be obtained mathematically as follows:

1. $V_{avg} = \dfrac{1}{T}\displaystyle\int_0^T V(\omega t)d(\omega t) = \dfrac{1}{2\pi}\left(\displaystyle\int_\alpha^\pi 240\sin(\omega t)d(\omega t) - \displaystyle\int_\pi^{\pi+\beta} 240\sin(\omega t)d(\omega t)\right)$

$= 71.08$ V

2. $V_{rms} = \dfrac{1}{T}\left\{\displaystyle\int_0^T V^2(\omega t)d(\omega t)\right\}^{0.5} = \left[\dfrac{1}{2\pi}\left\{\displaystyle\int_\alpha^\pi (240\sin(\omega t))^2 d(\omega t) - \displaystyle\int_\pi^{\pi+\beta} (240\sin(\omega t))^2 d(\omega t)\right\}\right]^{0.5}$

$= 118.4$ V

3. $I_{avg} = \dfrac{1}{T}\displaystyle\int_0^T I(\omega t)d(\omega t) = \dfrac{1}{2\pi}\left\{\displaystyle\int_\alpha^{\pi+\beta} 240\sin(\omega t)d(\omega t)\right\}$

$= 1.653$ A

4. $I_{rms} = \dfrac{1}{T}\left\{\displaystyle\int_0^T I^2(\omega t)d(\omega t)\right\}^{0.5} = \left[\dfrac{1}{2\pi}\left\{\displaystyle\int_\alpha^{\pi+\beta} 240\sin(\omega t)d(\omega t)\right\}\right]^{0.5}$

$= 2.674$ A

So, for this rectifier FF = 1.66 and RF = 1.32. As the form factor and the ripple factor also depend on the firing angle α and the delay angle β, they vary with the values of α and β. The delay angle is given by

$$\beta = (\pi/0.01) \times \tau$$

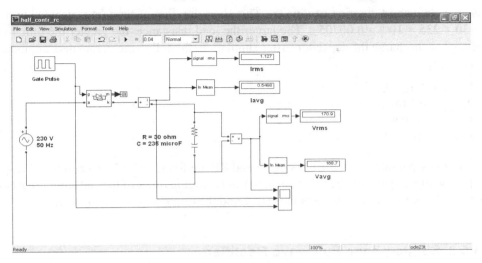

Fig. 5.23   Single-phase half-wave controlled rectifier with R-C load

Figure 5.23 illustrates a model of a single-phase half-wave controlled rectifier with a R-C load. This model is also same in principle as the model shown in Fig. 5.19 expect that the load is capacitive in this model. The supply voltage is of magnitude 230 V, 50 Hz frequency. The load applied to this rectifier is 30 Ω and 235 μF. The thyristor is ON only when the supply voltage is positive and the control signal is present at the gate terminal of the thyristor. The gate pulse is applied with a phase delay time of 0.0015 s. Figure 5.24 shows load voltage and current along with the control signal. As the voltage cannot change instantaneously across a capacitor, so even if the supply voltage goes negative, the load voltage remains positive as the capacitor

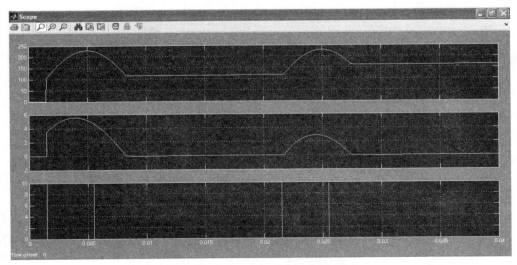

Fig. 5.24   Voltage and current waveforms of load and gate pulse of the thyristor

charges in a short duration of time (τ). This short duration of time (τ) is decided by $\tau = R \times C$ = $30 \times 235 \times 10^{-6} = 0.00705$ s. The larger the load resistance, the greater the time required to charge or discharge the capacitor. After simulation, the following parameters are measured for this model:

1. $I_{rms} = 1.127$ A
2. $I_{avg} = 0.5498$ A
3. $V_{rms} = 170.9$ V
4. $V_{avg} = 168.7$ V

From these parameters, the values of FF and RF are found to be 1.013 and 0.162 respectively. These values depend on the firing angle (α) and charging times of the capacitor (τ). Also, in case of R-C load, the form factor was close to 1 and the ripple factor is also lower.

5.5.2 Single-phase Full-wave Rectifier

A single-phase full-wave controlled rectifier uses four controlled switches in order to provide full wave at the load terminals. One pair of controlled switches are turned ON when the input is positive and the other pair of switches are turned ON when the input is negative. Figure 5.25 illustrates a model of a single-phase full-wave rectifier containing four thyristors and feeding a resistive load. The input supply has a magnitude of 235 V, 50 Hz frequency. The load resistance is of 45 Ω and the control pulses are applied after a delay of 0.0025 s. The frequency of the

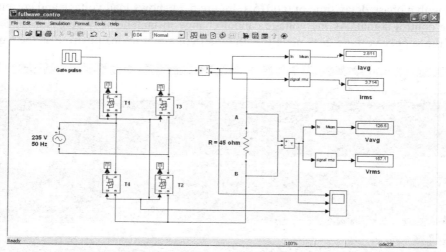

Fig. 5.25   Single-phase full-wave controlled rectifier

firing pulses is 100 Hz. When the supply voltage is positive, thyristors T1 and T2 are forward biased as anode of T1 is positive and cathode of T2 is negative. Now when the gate pulse arrives at the gate terminals, T1 and T2 start conducting and supply current to the load from A to

B. In other case, when the supply voltage is negative, T3 and T4 become forward biased as the anode of T3 is positive and the cathode of T4 is negative. When the gate pulse arrives, T3 and T4 start conducting and supply current to the load from A to B. Load voltage and current are in phase as shown in Fig. 5.26. After simulating this model, the following readings are noted:

1. $I_{avg} = 2.811$ A
2. $I_{rms} = 3.714$ A
3. $V_{avg} = 126.5$ V
4. $V_{rms} = 157.1$ V

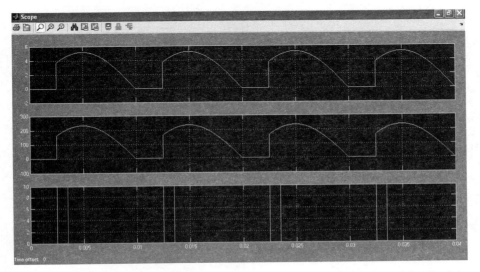

Fig. 5.26 Load current and voltage waveforms and gate pulse of the thyristor

These parameters can be evaluated mathematically as follows:

1. $V_{avg} = \dfrac{1}{T}\displaystyle\int_0^T V(\omega t)d(\omega t) = \dfrac{1}{2\pi}\displaystyle\int_\alpha^\pi 235\sin(\omega t)d(\omega t)$

$= 126.5$ V

2. $V_{rms} = \left\{\dfrac{1}{T}\displaystyle\int_0^T V^2(\omega t)d(\omega t)\right\}^{0.5} = \left\{\dfrac{1}{2\pi}\displaystyle\int_\alpha^\pi (235\sin(\omega t))^2 d(\omega t)\right\}^{0.5}$

$= 151.7$ V

3. $I_{avg} = V_{avg}/R = 126.5/45 = 2.811$ A
4. $I_{rms} = V_{rms}/R = 157.1/45 = 3.714$ A

So, for $\alpha = \pi \times 0.25$ rad, FF = 1.24, and RF = 0.733. The form factor and the ripple factor vary with the firing angle and can be verified by varying the phase delay of the gate pulse in the circuit model.

Figure 5.27 illustrates a circuit model of a single-phase full-wave controlled bridge-type rectifier feeding an inductive load. The supply voltage has a magnitude of 235 V and frequency of 50 Hz. The load resistance and inductance are 45 Ω and 30mH respectively. The magnitude of the gate pulse is 10 V and with a frequency of 100 Hz with a phase delay time of 0.0025 s. The rectifier bridge consists of four thyristors T1, T2, T3, and T4. When the supply voltage is

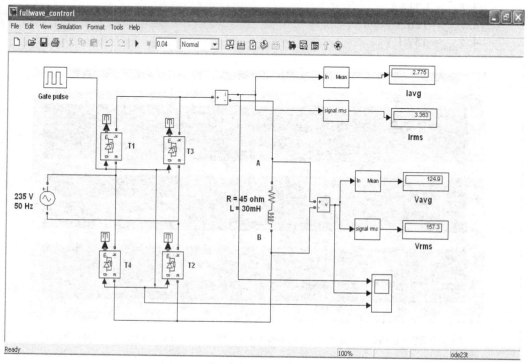

Fig. 5.27   Single-phase full-wave rectifier with inductive load

positive, T1 and T2 are forward biased and conduct when a gate pulse is present. In the other case, when the supply voltage is negative, T3 and T4 are forward biased and conduct when the gate pulse is present. In both these cases, the current through the load flows from A to B. As current flowing through an inductor cannot change instantaneously, the load current continue to flow for a short duration of time even if the supply voltage goes negative. The time for which the load current flows due to the load inductance is decided by $\tau = L/R = 0.03/45 = 0.00067$ s. Load current, voltage, and gate pulse are shown in Fig. 5.28. This model is simulated and the following parameters are measured for this model:

1. $I_{avg} = 2.775$ A
2. $I_{rms} = 3.363$ A
3. $V_{avg} = 124.9$ V
4. $V_{rms} = 157.3$ V

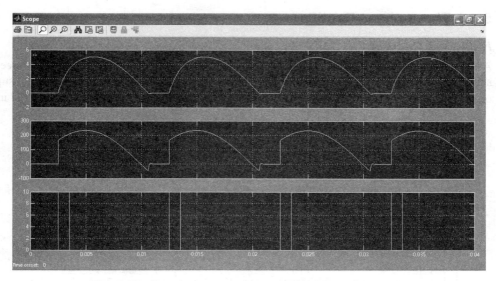

Fig. 5.28 Load current, voltage, and thyristor firing pulse

These parameters can be mathematically evaluated as follows:

1. $V_{avg} = \dfrac{1}{T}\displaystyle\int_0^T V(\omega t)d(\omega t) = \dfrac{1}{2\pi}\left\{\displaystyle\int_\alpha^\pi 235\sin(\omega t)d(\omega t) - \displaystyle\int_\pi^{\pi+\beta} 235\sin(\omega t)d(\omega t)\right\}$

$= 124.9\ \text{V}$

2. $V_{rms} = \left\{\dfrac{1}{T}\displaystyle\int_0^T V^2(\omega t)d(\omega t)\right\}^{0.5} = \left[\dfrac{1}{2\pi}\left\{\displaystyle\int_\alpha^\pi (235\sin(\omega t))^2 d(\omega t) - \displaystyle\int_\pi^{\pi+\beta}(235\sin(\omega t))^2 d(\omega t)\right\}\right]^{0.5}$

$= 157.3\ \text{V}$

3. $I_{avg} = \dfrac{1}{T}\displaystyle\int_0^T I(\omega t)d(\omega t) = \dfrac{1}{2\pi}\left\{\displaystyle\int_\pi^{\pi+\beta} 235\sin(\omega t)d(\omega t)\right\}$

$= 2.775\ \text{A}$

4. $I_{rms} = \left\{\dfrac{1}{T}\displaystyle\int_0^T I^2(\omega t)d(\omega t)\right\}^{0.5} = \left\{\dfrac{1}{2\pi}\left(\displaystyle\int_\pi^{\pi+\beta} 235\sin(\omega t)d(\omega t)\right)\right\}^{0.5}$

$= 3.363\ \text{A}$

Hence, for this rectifier, FF = 11.26 and RF = 0.766. As the form factor and the ripple factor depend also on the firing angle α and the delay angle β, they vary with the values of α and β. The delay angle is given by

$$\beta = (\pi/0.01)\times\tau$$

5.5.3 Three-phase Rectifiers

Half-wave Rectifier

A three-phase half-wave rectifier model is shown in Fig. 5.29. This model consists of a three-phase star-connected input supply, three thyristors and is feeding a load resistance. The input supply is of magnitude 415 V and frequency 50 Hz. Three thyristors T1, T2, and T3 are connected in each

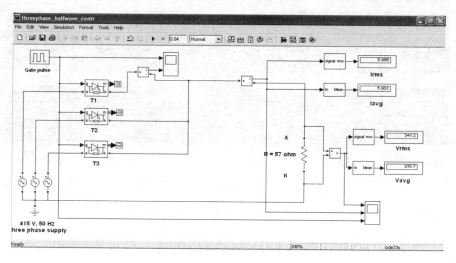

Fig. 5.29 Three-phase half-wave controlled rectifier

phase. The load is purely resistive and is of 57 Ω. The gate pulse is applied at the gate terminals of each thyristor and is of 10 V, frequency 150 Hz, and a phase delay time of 0.0025 s. Thyristor T1 is forward biased when the phase in which it is connected is most positive and similarly thyristors T2 and T3 are forward biased when their resistive phases are most positive. To turn ON these thyristors, a gate pulse is required when the thyristor is forward biased. In this model, the gate pulse applied at a phase delay time of 0.0025 s. Figure 5.30 shows the current flowing through T1 and the gate pulse and Fig. 5.31 shows load current, voltage, and gate pulse. It can be observed from Fig. 5.31 that the frequency of the output ripple is 150 Hz. After simulating this model, the following parameters are obtained:

1. $I_{rms} = 5.986$ A
2. $I_{avg} = 5.802$ A
3. $V_{rms} = 341.2$ V
4. $V_{avg} = 341.2$ V

These parameters can be evaluated mathematically as follows:

1. $V_{avg} = \dfrac{1}{T} \int_0^T V(\omega t)d(\omega t) = \dfrac{3}{\pi} \int_\alpha^{\pi/3} 415 \sin(\omega t)d(\omega t) = 330.7$ V

2. $V_{rms} = \left\{ \dfrac{1}{T} \int_0^T V^2(\omega t)d(\omega t) \right\}^{0.5} = \left\{ \dfrac{3}{\pi} \int_\alpha^{\pi/3} (415 \sin(\omega t))^2 d(\omega t) \right\}^{0.5} = 341.2$ V

3. $I_{avg} = V_{avg}/R = 94.25/54 = 5.802$ A

4. $I_{rms} = V_{rms}/R = 154.8/54 = 5.8$ A

Hence, for this rectifier model, FF = 1 and RF = 0. So in case of three-phase rectifiers, the values of RF and FF are almost equal to the ideal values.

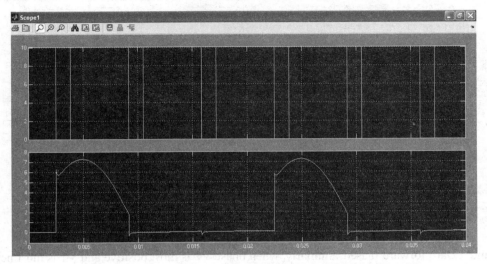

Fig. 5.30   Firing pulse and current waveform for thyristor T1

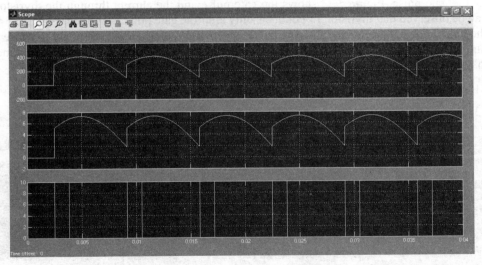

Fig. 5.31   Load voltage and current waveforms along with the firing pulses

Full-wave Rectifier

A three-phase full-controlled rectifier bridge-type model is shown in Fig. 5.32. In this model, the input supply is three-phase star-connected, and its magnitude is 415 V and frequency of 50 Hz. Six thyristors are used to form the rectifier bridge as shown in Fig. 5.32. The load is purely resistive and is of 57 Ω. The gate pulse has a magnitude of 10 V, frequency 150 Hz, and phase delay time of 0.0025 s. When phase V1 is most positive and V3 is most negative, thyristors T1

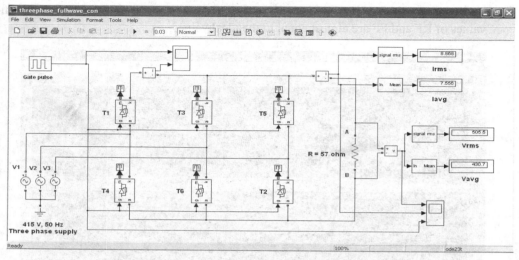

Fig. 5.32 Three-phase full-wave controlled rectifier

and T2 are forward biased. Similarly, when phase V2 is most positive and V1 is most negative, thyristors T3 and T4 are forward biased; and when phase V3 is most positive and V2 is most negative, thyristors T5 and T6 are forward biased. These forward-biased thyristors are turned ON by the gate pulse. Figure 5.33 shows the gate pulse and the current through the thyristor T1. It can be observed that thyristor T1 conducts only after the gate pulse is applied. Figure 5.34 shows the load voltage, current, and the gate pulse. The output voltage and current are in phase and the output ripple frequency is 300 Hz. After simulating this model, the following measurements are taken:

1. $I_{\text{rms}} = 8.68$ A

2. $I_{\text{avg}} = 7.556$ A

3. $V_{\text{rms}} = 505.5$ V

4. $V_{\text{avg}} = 505.5$ V

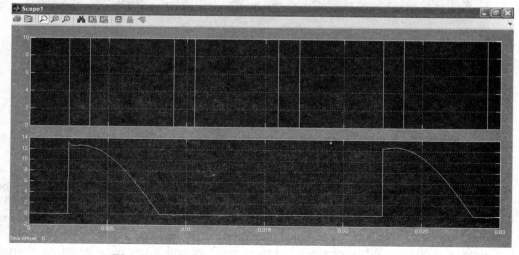

Fig. 5.33 Gate pulse and current of thyristor T1

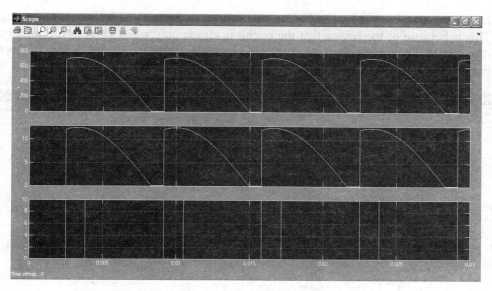

Fig. 5.34   Load voltage, current, and gate pulse

These parameters can be evaluated mathematically as follows:

1. $V_{avg} = \dfrac{1}{T} \displaystyle\int_0^T V(\omega t)\, d(\omega t) = \dfrac{1}{3\pi} \displaystyle\int_\alpha^{\pi/6} 415 \sin(\omega t)\, d(\omega t) = 430.7 \text{ V}$

2. $V_{rms} = \left\{ \dfrac{1}{T} \displaystyle\int_0^T V^2(\omega t)\, d(\omega t) \right\}^{0.5} = \left\{ \dfrac{1}{3\pi} \displaystyle\int_\alpha^{\pi/6} (415 \sin(\omega t))^2 d(\omega t) \right\}^{0.5} = 505.5 \text{ V}$

3. $I_{avg} = V_{avg}/R = 505.5/57 = 7.556 \text{ A}$

4. $I_{rms} = V_{rms}/R = 154.8/54 = 8.86 \text{ A}$

The difference between the measured and calculated values might be due to the calculation errors of the software. Hence, for this rectifier model, FF = 1.17 and RF = 0.6144. So, in the case of three-phase rectifiers, the values of RF and FF are near to the ideal values.

5.6 SIMULATION PROJECTS

Project 1

The rectifiers made up of a combination of controlled and uncontrolled switches are called semi-controlled rectifiers. For instance, if a single-phase full-wave rectifier uses two SCRs and two diodes for rectification, it is a semi-controlled rectifier. A model of a semi-controlled single-phase full-wave rectifier is shown in Fig. 5.35. This is same as the full-wave rectifiers discussed in earlier sections except that the semi-controlled rectifiers use a combination of SCRs (T1 and T3) and diodes (D3 and D4). The output of this rectifier is same as that of a controlled rectifier using four SCRs. When the input supply is positive, T1 and D2 are forward biased but the current is supplied to the load only when the SCR T1 is triggered by a gate pulse. This is because of the reason that the circuit is closed only when T1 is ON. Similarly, when the input supply is negative, T3 and D4 conduct when T3 is ON. The advantage of this type of rectifier is that it can be fully controlled by using only two SCRs instead of four.

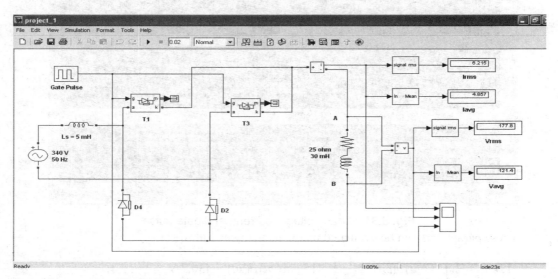

Fig. 5.35   Semi-controlled single-phase full-wave rectifier

In the model shown in Fig. 5.35, the input supply is 340 V peak at 50 Hz. The load parameters are $R = 25 \ \Omega$ and $L = 30$ mH. The control gate pulse is of 5 V, 100 Hz, pulse width of 10% at 0.0045 s time delay. Figure 5.36 shows the load current, voltage, and control pulse. It is clear

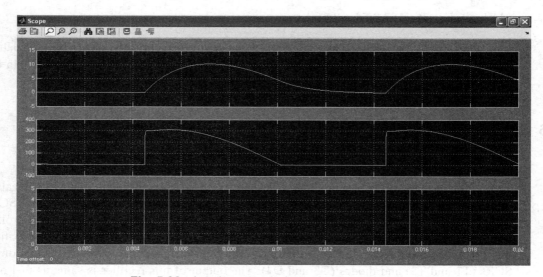

Fig. 5.36   Load current, voltage, and control pulse

from the output that it functions similar to that of a controlled rectifier. After simulating this model, the following readings are observed:

1. $I_{rms} = 6.215$ A
2. $I_{avg} = 4.857$ A
3. $V_{rms} = 177.8$ V
4. $V_{avg} = 121.4$ V

These values can be verified mathematically like previous models.

Project 2

Figure 5.37 shows another model of a single-phase semi-controlled rectifier. In this model, the SCR pair T1 and T2 conducts when the supply is positive. If the supply is negative, diodes D3 and D4 conduct. In this case the positive half of the input supply can be controlled by the control signal but the negative half is uncontrolled. For this model, the supply and control pulse has the same specifications as used in model of Fig. 5.35. The load is of 40 Ω, 20 mH, and a source inductance of 1 mH is also taken. The source inductance helps in freewheeling the

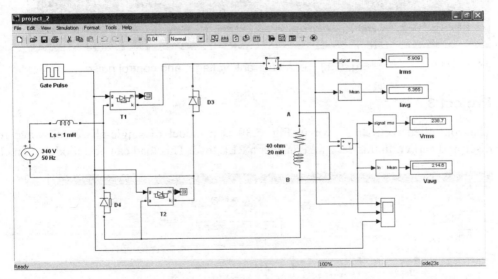

Fig. 5.37 Semi-controlled single-phase full-wave rectifier

current through freewheeling diodes used with the switches. It also increases the effective load resistance. These can be observed by simulating this model for various values of L_s. One can do this exercise with this model. The output current, voltage, and control pulses are shown in Fig. 5.38. It can be observed from the figure that in the positive cycle, the output starts commencing after the firing angle (α) but in the negative input cycle, the output exists for complete half cycle. The following observations, which can be verified mathematically, are taken:

1. $I_{rms} = 5.909$ A
2. $I_{avg} = 5.363$ A
3. $V_{rms} = 238.7$ V
4. $V_{avg} = 214.6$ V

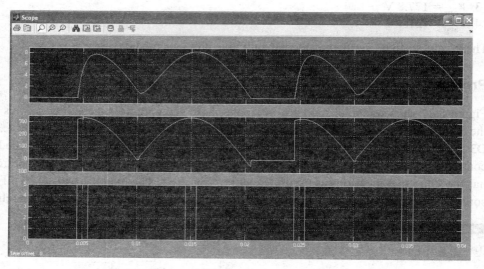

Fig. 5.38   Load current, voltage, and control pulse

Project 3

The controlled rectifier shown in Fig. 5.39 is a model of single-phase full-wave rectifier discussed earlier. In this case we consider a RLE load. This load can be a model of an electric

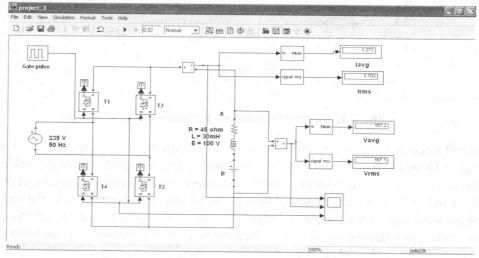

Fig. 5.39   Single-phase full-wave controlled rectifier with RLE load

motor or a storage system. The functions of this rectifier are similar to that of a controlled rectifier. At load terminals, the output voltage cannot fall below the battery voltage. When the load voltage is more than the battery terminal voltage, the battery is charged; otherwise, it discharges through the rectifier circuit and the R–L load. The output load current and voltage are shown in Fig. 5.40. The observations taken for this model, which can be mathematically calculated, are as follows:

1. $I_{rms} = 1.702$ A
2. $I_{avg} = 1.272$ A
3. $V_{rms} = 167.1$ V
4. $V_{avg} = 157.2$ V

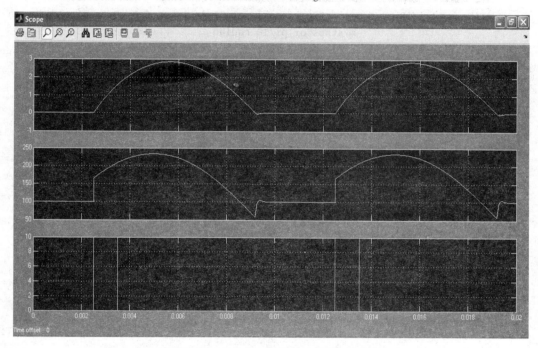

Fig. 5.40 Load current, voltage, and firing pulse

Project 4

```
% This program plots graph between Ripple Factor and firing angle for a
% single phase half wave and full wave rectifier in case of a resistive
% load

clc
clear all
clf

%Input Supply
```

```
%%%%%%%%%%%%%%%%%%%%%%%%%%%%%%%%%%%%%%%%%%%%%%%%%%%%%%%%%%%%%%%%%%%%%%%%%
t = 0:10e-6:.04;  %    Time  is  taken  from  0  sec  to  0.04  sec  in  steps  of
10 micro sec.
f = 50;           %    Frequency is taken as 50 Hertz.
w = 2 * pi * f;   % Angular frequency in rad/sec.
Vm = 240;         %    Peak amplitude 240

%Average & RMS output voltage values for half wave
%%%%%%%%%%%%%%%%%%%%%%%%%%%%%%%%%%%%%%%%%%%%%%%%%%%%%%%%%%%%%%%%%%%%%%%%%
alpha = 0:pi/10: 2 * pi; % Firing angle alpha is taken from 0 to pi radians in
                         % steps of pi/10 radian
Vdc_half = (Vm / pi) .* (1 + cos(alpha)); % Average output voltage
Vrms_half = (Vm / 2) .* ((1 / pi)* (pi - alpha + (sin(2 .* alpha) / 2)));
                         % RMS output voltage

% Average & RMS output voltage values for full wave
%%%%%%%%%%%%%%%%%%%%%%%%%%%%%%%%%%%%%%%%%%%%%%%%%%%%%%%%%%%%%%%%%%%%%%%%%
Vdc_full = ((2 * Vm) / pi) .* cos(alpha);
Vrms_full = Vm / sqrt(2);

% Ripple factor

FF_half = Vrms_half ./ Vdc_half;
RF_half = ((FF_half .^ 2) - 1).^0.5;

FF_full = Vrms_full ./ Vdc_full;
RF_full = ((FF_full .^ 2) - 1).^0.5;

% Plot alpha versus RF and aplha versus FF

figure(1)
plot(alpha, RF_half, 'k')
title('\Variantion in ripple faction with firing angle for half wave rectifier');
xlabel('\bf Firing Angle');
ylabel('\bf Ripple Factor');

figure(2)
plot(alpha, RF_half, 'k')
title('\Variantion in ripple faction with firing angle for full wave rectifier');
xlabel('\bf Firing Angle');
ylabel('\bf Ripple Factor');
```

The output plots of this program are shown in Figs 5.41 and 5.42.

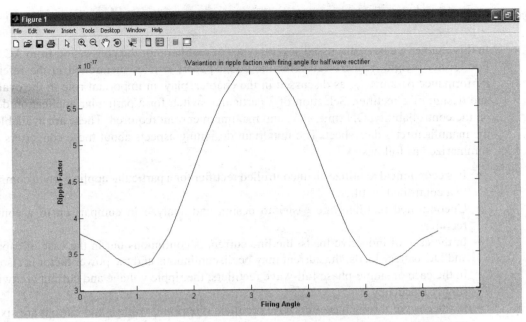

Fig. 5.41   Ripple factor variation with α in case of half-wave controlled rectifier

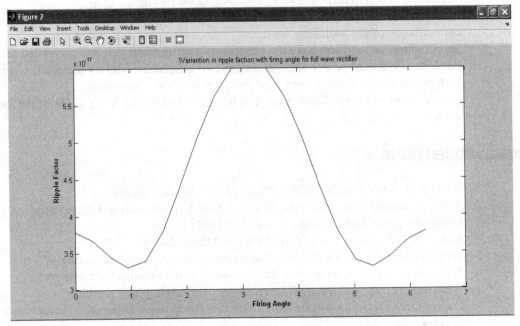

Fig. 5.42   Ripple factor variation with α in case of full-wave controlled rectifier

SUMMARY

In this chapter we have discussed several types of rectifiers for the conversion from AC to DC. For further insight, readers can refer the books/papers/articles mentioned in the references. Performance parameters, as discussed in the chapter, play an important role in the evaluation and design of a rectifier. Selection of a particular switch for a particular application depends on the controllability, ON time, PIV, and maximum current required. These are available from the manufacturer's data sheet. The important designing aspects about these converters can be summarized as follows:

- It is economical to utilize an uncontrolled rectifier for a particular application in comparison to a controlled rectifier.
- Uncontrolled rectifiers are easier to design and analyze in comparison to a controlled rectifier.
- In the case of inductive loads, the line current is continuous but in the case of capacitive and DC battery loads, the current may be discontinuous and the power factor is also poor.
- In the case of single-phase full-wave rectifiers, the ripple voltage and current are twice the supply frequency.
- In the case of three-phase full-wave rectifiers, the ripple voltage and current are six times the supply or line frequency.
- The source inductance L_s reduces the output voltage as it gives rise to an effective source resistance. It is also responsible for notching in the AC line.
- All techniques, except firing angle control as used in this chapter required forced commutation of the switching device.
- Semi-controlled rectifiers have the benefits of both uncontrolled and controlled rectifiers.
- For low to medium power applications, the single-phase rectifiers are used; whereas, for high power applications, three-phase rectifiers are more convenient.
- Ripple factor of controlled rectifier can be controlled by the firing angle of the switching device.

REVIEW QUESTIONS

1. What is the basic function of a rectifier? Define an ideal rectifier.
2. What are controlled and uncontrolled switches? Explain with suitable examples.
3. Define the performance parameters for a rectifier circuit.
4. Elaborate the significance of form factor and ripple factor.
5. What are the advantages of controlled rectifier over uncontrolled rectifiers?
6. What are the advantages of uncontrolled rectifiers over controlled rectifiers?
7. What are the merits of three-phase rectifiers over single-phase rectifiers?
8. Define the SCR firing angle. How does this firing angle affect the performance of the rectifier?
9. How is the ripple factor affected by changing the load inductance in case of a single-phase rectifier?

10. Mention the applications of three-phase rectifiers.
11. Why are transformers used in rectifiers?
12. What is a harmonic factor?
13. Discuss the effects of source inductance on rectifier output.
14. Which of the rectifier is more efficient, full wave or half wave, and why?
15. What is the output frequency in case of a single-phase full-wave rectifier if the input frequency is f?
16. What will be the output frequency in case of a three-phase full-wave rectifier if the input supply frequency is f?
17. Discuss the problems of electromagnetic interference.

SIMULATION EXERCISES

1. Construct a model of a half-wave uncontrolled rectifier having an input supply of 220 V rms, 50 Hz. Calculate the voltage form factor and ripple for load resistance of 10,100, and 1,000 Ω. Determine the average power delivered and plot the output voltage, current I_d through the diode, and the source voltage V_s waveforms.

2. Construct a rectifier model to produce 100 V $\pm$ 5% at power levels up to 1 kW. The input supply is of 240 V rms and 60 Hz.

3. What are the benefits of using SCRs in place of diodes in the rectifier circuit? Can SCRs provide control over the output voltage? If so, plot the average output voltage as a function of SCR phase delay angle.

4. An SCR bridge takes input from a 210 V rms, 50 Hz source and delivers it to a load at an average of 72 V. Construct a model/program for this circuit and determine the value of α in order to meet this requirement. Plot the gate pulse, input supply, and voltage across one thyristor. Also determine the ripple and form factor for this circuit.

5. Develop a single-phase rectifier as shown in Fig. 5.11. If the peak voltage of the input supply is 314, the supply frequency $f = 50$ Hz, and the load resistance $R = 50$ Ω, determine the value of the load inductance in order to limit the load current harmonics to 3% of the average values I_{dc}. Plot this load current and voltage and verify the result obtained.

6. A single-phase bridge rectifier as shown in Fig. 5.9 has a load resistance of 150 Ω. The supply peak voltage is of 180 V and frequency of the supply is 50 Hz. Determine the average and rms values of the current through each diode. Also plot the voltage across each diode and determine their peak inverse voltages.

7. A single-phase bridge rectifier as shown in Fig. 5.9 is designed using a capacitor C so that the ripple factor of the load voltage is 5%. Estimate the value of this capacitor and plot the load voltage after applying this capacitor at the load terminals. Also determine the PF of this circuit.

8. Develop a three-phase bridge type rectifier feeding an inductive load. The input supply is star connected with peak phase voltage of 170 V at 50 Hz. The load resistance is of 170 Ω. Estimate the value of load inductance to limit the load current to 1% of the average value I_{dc}. Plot the load current and voltage for this model.

9. Develop a three-phase bridge-type controlled rectifier feeding R-L load. The input supply is 415 V peak at 50 Hz and the load resistance is of 200 Ω. Plot the load current and voltage for $L = 10$ mH, 100 mH, and $1,000$ mH. Also vary the firing angle of the SCRs used and determine the values of average and rms load voltage. Take $\alpha = 10°$, 25°, and 30°. Analyze the results obtained.

10. Develop a three-phase full-wave bridge-type uncontrolled rectifier feeding a R-L load of 40 Ω, 70 mH. The input supply is of 340 V peak at 50 Hz and has an inductance L_s. Determine the load average and rms current values for $L_s = 5$ mH, 10 mH, and 50 mH.

SUGGESTED READING

Agrawal, J.P., *Power Electronic Systems: Theory and Design*, Pearson Education, New Delhi, 2006.

Davis, T.A., *MATLAB Primer*, 8th Ed., Chapman & Hall/CRC, Florida, 2011.

Driscoll, T.A., *Learning MATLAB*, SIAM, Philadelphia, PA, 2009.

Dubey, G.K., S.R. Doradla, A. Joshi, and R.M.K. Sinha, *Thyristorised Power Controllers*, Wiley, New York, 1986.

Kassakian, J.G., M.F. Sclecht, and G. Vergassian, *Principles of Power Electronics*, Addison-Wesley, Boston, 1991.

Kelley, A.W., and W.F. Yadusky, 'Rectifier design for minimum line-current harmonics and maximum power factor', *IEEE Transactions on Power Electronics*, Vol. 7, No. 2, pp. 332–341, April 1992.

Krein, P.T., *Elements of Power Electronics*, Oxford University Press, New York, 2009.

Mohan, N., T.M. Undeland, and W.P. Robbins, *Power Electronics: Converters, Applications and Design*, 2nd Ed., John Wiley & Sons, Inc., New York, 1995.

Pisctaway, NJ, 'Practices and requirements for general purpose thyristor drives', *IEEE Standard 597*, 1983.

Rashid, M.H., *Power Electronics: Circuits, Devices and Applications*, 2nd Ed., Prentice-Hall Inc., New Jersey, 1993.

Schaefer, J., *Rectifier Circuits—Theory and Design*, John Wiley & Sons, New York, 1975.

Tse, C.K., S.C. Wong, and M.H.L. Chow, 'On lossless switched-capacitor power converters', *IEEE Transactions on Power Electronics*, Vol. 10, No. 3, pp. 286–291, May 1995.

Vithayathil, J., *Power Electronics: Principles and Applications*, McGraw-Hill Inc., New York, 1995.

6

SIMULATION
OF INVERTERS

6.1 INTRODUCTION

In electrical engineering, the device which converts the DC power into AC power at a desired voltage and frequency is known as an inverter. The required DC input is taken from an existing power supply network or from a rotating alternator through a rectifier or a battery. The output voltage could be fixed or variable at a fixed or variable frequency. A variable output voltage can be obtained by varying the input DC voltage and maintaining the gain of the inverter constant. Also, the gain of the inverter can be varied by keeping the DC input voltage constant, which can be accomplished by pulse width modulation (PWM) control. The output of an ideal inverter should be constant in magnitude and should be sinusoidal (free of harmonics). In practical cases, however, the output is non-sinusoidal and contains harmonics of higher and/or lower order. The square or quasi-square wave output of the inverter is acceptable in low or medium power (lighting and domestic) applications. For high power (industrial) applications, sinusoidal output with low distortion is required. The harmonic content of the output wave can be reduced by high frequency switching techniques. The applications of inverters in industries are for adjustable speed AC drives, induction heating, uninterrupted power supplies (UPS), HVDC transmission lines, and standby aircraft power supplies.

The inverters can be broadly classified into voltage source inverters (VSI) and current source inverters (CSI), which can further be divided as single-phase inverters and three-phase inverters. The CSIs use DC source which tends to maintain a fixed current and they usually look like the controlled bridge rectifiers discussed in Chapter 5. The voltage source inverters convert the energy from a battery or other fixed voltage DC source into AC. Each type can use controlled switches like BJTs, MOSFETs, IGBTs, MCTs, static induction transistors (SITs), GTOs, and TRIACs. If the output voltage or current of the inverter is forced to cross zero by using a LC resonant circuit, the inverter is called as resonant pulse inverter. These inverters find application in high power induction heating.

6.2 PERFORMANCE PARAMETERS

Harmonics are generated by all practical inverters circuits as they utilize nonlinear devices for switching. The performance of an inverter can be evaluated in terms of the following parameters:

1. **Total harmonic distortion (THD)** The THD which tells the amount of harmonics present is defined as

$$\text{THD} = 1/V_{01} \{ \Sigma_{n=2,3,...\infty} V_{0n}^2 \}^{0.5}$$

where $n > 1$; $n < 1$ means sub-harmonics (harmonics having frequency less than the fundamental frequency by a fraction, i.e., 0.2 and 0.3).

2. **Distortion factor (DF)** The DF for an inverter is defined as

$$\text{DF} = 1/V_{01} [\Sigma_{n=2,3,...\infty} (V_{0n}/n^2)^2]^{0.5}$$

where V_{01} is the RMS value of fundamental harmonic, n is the index of the nth harmonic, and V_{0n} is the RMS value of the nth harmonic. In case of THD, the total harmonic content in the output is obtained but still there is no idea about the magnitude of each harmonic present. If a filter is employed at the output, the highest order harmonic will be most attenuated. So, the DF is a measure of effectiveness in reducing the unwanted harmonics.

3. **Harmonic factor (HF)** The HF for nth harmonic is given by

$$\text{HF}_n = V_{on}/V_{01}$$

It clearly indicates the contribution of individual harmonic in the complex component of voltage wave.

4. **Lowest order harmonic** The lowest order harmonic has frequency nearest to the fundamental frequency and its magnitude is equal to or more than 3% of the fundamental component magnitude.

6.3 SINGLE-PHASE INVERTERS

This section discusses the single-phase bridge-type inverters. These inverters are of two types—single-phase half-bridge inverter and single-phase full-bridge inverter. Both these types of inverters are modeled and simulated in this section.

6.3.1 Half-wave Inverter

A single-phase 50 Hz half-wave VSI model is shown in Fig. 6.1. In this model, two DC voltage sources of amplitude 110 V are taken for input supply, two MOSFETs are taken for switching, two pulse generators are selected for generating gate pulses for firing the two switches, a resistance of 50 Ω is taken as load, and a multimeter is used for measurements. Switch M1 and M2 are forward biased but can be turned ON by the gate pulse. The duration for which the gate pulse is high, these switches conducts. The *Gate Pulse 1* is of amplitude 5 V, time period 0.02 s, pulse width 50%, and a phase delay of zero degree. In case of *Gate Pulse 2*, all the parameters are same as that of *Gate Pulse 1* except the phase delay, which is 0.01 s. Now, when the gate

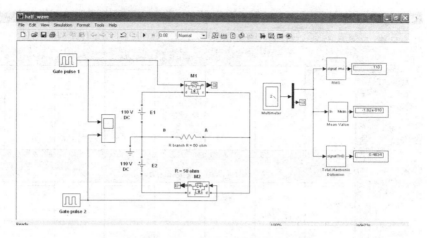

Fig. 6.1   Single-phase half-wave inverter

pulse is applied to the switch M1, it conducts and the current flows in resistance from A to B through E1. Similarly, M2 conducts when a gate pulse is applied to it but the current flows in the resistance from B to A through E2. In order to perform the inverter operation, gate pulse is first applied to M1, then to M2, and M1 and M2 are operated alternatively as shown in Fig. 6.2. The branch voltage and current across the load resistance are shown in Fig. 6.3. The output waveform is a square wave of 50 Hz. This model is simulated from 0 to 0.08 s with a load resistance of 50 Ω by using variable step ode23s (stiff/Mod. Rosenbrock) solver. In this circuit, both M1 and M2 should not be conducting at any point of time as this would create a short circuit at the output. The rms and average values of the output voltage measured by the multimeter are 110 V and 1.17e–12 V (almost zero) and the total harmonics distortion for fundamental frequency of 50 Hz is 0.04826. These parameters are observed on the display blocks used in this model.

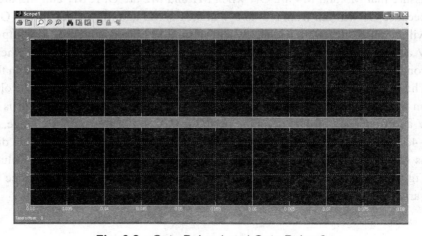

Fig. 6.2   Gate Pulse 1 and Gate Pulse 2

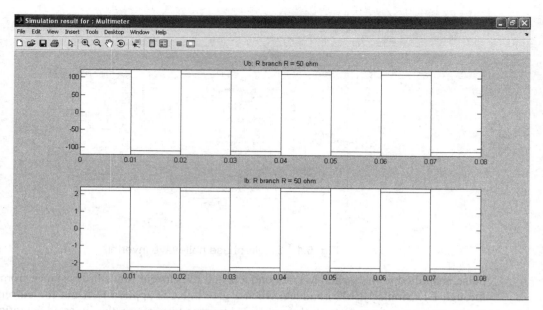

Fig. 6.3   Output voltage and current waveforms

6.3.2 Full-wave Inverter

The Simulink model of a single-phase 50 Hz full-wave inverter consisting of four MOSFETs, DC voltage source of 300 V, a resistive load of 60 Ω, gate pulses G1 and G2, and measurement and display devices is shown in Fig. 6.4. Gate pulse G1 is of amplitude 5 V, period 0.02 s, pulse width 50%, and a phase delay of 0 s. This pulse is applied to switches M1 and M2, which form one arm of the inverter. Gate pulse G2 has a phase delay of 0.01 s. This delay is provided to assure that M3 and M4 are ON when M1 and M2 are OFF. M1 and M2 are OFF after 0.01 s because G1 will fall from 5 V to 0 V at 0.01 s. M3 and M4 will be turned ON after 0.01 s and will conduct till 0.02 s. Hence, when M1 and M2 are conducting, the output voltage is –300 V and the current flows from B to A in the load resistance. Similarly, when M3 and M4 are conducting, the output voltage is 300 V and the current flows from A to B in the load resistance. The waveforms of gate pulse G1, current through M1, and the output voltage are observed on the Scope and are shown in Fig. 6.5. The output voltage waveform is of amplitude 300 V and frequency 50 Hz, i.e., time period is 0.02 s. For this output voltage, the mean value is –4.136e–12 (almost zero). the rms value is 300 V, and the total harmonic distortion is 0.4835 as observed on the display blocks after simulating the model. These values can be seen in Fig. 6.4. This model is simulated from 0 to 0.08 s with a load resistance of 60 Ω by using variable step ode23s (stiff/Mod. Rosenbrock) solver.

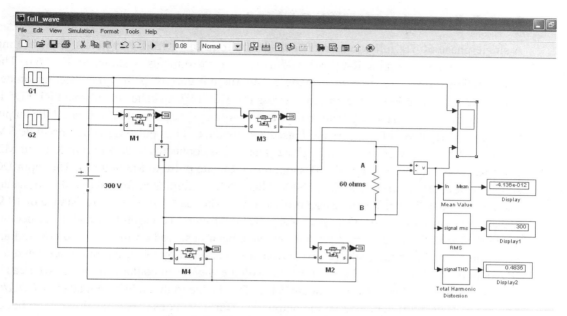

Fig. 6.4 Single-phase full-wave inverter

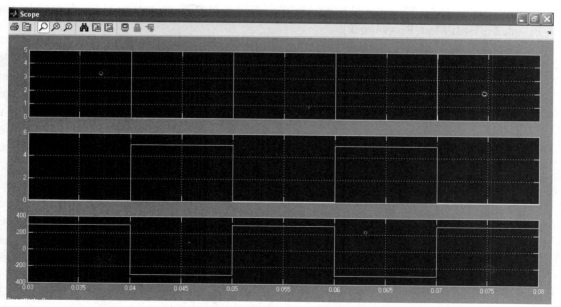

Fig. 6.5 Waveforms of gate pulse G1, current in switch in M1, load voltage

6.3.3 Full-wave Inverter with R-L Load

A single-phase 50 Hz full-wave inverter model built up of MOSFET switches, freewheeling diodes, DC voltage supply, R-L load, and measuring instruments is shown in Fig. 6.6. The freewheeling diodes are used to provide an alternating current path. Freewheeling diodes FD1, FD2, FD3, and FD4 are connected across the MOSFET switches in anti-parallel. FD5 is connected in parallel to the R-L load so as to avoid any negative pulse at the output. If the output voltage goes negative, FD5 is forward biased and conducts. Gate pulse G1 is of amplitude 5 V, period 0.02 s, pulse width 50%, and a phase delay 0 s is connected to gate port of M1 and M2, whereas G2 has a phase delay of 0.01 s and provides gate pulse to M3 and M4. The input DC supply is of constant amplitude, i.e., 250 V. This model is simulated from 0 to 0.06 s by using variable step ode23s (stiff/Mod. Rosenbrock) solver. The load consists of a resistance of 10 Ω and an inductance of 20 mH. Gate pulse G1, current through M1, current through freewheeling diode FD5, and the load current waveforms have been observed on the Scope block and are shown in Fig. 6.7. Output voltage and current are measured by using the multimeter and are shown in Fig. 6.8. It can be observed that the voltage waveform contains impulses whereas in the case of resistive load, it was a square wave. This is due to the inductive load ($V = L \, di/dt$). The R-L acts as a differentiator, so the output waveform contains spikes (as the differential of a square wave is an impulse train). The load current is continuous as can be seen in Fig. 6.8. The output voltage parameters measured after simulating the model are average voltage 85.82 V, rms voltage 185.5 V, and total harmonic distortion 0.734. The solver used for this model was ode23s (stiff/Mod. Rosenbrock).

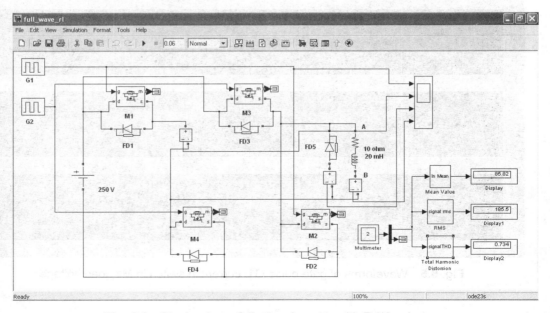

Fig. 6.6 Single-phase full-wave inverter with R-L load

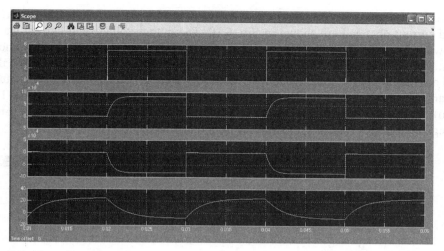

Fig. 6.7 Waveforms of gate pulse G1, current through M1 and FD5, and load current

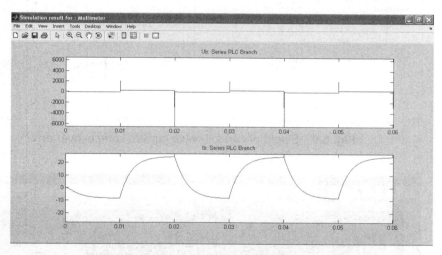

Fig. 6.8 Output voltage and current waveforms

6.4 CURRENT SOURCE INVERTER

A current source inverter (CSI) uses a constant current source instead of a voltage source at the input. A model of a single-phase CSI is shown in Fig. 6.9. It contains pulse generator, MOSFETs, DC voltage source of 250 V, source inductance of 50 mH, load resistance of 50 Ω, and various measuring instruments. The source inductance keeps the source current constant. The operation of this inverter is same as that of a VSI. Gate pulse G1 is applied to M1 and M2 and G2 is applied to M3 and M4. When G1 is positive, M1 and M2 are ON and the current flows from B to A in the load resistance. When G2 is positive, M3 and M4 are ON and the load current flows from A to B in the load resistance. The waveforms of G1, source current, and the load current obtained are shown in Fig. 6.10. The output voltage and current waveforms are shown in Fig. 6.11. It can be observed from Fig. 6.10 that the source current is almost constant, i.e., approximately 5 A. The downward spikes observed in the waveform show the charging and discharging of the source inductor. The load current is a square wave of amplitude 5 A. The parameters measured for the load current are as follows: average value—2.275e–12 A (almost zero), rms value—4.998 A and THD—0.4826. This model is simulated from 0 to 0.06 s by using variable step ode23s (stiff/Mod. Rosenbrock) solver.

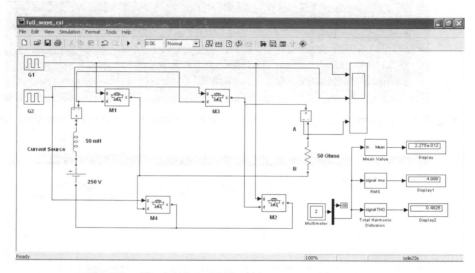

Fig. 6.9 Single-phase full-wave current source inverter

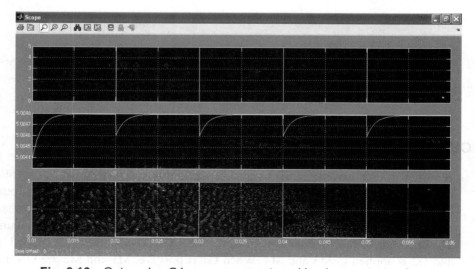

Fig. 6.10 Gate pulse G1, source current, and load current waveforms

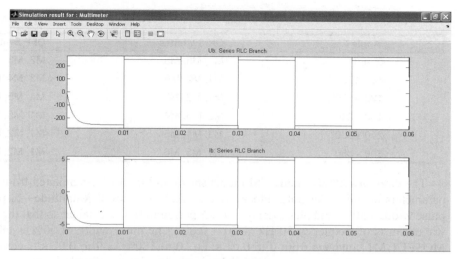

Fig. 6.11 Output voltage and current waveforms

6.5 THREE-PHASE INVERTERS

The three-phase inverters are used in industrial applications for providing variable frequency AC supply. The input is taken from a battery or DC supply and the output is a three-phase AC supply of desired frequency. An elementary three-phase inverter is of bridge type and consists of six controlled switches. In inverters, a step is defined as the change of conduction from one controlled switch to another. For example, in one cycle of 360°, each step will be of 60° for a six-step inverter. This implies that each controlled switch of a six-step inverter will get a control pulse at intervals of 60° in a proper sequence so that a three-phase AC voltage is synthesized at the output terminals of the inverter. There are two possible patterns for supplying the controlled pulses to the inverter switches. In one pattern each switch conducts for 180° and in the other, it conducts for 120°. In both these cases the control pulse is applied at an interval of 60° and these modes require six controlled switches.

6.5.1 Three-phase 180° Conduction Mode Inverter

A three-phase 180° conduction mode bridge-type inverter model built up of six MOSFETs, 450 V DC supply, star-connected load of 300 Ω, six pulse generators, and measurement blocks is shown in Fig. 6.12. In this inverter, each MOSFET conducts for 180° and hence it is known as three-phase 180° voltage source inverter. The MOSFET pair in each arm, i.e., M1 and M4, M3 and M6, and M5 and M2 conducts for a time interval of 180°. It implies that M1 conducts for 180° and then M4 conducts for 180°, and so on. Table 6.1 illustrates the switching pattern for 180° mode inverter. It can be seen from Table 6.1 that in every step of 60° duration, only three switches are conducting, i.e., one from the upper group and two from the lower group or two from the upper group and one from the lower group.

Table 6.1 Switching pattern for 180° mode three-phase VSI

Angle	ON Switches	OFF Switches
0°–60°	M1, M5, M6	M2, M3, M4
60°–120°	M1, M2, M6	M3, M4, M5
120°–180°	M1, M2, M3	M4, M5, M6
180°–240°	M2, M3, M4	M1, M5, M6
240°–300°	M3, M4, M5	M1, M2, M6
300°–360°	M4, M5, M6	M1, M2, M3

The three-phase 180° mode VSI model shown in Fig. 6.12 is simulated from 0 to 0.1 s. The parameters of Pulse Generator block G1 are taken as follows: Amplitude—5, period—0.02 s, pulse width—50%, and phase delay—0 s. A period of 0.02 s is taken so that the output voltage waveform is of frequency 50 Hz (1/50 = 0.02). The pulse width of 50% is taken so that the MOSFET M1 conducts from 0 to 0.01 s, i.e., 0 to 180°. As 0 to 0.02 s is equivalent to 0° to 360°, the delay of 0.01 s means 180° delay in electrical angle. Similarly, the phase delay for G2 is 0.01/3 s, for G3 it is 0.02/3 s, for G4 it is 0.01 s, for G5 it is 0.04/3 s, and for G6 it is 0.05/3 s, with all other parameters kept same as of G1. The gate pulses of switches M1, M3, and M5 are shown in Fig. 6.13. It can be observed from these gate pulses that each switch conducts for 0.01 s, i.e., 180°. The output voltage waveform of the three phases obtained after simulating the model is shown in Fig. 6.14. It can be observed from Fig. 6.14 that the first phase has phase angle of 0°, the second phase –120°, and third phase 120°. The following parameters are measured for the first phase as can be seen in Fig. 6.12:

1. $V_{avg} = -5.884 \times 10^{-12}$ V, i.e., almost zero
2. $V_{rms} = 212.1$ V
3. THD = 0.3108

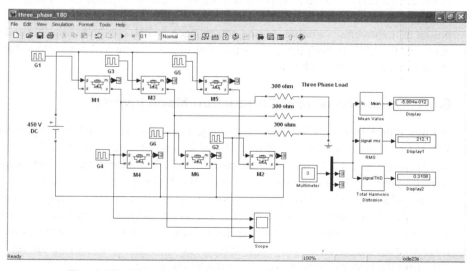

Fig. 6.12 Three-phase inverter with 180° conduction mode

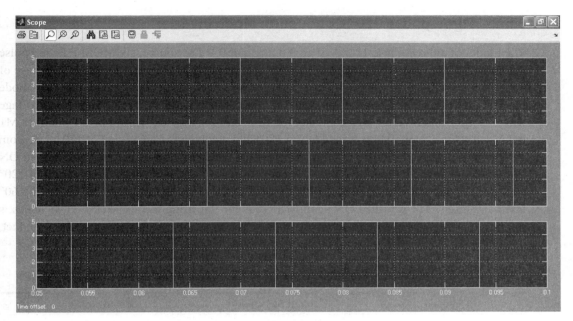

Fig. 6.13 Gate pulses of M1, M3, and M5

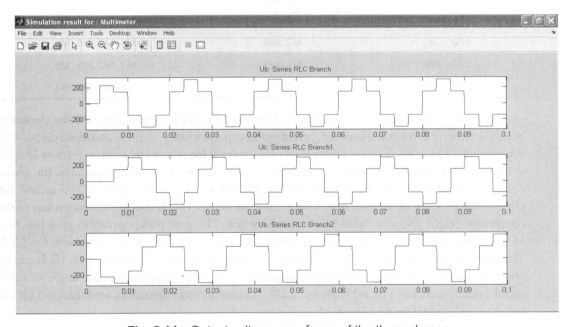

Fig. 6.14 Output voltage waveforms of the three phases

6.5.2   Three-phase 120° Conduction Mode Inverter

The model of a three-phase 120° VSI containing a 450 V DC source, six MOSFETs and pulse generators, 300 Ω star-connected load, and measuring blocks is shown in Fig. 6.15. In case of a 120° modeVSI, each switch conducts for 120° of a cycle. Like 180° mode, the 120° mode inverter also requires six steps of 60° duration for completing one cycle of output AC voltage waveform. In this inverter, M1 conducts for 120° and for the next period of 60°, neither M1 nor M4 conducts. Now, M4 is turned ON at 180° and it conducts for a period of 120°, i.e., from 180° to 300°. At 300°, M4 stops conducting and 60° interval elapses before M1 is turned ON again at 360°. In the bottom row, M3 is turned ON at 120° and it conducts for a period of 120° and then it is turned OFF. At 300°, M6 is turned ON and it conducts for 120°. After the 60° interval elapses, M3 is turned ON again. The sequence of firing the six MOSFET switches is same as that for the 180° mode inverter. In this case, in each step only two switches conduct, i.e., one from the upper group and one from the lower group.

Table 6.2   Switching pattern for 120° mode three-phase VSI

Angle	ON Switches	OFF Switches
0°–60°	M1, M6	M2, M3, M4, M6
60°–120°	M1, M2	M2, M3, M4, M5
120°–180°	M2, M3	M1, M4, M5, M6
180°–240°	M3, M4	M1, M2, M5, M6
240°–300°	M4, M5	M1, M2, M3, M6
300°–360°	M5, M6	M1, M2, M3, M4

The model shown in Fig. 6.15 is simulated from 0 to 0.1 s. The parameters set for Pulse Generator block G1 are as follows: Amplitude—5, period—0.02 s, pulse width—33.34%, and phase delay—0 s. A period of 0.02 s is taken so as to get a 50 Hz waveform at the output and a pulse width of 33.34% is taken for G1 because M1 conducts from 0° to 120°, i.e., 0 to 0.0067 s. Similarly, the phase delay for G2 is 0.01/3 s, for G3 is 0.02/3 s, for G4 is 0.01 s, for G5 is 0.04/3 s, and for G6 it is 0.05/3 s, all other parameters remaining the same. The delays are set according to the conduction angle given in Table 6.2 for each switch. The gate pulses given to switches M1, M3, and M5 are shown in Fig. 6.16. It can be observed that each switch conducts for 120°. The output voltage and current waveforms of the three phases are shown in Fig. 6.17. It can be observed that the first phase has a phase angle of 0°, second phase has a phase angle of 120°, and the third phase has a phase angle of –120°. The following parameters are measured for the first phase as can be seen in Fig. 6.15:

1. V_{avg}: -3.342×10^{-11} V, i.e., almost zero
2. V_{rms}: 183.7 V
3. THD: 0.3106

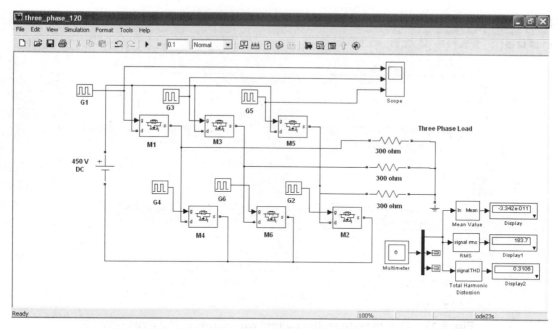

Fig. 6.15 Three-phase inverter with 120° conduction mode

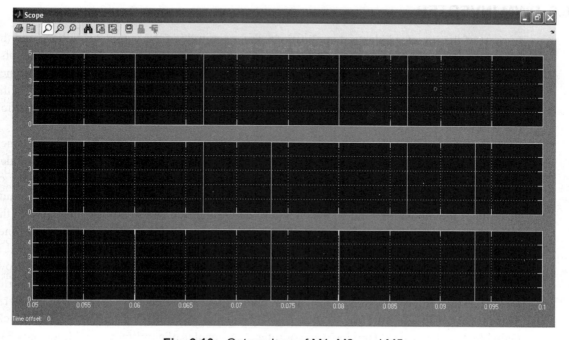

Fig. 6.16 Gate pulses of M1, M3, and M5

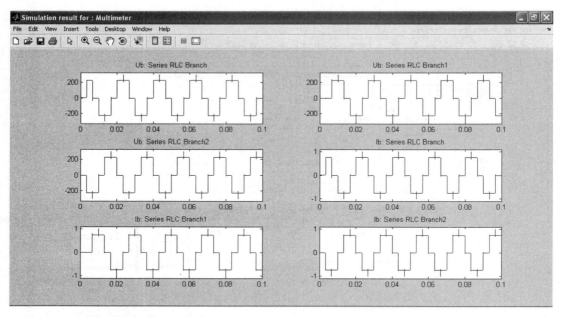

Fig. 6.17 Output voltage and current waveforms of the three phases

6.6 PWM INVERTER

In most of the inverter industrial applications, the magnitude of the output AC voltage and the input DC voltage varies continuously. This requires continuous variation of the control pulse width in order to achieve the desired output voltage waveform. In this method, a reference signal is compared with a high frequency triangular wave to generate the PWM signal. If the reference signal is a sine wave, the modulation is called sine wave pulse width modulation (SPWM).

The SPWM is realized by comparing a sine wave reference signal (V_r) with a high frequency triangular or sawtooth wave-carrier signal (V_c). The ratio of V_r/V_c is called modulation index (MI) and it determines the harmonic content in the inverter output voltage. The frequency of the reference signal determines the inverter output voltage frequency and its peak magnitude determines the modulation index, which in turn determines the RMS output voltage. When the reference signal (sine wave) is compared with the carrier signal, their intersection determines the switching instants of the inverter switches. When the magnitude of the reference signal is greater than the carrier signal, the output of the comparator is high; otherwise, the output is low. The output of the comparator is further processed such that the output pulse width of the inverter is same as that of the comparator. The circuit model of a PWM generator is shown in Fig. 6.18. This model consists of Sine Wave block, Signal Generator, Switch, Constant block, Sum block, Gain block, and a Scope. The sine wave or reference signal is of amplitude 5 and frequency 314 rad/s. The carrier wave is a sawtooth signal of amplitude 10 and frequency 1 kHz. The carrier signal is subtracted from the reference signal for their comparison. The output obtained is a PWM signal of varying magnitude. This signal is fed to input2 of the Switch block. Input1 of the Switch block is connected to a constant value 5 and input3 of the Switch is connected to the constant value 0 as shown in Fig. 6.18. The switch passes the input1, i.e., 5, if the control signal, i.e., input2, satisfies the selection criterion, else input3, i.e. 0, is passed. The selection criterion is taken as u2 >= threshold and threshold values is zero. Hence, when the input PWM signal is greater than zero, the output of the switch is 5, otherwise zero. Thus, a PWM switching signal for switches M1 and M2 is obtained. This PWM signal is multiplied by –1 (inversion) and then added to a constant value 5 in order to obtain the PWM switching waveform for switches M3 and M4. Thus, the PWM signals for a single-phase inverter are obtained.

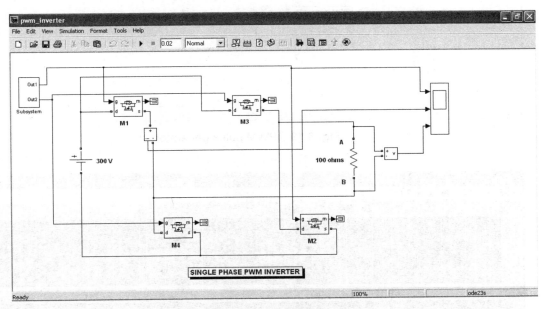

Fig. 6.18   PWM control inverter

A block diagram of a single-phase PWM generator is shown in Fig. 6.19. The inverter consists of four MOSFET switches M1, M2, M3, and M4 connected to DC supply of 300 V at the input side and a load of 100 Ω at the output side. A subsystem of the PWM pulse generator circuit is created in order to simplify the view of the model. The out1 and out2 signals of the subsystems are the two PWM control signals. When out1 is high, switches M1 and M2 are ON and the output is 300 V; otherwise, switches M3 and M4 are ON and the output is –300 V as shown in Fig. 6.20. It can be observed from Fig. 6.20 that the pulse width of the PWM signal and the output voltage is same. The output voltage waveform contains harmonics of higher order.

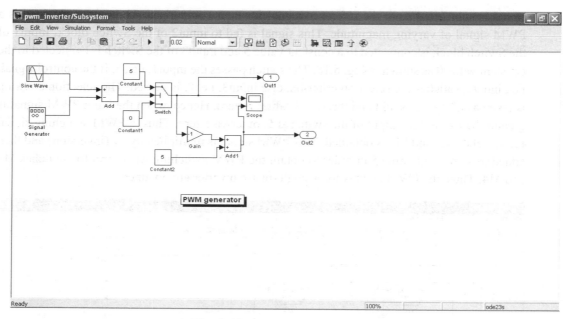

Fig. 6.19   PWM pulse generator

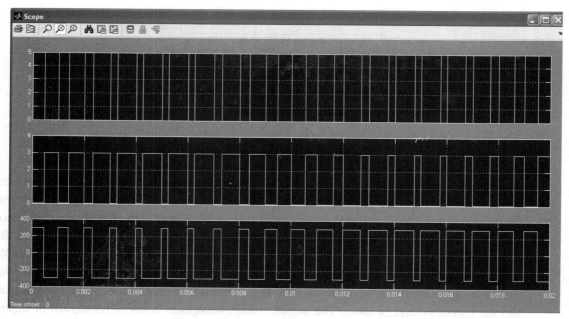

Fig. 6.20   Waveforms of gate pulse, current through M1, and output voltage

6.7 DUAL CONVERTERS

A rectifier circuit converts AC into DC and an inverter converts DC back into AC. A full converter can operate as a rectifier in the first quadrant for firing angle 0° to 90° and as an inverter in the fourth quadrant for firing angle 90° to 180°. In case a four-quadrant operation is required, the rectifier and inverter can be connected to the same load back to back. In this case, the rectifier converts AC supply into DC and feed it to the load and the load DC voltage acts as a DC input to the inverter which converts it into AC voltage again. In this manner, a four-quadrant operation is achieved. Such an arrangement of using rectifier and inverter in anti-parallel and connecting them to the same load is known as dual converter.

In an ideal dual converter, both the converters are controlled in such a manner that their average output voltages are equal in magnitude and have the same polarity, which is possible only when one converter is operating as a rectifier and the other as an inverter. This is possible if the firing angles to both the converters meet the requirement $\alpha_1 - \alpha_2 = 180°$, where α_1 is the firing angle of the first converter, i.e., rectifier, and α_2 is the firing angle of the second converter, i.e., inverter.

Practically, the firing angles of both the converters are controlled such that their output voltages are equal in magnitude and phase. In case the output voltages of the two converters are slightly different, the current will flow from one converter to another. This current which is caused by the instantaneous voltage difference of the two converters is called circulating current as it circulates between the two converters. The circulating current does not flow through the load and therefore causes heating in the converter switches. This current can be avoided in two ways. The first way is to operate only one converter at a time, which can be achieved by firing only one converter at a time. Thus, only one converter is operating at a time while the other one is idle. The second way is to insert an inductor between the two converters. As the current cannot build up suddenly inside an inductor, this inductor limits the magnitude of the circulating current to a reasonable value. The size and cost of this inductor can be significant in high power applications.

A single-phase dual converter consisting of a controlled rectifier and an inverter is shown in Fig. 6.21. The rectifier circuit consists of MOSFET switches S11, S13, S13, and S14 and pulse generators G1 and G2. This rectifier circuit is supplied by a 314 V, 50 Hz voltage source. The inverter circuit consists of MOSFET switches S22, S24, S21, and S23 and pulse generators G3 and G4. This inverter gets input from the load resistance of 100 Ω connected between the rectifier and inverter circuits. The output of the inverter is connected to a resistive load of 100 Ω. Now the firing of the rectifier and inverter switches should be done such that the inverter output voltage is similar to that of input supply voltage. For accomplishing this task, the parameters selected for the pulse generators are as follows: for G1: Amplitude—5, period—0.02 s, pulse width—50%, and phase delay—0; for G2: Amplitude—5, period—0.02 s, pulse width—50%, and phase delay—0.01 s; for G3: Amplitude—5, period—0.02 s, pulse width—50%, and phase delay—0.01 s; and for G4: Amplitude—5, period—0.02 s, pulse width—50%, and phase delay—0 s. Thus, the firing angle of the inverter switches is 180° out of phase with firing angle of rectifier switches. The waveforms obtained after simulating the model are shown in Fig.

6.22. It can be observed that the output of the inverter is same as the input supply. Also, the waveform of the voltage at the load resistance connected between the rectifier and the inverter is that of pulsating DC.

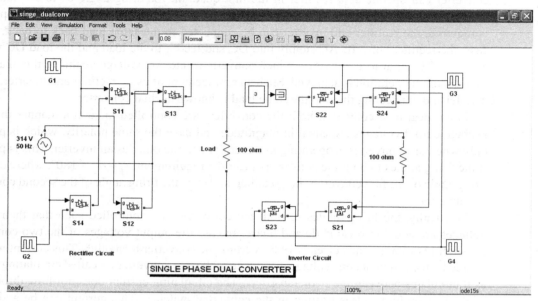

Fig. 6.21 Single-phase dual converter

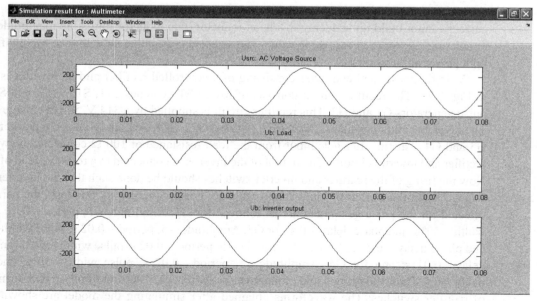

Fig. 6.22 Supply, rectifier output, and inverter output voltage waveforms

6.8 SIMULATION PROJECTS

Three simulation projects are elaborated in this section. The first one is of a parallel inverter, the second one is of a Mc Murray Bedford full-wave inverter, and the third is of a three-phase current source inverter.

Project 1

Design and simulation of a parallel inverter A circuit model of a single-phase parallel thyristor inverter is shown in Fig. 6.23. It consists of two thyristors T1 and T2, an inductor L of 1 mH, an output three-winding transformer, and a commutating capacitor C of 22 µF. The transformer turns ratio from each primary half to secondary winding are taken as unity. The source inductor L_s is placed to keep the source current constant. During the operation of this inverter, capacitor C is parallel with the load through the transformer. This is the reason why this inverter is being called a parallel inverter.

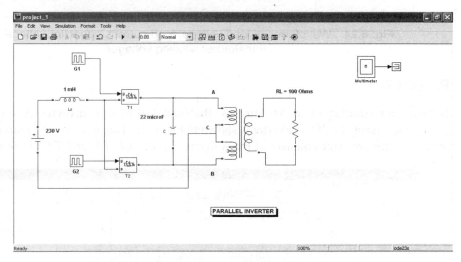

Fig. 6.23   Single-phase thyristor parallel inverter

This inverter is operated at a switching frequency of 50 Hz, i.e., time period is 0.02 s. Now, when thyristor T1 is triggered, the current flows through the upper primary winding (A-C) of the transformer. Due to this, the current induces magnetic flux in the transformer that links both the halves of the primary winding. So the total voltage across the capacitor becomes 460 V as it is connected in parallel to both the primary windings, i.e., points A and B. The upper plate of the capacitor is positive and the lower plate is negative. Now at $t = 0.01$ s, thyristor T2 is triggered and the capacitor voltage 460 V appears as a reverse bias across the conducting thyristor T1 and it is turned OFF. Now, the current starts flowing through the lower winding (C to B) of the transformer. This time the capacitor gets negatively charged and the voltage across it is –460 V. Now, when T1 is again triggered at $t = 0.02$ s, the capacitor voltage reverse biases T2 and the current again flows through the upper primary winding. Thus the cycle repeats, and the output AC voltage is obtained at the secondary winding of the transformer. The various waveforms obtained after simulating the model are shown in Fig. 6.24.

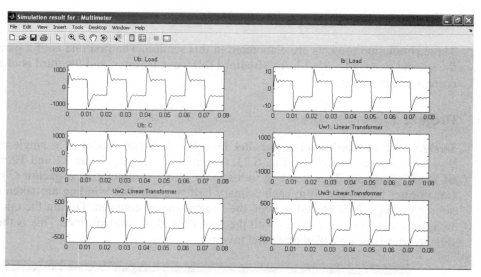

Fig. 6.24 Waveforms of load voltage, load current, capacitor voltage, and transformer winding voltages

Project 2

Design and simulation of McMurray Bedford full-bridge inverter A single-phase full bridge Mc Murray Bedford inverter is shown in Fig. 6.25. This is a forced commutated thyristor inverter. This inverter consists of main thyristors T1, T2, T3, and T4 and feedback diodes

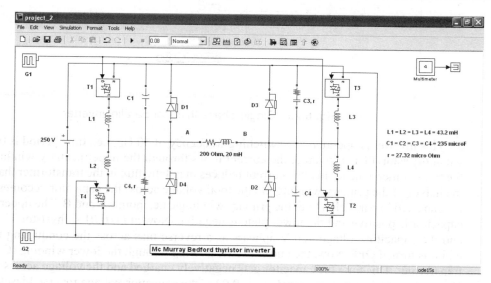

Fig. 6.25 McMurray Bedford full-bridge inverter

D1, D2, D3, and D4. The commutation circuit consists of capacitors C1, C2, C3, and C4 and coupled inductors L1, L2 and L3, L4. In a branch containing two inductors in series with two thyristors, if one thyristor is ON the other one gets turned OFF. This method of commutation in which only one thyristor conducts is called complementary commutation.

In the model of Fig. 6.25, the values of the inductors L1, L2, L3, and L4 is 43.2 mH, that of capacitors C1, C2, C3, and C4 is 235 µF, and that of capacitive resistance r is 27.32 µΩ. The value of capacitive resistance is taken quite low as it cannot be set to zero. The input supply of the inverter is of 250 V and it supplies to a 200 Ω, 20 mH load. Now, at $t = 0$ s, thyristors T1 and T2 are triggered. The load current starts flowing through T1, L1, Load, L2, and T2. Now as the load capacitor C4 and V3 get charge up to the supply voltage through the load. Capacitor C4 is connected to the negative terminal of the DC supply and capacitor C3 is connected to the positive terminal of the DC supply. Now, at $t = 0.01$ s, when T4 and T3 are triggered, the voltage of C4 reverse biases thyristor T1 and the voltage of C3 reverse biases T2, as the voltage drop across L1 and L2 is negligible. Now, capacitors C1 and C2 get charged up to the supply voltage and the voltage drop across L4 and L3 is negligible, as $V = L \, di/dt \approx 0$. Now, at $t = 0.02$ s, when T1 and T2 are triggered, the potential of C1 reverse biases T4 and the potential of C2 reverse biases T3. Like this, the output voltage is obtained at the output at a frequency of 50 Hz. The energy lost in the commutation and the voltage drop across the capacitive resistance reduces the output power considerably if the circuit is not designed properly. The waveforms obtained after simulating the model are shown in Fig. 6.26.

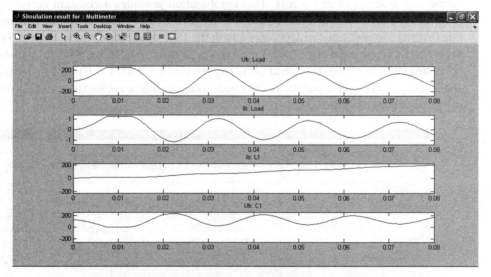

Fig. 6.26 Load voltage, load current, inductor L1 current, and capacitor
C1 current waveforms

Project 3

Design and simulation of a three-phase current source inverter A model of a three-phase current source inverter is shown in Fig. 6.27. This inverter consists of MOSFET switches S1, S2, S3, S4, S5, and S6. The input current source is built by connecting an inductance in series

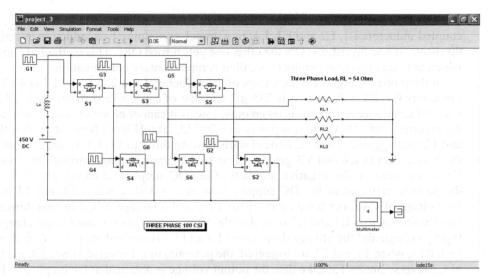

Fig. 6.27 Three-phase current source inverter

with the DC voltage source. The input DC supply of the inverter is 450 V and the inductor connected in series is of 1.5 mH. This combination provides a constant current source. This inverter supplies power to a star-connected resistive load of 54 Ω per phase. The firing pulses of this inverter are generated in accordance with the 180° conduction mode. Switch S1 is triggered at $t = 0$ s, S2 is triggered at $t = 0.01/3$ s, S3 is triggered at $t = 0.02/3$ s, S4 is triggered at $t = 0.01$ s, S5 is triggered at $t = 0.04/3$ s, and S6 is triggered at $t = 0.05/3$ s. By following this switching sequence, a three-phase output is obtained. The various waveforms obtained for the circuit model are shown in Fig. 6.28.

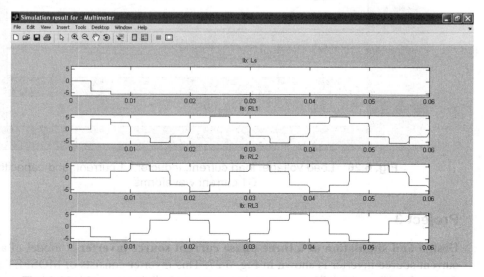

Fig. 6.28 Source inductor, phase 1, phase 2, and phase 3 current waveforms

SUMMARY

The power electronic device that converts DC voltage into AC voltage is called an inverter. The inverters can provide single-phase or three-phase AC voltages from a DC voltage source. The main points of this chapter can be summarized as follows:

- Voltage source inverters (VSI) have a DC voltage source at the input and provide AC voltage at the output.
- Current source inverters (CSI) have a DC current source at the input and provide constant AC current at the output. The DC current source is made by a using a constant DC voltage source in parallel with a resistance.
- Feedback diodes are not used in anti-parallel with the switches in CSI.
- A half-bridge inverter requires a center-tapped DC source which is not required in a full-bridge inverter.
- The square wave switching pulses controls the frequency of the output AC voltage while the magnitude is governed by the input DC voltage.
- The PWM switching technique permits the control of output voltage magnitude as well as frequency. The harmonics generated in PWM control are near to the switching frequency and can be easily removed by filtering.
- The phase inverters follow two conduction strategies, namely 120° conduction mode and 180° conduction mode. In 180° conduction mode, each switch conducts for 180° while in 120° conduction mode, each switch conducts for 120° during each cycle.
- Dual converts are made up of a rectifier and an inverter connected in anti-parallel. The DC output of the rectifier acts as an input to the inverter. A dual converter can operate in all the four quadrants.

REVIEW QUESTIONS

1. Explain the function of an inverter and mention its important industrial applications.
2. Classify the different types of inverters.
3. Define the performance parameters of an inverter.
4. Explain the working of single-phase bridge-type inverter.
5. Explain the working of a single-phase full-wave bridge inverter.
6. Mention the advantages of a three phase 120° inverter over a three-phase 180° conduction mode inverter.
7. Mention the important applications of three-phase inverters.
8. What is the purpose of modulation?
9. What is the purpose of feedback diodes in inverters?
10. What are the merits of sinusoidal pulse width modulation?
11. Mention the merits and demerits of current source inverters.
12. Write the techniques of harmonic reductions.

13. Mention the different types of modulation techniques used for inverter control.
14. Mention the advantages and disadvantages of current source inverter.
15. Explain the difference between MOSFET and thyristor-based inverter.
16. Explain the functioning of a dual converter and its applications.

SIMULATION EXERCISES

1. Develop a model of a half-bridge inverter which supplies power to a resistive load of 30 Ω by a DC source of $V/2 = 125$ V. In this model, determine the rms output voltage, power delivered to the load, peak current of each switch, and plot the output voltage of the inverter. The switching frequency of the inverter is 100 Hz.

2. Repeat Problem 1 for a full-wave inverter. Mention the differences observed in this case.

3. A voltage source inverter is connected to a large battery storage system in order to supply power to a utility grid. The DC voltage of the battery is 2,500 V. Construct a model of the system and measure the power supplied per phase to the utility grid. The load connected to the grid per phase is 30 Ω.

4. Design a dual converter of a particular home application. The supply coming to the home is 220 V rms, 50 Hz. If the power supply is interrupted, the power is taken from a 110 V DC battery. When the power is restored, the battery is again charged by a rectifier. The total load of the home is 1 kW. Also plot the input AC supply and inverter output voltage waveforms.

5. The PWM can be performed based on a saw-tooth carrier waveform also. Consider a system with 170 V input DC voltage which provides a 120 V RMS output at 50 Hz. For a switching frequency of 750 Hz, plot the inverter output voltage waveforms for both triangle and saw-tooth carrier signal. Compare the results obtained.

6. A square wave DC to AC converter is powered by a 400 V source and provides the power to a load of 10 Ω in series with a 10 mH inductor. The switching frequency of the inverter is 400 Hz. Develop a model of this converter and plot the output voltage and current waveforms. Also determine the power delivered to the load.

7. A three-phase full-wave inverter is supplied by a 150 V DC battery. The three-phase load is balanced star connected and the output frequency is 50 Hz and the load per phase is 2 Ω, 10 mH. Plot the voltage and current of any one phase and determine the total harmonic distortion of the output current.

8. A PWM controlled inverter is loaded by a 2 Ω resistance in series with a 2 mH inductance. The inverter is powered by 170 V battery and is switched at 50 Hz. Determine the harmonic profile of the output current of this inverter up to 7th harmonic.

SUGGESTED READING

Almazan, J., N. Vazquez, C. Hernandez, J. Alvarez, and J. Arau, 'Comparison between the buck, boost and buck-boost inverters', *International Power Electronics Congress,* Acapulco, Mexico, pp. 341–346, October 2000.

Agrawal, J.P., *Power Electronic Systems: Theory and Design*, Chapter 7, Pearson Education, Delhi, 2006.

Bedford, B.D. and R.G. Hoft, *Principle of Inverter Circuits*, John Wiley & Sons, New York, 1964.

Bimbhra, P.S., *Power Electronics*, Chapter 8, Khanna Publishers, New Delhi, 2009.

Boost, M.A. and P.D. Ziogas, 'State-of-the-art carrier PWM techniques: A critical evaluation', *IEEE Transactions on Industry Applications*, Vol. IA24, No. 2, Corcodia Univ., Montreal Que, pp. 271–279, 1988.

CaCeres, R.O. and I. Barbi, 'A boost DC-AC converter: Operation, analysis, control and experimentation', *Industrial Electronics Control and Instrumentation Conference*, Los Andes Univ., Merida, pp. 546–551, November 1995.

Chen, T.P., Y.S. Lai, and C.H. Liu, 'New space vector modulation technique for inverter control', *IEEE Power Electronics Conference*, Vol. 2, Dept. of Elect., Engg., Nat. Taiwan Univ, 1999, pp. 777–782.

Iwaji, Y. and S. Fukuda, 'A pulse frequency modulated PWM inverter for induction motor drives', *IEEE Transactions on Power Electronics*, Vol. 7, No. 2, Dept. of Electr. Engg., Hokkaido Univ., Sapporo, April 1992, pp. 404–410.

Jacobina, C.B., A.M.N. Lima, E.R. Cabral da Silva, R.N.C. Alves, and P.F. Seixas, 'Digital scalar pulse-width modulation: A simple approach to introduce non-sinusoidal modulating waveforms', *IEEE Transactions on Power Electronics*, Vol. 16, No. 3, Dept. de Engenharia Electrica Univ. Federal da Paraiba, pp. 351–359, May 2001.

Krein, P.T., *Elements of Power Electronics*, Chapter 6, Oxford University Press, New York, 2009.

Patel, H.S. and R.G. Hoft, 'Generalized techniques of harmonics elimination and voltage control in thyristor converter', *IEEE Transactions on Industrial Applications*, Vol. IA9, No. 3, General Electric Company, Philadelphia, PA, pp. 310–317, 1973.

Rashid, M.H., *Power Electronics: Circuits, Devices, and Applications*, Chapter 6, Pearson Education, New Delhi, 2005.

Taniguchi, K. and H. Irie, 'PWM technique for power MOSFET inverter', *IEEE Transactions on Power Electronics*, Vol. PE3, No. 3, Osaka Inst. of Technol, pp. 328–334, 1988.

Van dar Broek, H.W., H.C. Skudelny, and G.V. Stanke, 'Analysis and realization of a pulse-width modulator based on voltage space vectors', *IEEE Transactions on Industry Applications*, Vol. 24, No. 1, Philips Forschungs Pab., Aacha, pp. 142–150, January/February, 1988.

Ziogas, P.D., 'The delta modulation techniques in static PWM inverters', *IEEE Transactions on Industry Applications*, Corcordia University, Montreal, pp. 199–204, March/April 1981.

7

SIMULATION OF CHOPPERS AND CYCLOCONVERTERS

7.1 INTRODUCTION

A chopper is a static power electronic device that converts the fixed input DC voltage to a variable DC output voltage in a single stage. These devices are DC equivalent of an AC transformer which converts an input AC voltage of a given magnitude to an AC voltage of another magnitude depending on the type of transformer (step up or step down). Similar to a transformer, a chopper can be used to step down or step up the fixed DC input voltage. As choppers perform DC to DC conversion in a single stage, i.e., DC is not converted into AC and then again into DC, they are more efficient and compact. The conversion of a fixed DC voltage to an adjustable DC output voltage by using a controlled switch is known as chopping. The chopper systems offer smooth control, high efficiency, fast response, and regeneration.

Conventionally, a variable DC voltage was obtained from a fixed DC voltage by resistance control (potentiometer) method or by motor-generator set method. In the former method, a variable resistance (potentiometer) was used in between the fixed voltage DC source and the load. The input DC voltage is connected at the two fixed terminals of the potentiometer and the load is connected at the fixed end from one side and to the movable terminal of the potentiometer from another side. This method is inefficient because of high-energy losses in the resistance as continuous current flows in the variable resistance and it consumes supply power. In motor-generator set method, a variable DC output voltage is obtained by controlling the field current of the DC generator. Although this method is still used in industrial drives, the system is bulky, costly, slow in response, and less efficient.

Nowadays, high power-controlled semiconductor switches like thyristors, power MOSFETs, and IGBTs are available for constructing DC power converters. These converters are of two types—inverter–rectifier system and DC chopper system. In the inverter–rectifier system, the DC is first converted into AC by using an inverter, which is then stepped up or down by using a step-up or step-down transformer and then rectified back to DC by using a rectifier. As conversion is done in two stages, DC to AC (inverter system) and AC to DC (rectifier system), this system is costly, bulky, and less efficient. The DC chopper system converts directly from DC to DC and is a latest technology with good efficiency.

As the conversion in a chopper circuit is performed by switching action, the power is lost only in turning the switch ON/OFF.

The DC chopper finds tremendous applications in drives and control. The variable resistance commonly used in series with the armature of the DC motors for speed control can be replaced by a chopper resulting in better efficiency. Therefore, it can be utilized in battery-operated vehicles and in applications where saving energy is a prime consideration. Tunnel heating can be greatly reduced by using choppers in subway cars. They can also provide regenerative braking of the DC motor and return the energy back to the supply (or battery) which results in energy saving for transportation systems with frequent stops. They are also used in other applications such as inter-urban and trolley cars, marine hoists, forklifts trucks, mine haulers, and in metro rails. They are finding applications in continuous process plants like glass, fertilizers, and tyre manufacturing industries.

7.2 PRINCIPLE OF CHOPPER OPERATION

A chopper is like an ON/OFF switch which connects and disconnects the load from the DC power supply and produces a chopped load voltage. When the switch is ON, the load is connected to the supply and when it is OFF, the load is disconnected from the supply. An elementary chopper circuit having constant input supply of V_s and a load resistance of R is shown in Fig. 7.1. If the switch S is ON, the load terminals are connected to the supply and the load voltage is V_s; when the switch is OFF, the load is disconnected from the supply and the load voltage is zero. A chopped DC voltage is, thus, produced at the load terminals.

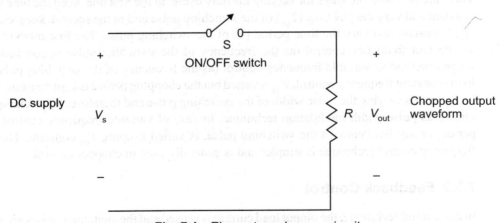

Fig. 7.1 Elementary chopper circuit

The average output voltage is given as follows:

$$V_{out} = T_{on}/(T_{on} + T_{off}) \times V_s = D \times V_s$$

The duty cycle of the switch is given as follows:

$$D = T_{on}/T$$

where

V_{out} is the average output voltage

T_{on} is the time for which the switch remains conducting

T_{off} is the time for which the switch does not conduct

T is the time period of the switching pulse $= T_{on} + T_{off}$

D is the duty cycle of the switch

Thus, the load voltage can be controlled by varying only the duty cycle of the switch. For instance, if the duty cycle of the switch is 40%, output voltage, $V_{out} = 0.4 \, V_s$. As the current flows only when the switch is in ON state, the power is consumed by the switch and the rest is delivered to the load resistance. The power consumed by the switch is quite low (few watts). Thus, the overall efficiency of the system is high.

7.3 CONTROL TECHNIQUES

There are broadly two types of control techniques used for chopper control—duty cycle control and feedback control. These techniques are discussed in this section.

7.3.1 Duty Cycle Control

As discussed earlier in the chapter, the output voltage can be controlled by the duty cycle of the switch. This technique controls the duty cycle in order to achieve the desired output voltage. There are two possible ways for varying the duty cycle. In the first one, keep the time period (T) constant and vary the ON time (T_{on}) of the switching pulse and in the second, keep the ON time (T_{on}) constant and vary the time period (T) of the switching pulse. The first method is known as constant frequency control (as the frequency of the switching pulse is constant) and the second method as variable frequency control (as the frequency of the switching pulse varies). In the constant frequency control, T_{on} is varied but the chopping period is kept constant. Variation in T_{on} means varying the pulse width of the switching pulse and therefore, this technique is also known as pulse width modulation technique. In case of variable frequency control, the time period, or say frequency of the switching pulse, is varied keeping T_{on} constant. The constant frequency control technique is simpler and is generally used in chopper control.

7.3.2 Feedback Control

In this control technique, the output load current is sensed and the switching pulses are generated accordingly. Lower and upper limits of the load current are set. When the load current reaches the upper limit, the chopper switch is opened, i.e., put OFF, so that the load current starts falling. When the load current falls to the lower limit, the chopper switch is closed, i.e., put ON, so that the load current starts building up. Hence, the load current fluctuates between the upper and the lower limits. This method requires feedback control and hence is more complex. However, it might be used if less ripples are desired in the load current.

7.4 STEP-DOWN OR BUCK CHOPPER

The chopper circuit in which the output voltage is lower than the input supply voltage is called step-down chopper or buck converter. A step-down chopper or buck converter model containing a DC supply of 200 V, controlled switch S1, freewheeling diode, inductance of 0.5 mH, pulse generator, load resistance of 10 Ω, and measurement blocks is shown in Fig. 7.2. When switch S1 is in ON state, the output voltage is equal to 200 V otherwise it is equal to 0 V. The parameters set for Pulse Generator block named Gate Pulse are as follows: Amplitude—5, period—0.01 s, pulse width—50%, and phase delay—0 s. Freewheeling diode is placed in order to avoid negative voltage spike across the load terminals and to provide alternating path for the inductor current. This model is simulated from 0 to 0.08 s. Figure 7.3 shows the voltage waveforms across the inductor, load voltage, and current. It can be observed that the load voltage varies from 0 to 200 V and load current varies from 0 to 20 A. The average output voltage measured for the circuit was 99.99 V, taking the load voltage frequency to be 100 Hz. The output voltage can be varied by changing the duty cycle of the switch S1, which can be accomplished by changing the pulse width of the Gate Pulse block. The average output voltage for this circuit is given as follows:

$$V_{out} = (T_{on}/T) \times V_s = (0.005/0.01) \times 200 = 100 \text{ V}$$

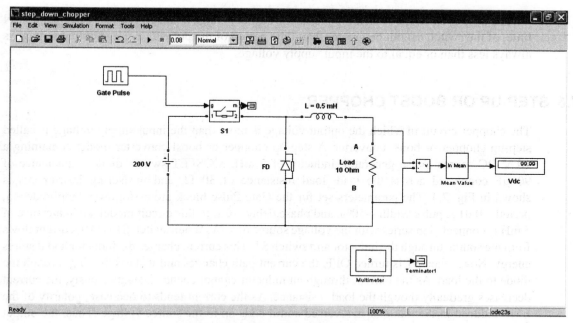

Fig. 7.2 Buck chopper

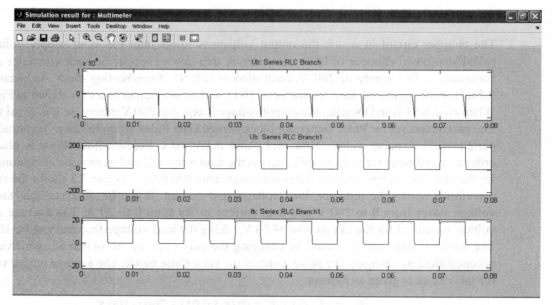

Fig. 7.3 Waveforms of voltage across inductor, output voltage, and current

The duty cycle of the switch is 50%, so the output voltage obtained is 100 V. As T_{on}, i.e., ON time, of the switch cannot be greater than time period of the pulse T, average output voltage is always less than or equal to the input supply voltage.

7.5 STEP-UP OR BOOST CHOPPER

The chopper circuit in which the output voltage is more than the input supply voltage is called step-up chopper or boost converter. A step-up chopper or boost converter model containing a 10 V DC source, pulse generator, inductor of 5 mH, MOSFET switch, diode, capacitance of 90 μF connected across the load, load resistance of 30 Ω, and measuring instruments is shown in Fig. 7.4. The parameters set for the Gate Pulse block are as follows: Amplitude—5, period—0.01 s, pulse width—50%, and phase delay—0 s. In this circuit model, an inductance of 5 mH is connected in series with the voltage source of 10 V. When switch S1 is ON, current flows from the source through the inductor and switch S1. This current charges the inductor and it stores energy. Now, when S1 is turned OFF, the current path changes and it starts flowing through the diode to the load. As the current through an inductor cannot change instantaneously, the current decreases gradually though the load resistance. As the current tends to decrease, polarity of the EMF induced in the inductor is reversed and it adds up to the source voltage. Thus, the output voltage increases, i.e., $V_{out} = V_s + L(di/dt)$. In this manner, this chopper acts as a step-up chopper and the energy stored in the inductor is released to the load. Figure 7.5 shows the inductor current, load voltage, and current waveforms obtained after simulating the circuit. The average output voltage measured was 19.94 V. The average output voltage of this chopper is given as follows:

$$V_{out} = (T/T_{off}) \times V_s = (T/(T - T_{on})) \times V_s = \{1/(1 - D)\} \times V_s$$

Therefore, for duty cycle of 50%, V_{out} for this circuit will be 10/0.5 = 20 V (V_s = 10 V). As duty cycle of the switch cannot be greater than 1, the output voltage is always greater than the input voltage. The output voltage can be varied by changing the duty cycle or T_{on} of the switch.

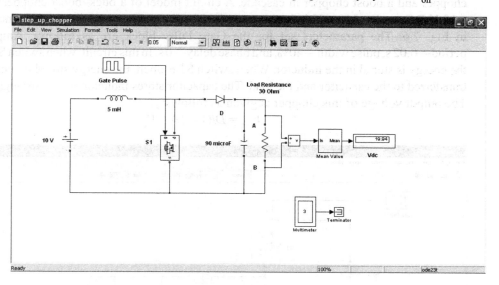

Fig. 7.4 Boost chopper

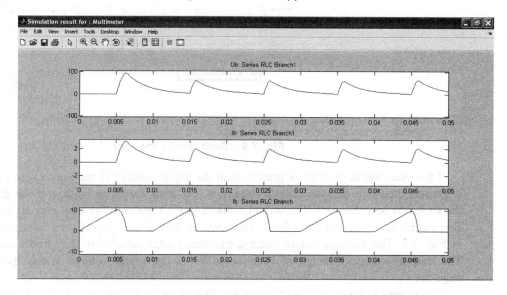

Fig. 7.5 Waveforms of the load voltage, load current, and current through the inductor

7.6 STEP-UP/STEP-DOWN OR BUCK–BOOST CHOPPER

The chopper circuit for which the output voltage can be greater or less than the input voltage is called as buck–boost chopper. A buck–boost chopper can be obtained by connecting a buck chopper and a boost chopper in cascade. A circuit model of a buck–boost chopper containing a 10 V DC supply, 150 mH inductor, a diode, 22 μF capacitor, and a load resistance is shown in Fig. 7.6. The parameters set for the Gate Pulse block are as follows: Amplitude—5, period—0.02 s, pulse width—40%, and phase delay—0 s. In this circuit, when switch S1 is closed, the energy is stored in the inductor. When switch S1 is open, the energy stored in the inductor is transferred to the capacitor and the load. The capacitor stores inductor energy through diode D1. The output voltage of this chopper is given as follows:

$$V_{out} = D/(1 - D) \times V_s$$

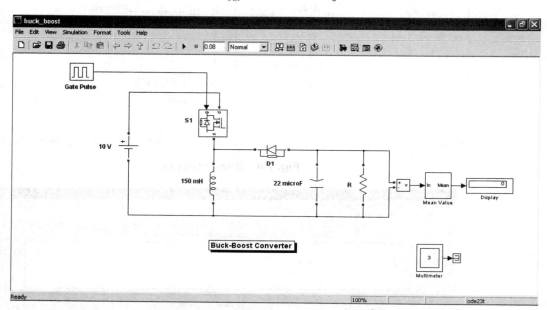

Fig. 7.6 Buck–boost chopper

Thus, the output voltage is greater than the input voltage if the duty cycle of the switch is greater than 0.5, and the output voltage is less than the input voltage if the duty cycle of the switch is less than 0.5. The output voltage also varies with the load current in this circuit. The circuit is simulated for a load resistance of 126 Ω and waveforms obtained are shown in Fig. 7.7. The average output voltage measured after simulating the circuit was −6.331 V. The output voltage is negative because as S1 is open, the voltage across the inductor is negative.

The output voltage of this chopper for 40% duty cycle can be calculated theoretically as follows:

$$V_{out} = 0.4/0.6 \times 10 = 6.67 \text{ V}$$

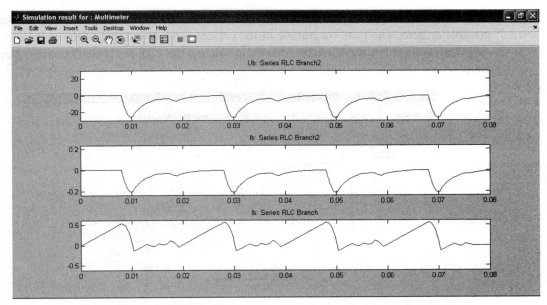

Fig. 7.7   Load voltage, load current, and inductor current waveforms

This type of circuit is again simulated for a load resistance of 230 Ω and taking duty cycle as 60%, the output voltage obtained after simulation was –15.16 V. Theoretically, the output voltage should be

$$V_{out} = 0.6/0.4 \times 10 = 60/4 = 15 \text{ V}$$

Thus, the chopper can supply an output voltage greater or less than the input supply voltage depending on the duty cycle of the switch and the load resistance.

7.7 TYPES OF CHOPPERS

The chopper circuits are classified according to the polarity of the output voltage and current. They are classified as Type-A or first quadrant, Type-B or second quadrant, Type-C or two-quadrant Type-A, Type-D or two-quadrant Type-B, and Type-E or four-quadrant chopper, depending on the quadrant in which the output voltage and the current of the chopper lie.

7.7.1 Type-A Chopper

In a Type-A or first quadrant chopper, the output voltage and current can be zero or positive. So, the power delivered to the load is either positive or zero. Thus, the power can flow only from the source to the load. Figure 7.8 shows the circuit model of a Type-A chopper containing a DC supply of 300 V, pulse generator, MOSFET switch, freewheeling diode, and resistive load. The parameters set for the Pulse Generator are as follows: Amplitude—5, period—0.02 s, Pulse width—50%, and phase delay—0 s. The output voltage is 300 V when switch is closed and is zero when the switch is open. The output voltage and current waveforms obtained after simulating the model are shown in Fig. 7.8. It can be observed that current and voltage are positive or zero. The peak magnitude of the current is 15 A and that of the voltage is 300 V. As output voltage is always less than the input voltage, this chopper is also known as step-down chopper.

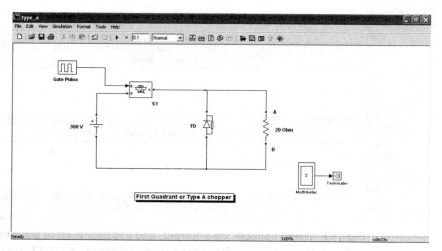

Fig. 7.8 Type-A chopper

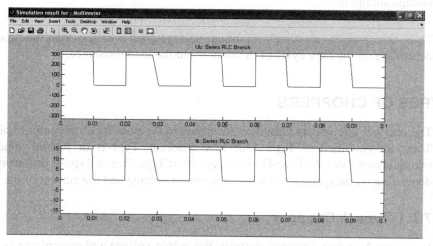

Fig. 7.9 Output voltage and current waveforms

7.7.2 Type-B Chopper

In a Type-B chopper, the output voltage is positive or zero, whereas the output current is zero or negative. The output power is negative in Type-B chopper. It implies that the power is delivered from the load to the source. The load must contain a DC source, i.e., a battery or a DC motor. The voltage across the load is $V_{out} = E + (iR\ L\ di/dt)$, which is always greater than the source voltage. So, this chopper is a step-up chopper. Figure 7.10 shows a model of a Type-B chopper containing a 120 V DC source, pulse generator, a diode, MOSFET switch, load inductance of 5 mH, resistance of 1 Ω along with a battery of 50 V, and measuring blocks. The voltage and current waveforms of the load inductor as obtained are shown in Fig. 7.11. The output voltage varies from 0 V to 50 kV and the output current varies from 0 to –100 A. Thus, the power delivered to the load is negative in this chopper circuit. Also, it is clear from the circuit that diode D will allow the current flow only from the load to the source. The average output voltage of the chopper depends on the duty cycle of the switch.

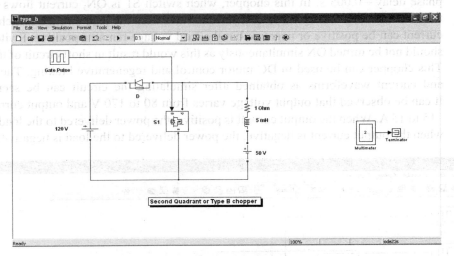

Fig. 7.10 Type-B chopper

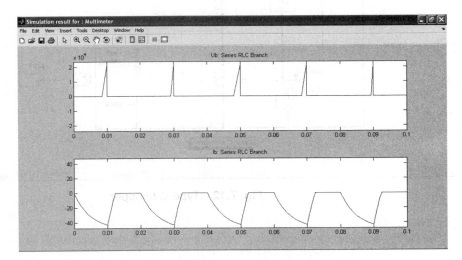

Fig. 7.11 Output voltage and current waveforms

7.7.3 Type-C Chopper

The Type-C chopper is a two-quadrant chopper. In this chopper, the output voltage is positive or zero whereas the output current can be either negative or positive. This type of chopper can be realized by connecting Type-A and Type-B choppers in parallel. A model of Type-C chopper is shown in Fig. 7.12. For this model, the input supply is of 250 V; load consists of 1 mH inductance, 5 Ω resistance, and 125 V battery; and two MOSFET switches S1 and S2 are used along with two diodes as can be observed in Fig. 7.12. The parameters set for the Gate Pulse blocks are as follows: For G1: Amplitude—5, period—0.01 s, pulse width—50%, and phase delay—0 s; and for G2: Amplitude—5, period—0.01 s, pulse width—50%, and phase delay—0.005 s. In this chopper, when switch S1 is ON, current flows from source to the load and when switch S2 is ON, current flows from the load through switch S2. Thus, load current can be positive or negative but the load voltage remains positive. Switches S1 and S2 should not be turned ON simultaneously as this would result in short circuit of the input supply. This chopper can be used in DC motor control and regenerative braking. The output voltage and current waveforms as obtained after simulating the circuit can be seen in Fig. 7.13. It can be observed that output voltage varies from 80 to 170 V and output current varies from −15 to 15 A. When the output current is positive, the power delivered to the load is positive and when the output current is negative, the power delivered to the load is negative.

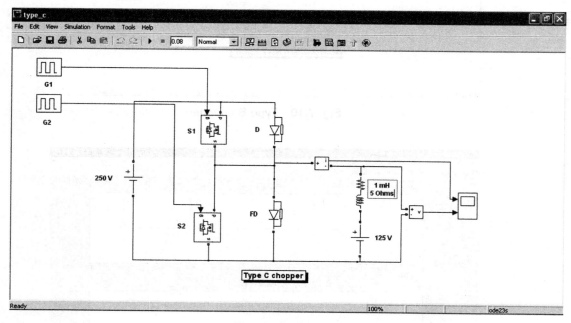

Fig. 7.12 Type-C chopper

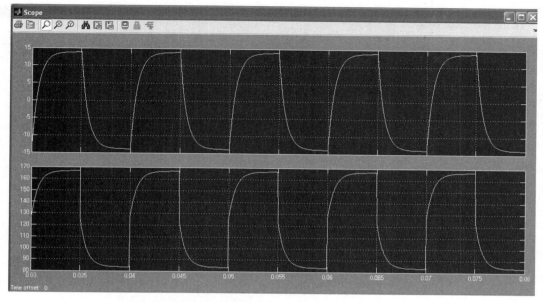

Fig. 7.13 Output voltage and current waveforms

7.7.4 Type-D Chopper

In Type-D or two-quadrant Type-B chopper, the output current is positive or zero whereas the output voltage can be either positive or negative depending upon the duty cycle of the chopper. A model of a Type-D chopper containing a DC source of 80 V, MOSFET switches S1 and S2, diodes D1 and D2, a load of 25 Ω, 40 V, pulse generator, and measurement blocks is shown in Fig. 7.14. The parameters set for Gate Pulse block are as follows: Amplitude—5, period—0.02 s, pulse width—80%, and phase delay—0 s. When switches S1 and S2 are ON, current flows from A to B and the output voltage is positive. When switches S1 and S2 are OFF and diodes D1 and D2 conduct, the output voltage becomes negative whereas the output current continues to flow from A to B as can be seen from Fig. 7.14. Now the average output voltage is positive when the duty cycle of S1 and S2 is greater than 0.5; in other words, $T_{on} > T_{off}$. If the duty cycle of S1 and S2 is less than 0.5, the average output voltage becomes negative. This model is simulated taking the duty cycle as 80% and the average output voltage obtained was 56.01 V. The output voltage and current waveforms are shown in Fig. 7.15. It can be observed that the output voltage is positive, i.e., 80 V, for a longer duration of time in one complete cycle. This model is simulated again taking the duty cycle as 20% and the average output voltage obtained was −15.97 V. The output voltage and current waveforms obtained in this case are shown in Fig. 7.16. It can be observed that the output voltage is negative, i.e., −40 V for a longer duration of time in one cycle. Mathematically, the average output voltage for this chopper can be given as follows:

$$V_{out} = (2 \times D - 1) \times V_s$$

where D is the duty cycle of the chopper and V_s is the average supply voltage.

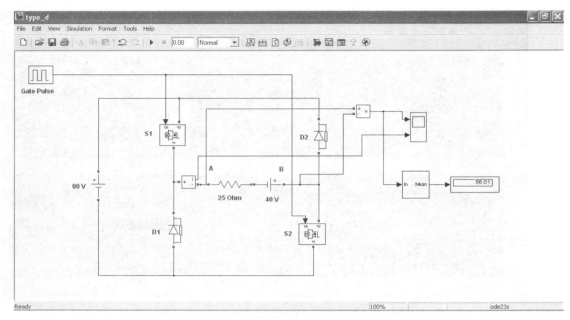

Fig. 7.14 Type-D chopper

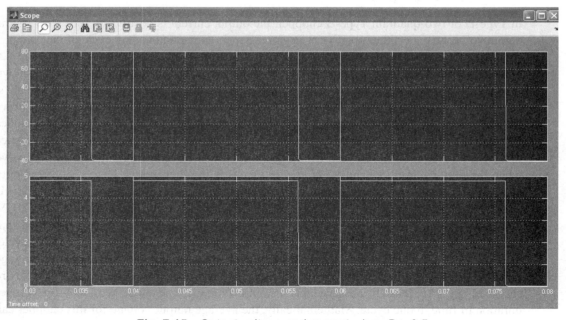

Fig. 7.15 Output voltage and current when $D > 0.5$

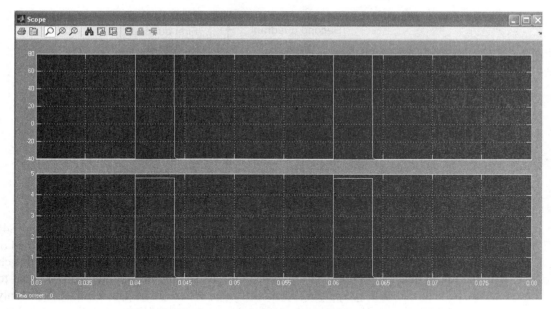

Fig. 7.16 Output voltage and current when $D < 0.5$

7.7.5 Type-E Chopper

The Type-E chopper is a four-quadrant chopper in which the output voltage as well as current can be either positive or negative. This chopper can operate in all the four quadrants. For the first-quadrant operation, switch S1 is operated, S4 is kept ON, and S3 and S2 are kept OFF. When S1 is ON, the output voltage is equal to the supply voltage and the load current increases. If S1 is OFF, then the load current decreases gradually due to the inductive load. For the second-quadrant operation, switch S2 is operated with all other switches, i.e., S2, S3, and S4, kept OFF. When S2 in ON, the current in the load flows in negative direction and the load inductor stores the energy, and when it is OFF, the energy stored in the load inductor is discharged through diodes D1 and D4. As the load voltage is positive and the current negative, so the chopper operates in the second quadrant. Also, in this case, the load voltage is higher than the supply voltage. For the third-quadrant operation, switch S3 is operated, S1 and S4 are kept OFF, and S2 is kept ON and the polarity of load battery is reversed. When S3 is ON, load is connected to the source leading to negative load voltage and current. When switch S3 is OFF, the negative current flows through switch S2 and diode D4. For the fourth-quadrant operation, switch S4 is operated and rest of the switches are turned OFF and the polarity of the load battery is reversed. When S4 is ON, positive current flows through the load and when it is OFF, the load discharges the current through diodes D2 and D3. In this case, the load voltage is negative whereas the load current is positive. Therefore, any type of chopper can be realized by Type-E chopper. Figure 7.17 shows the four quadrants and the operations of various choppers in these quadrants.

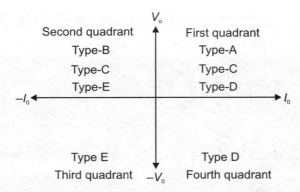

Fig. 7.17 Operation of various choppers in the four quadrants

A model of Type-E chopper containing four MOSFET switches (S1, S2, S3 and S4), four diodes (D1, D2, D3 and D4), 230 V DC supply, pulse generators (G1, G2), R-L-E load of 5 Ω, 5 H, and –190 V, and measurement blocks is shown in Fig. 7.18. In order to operate this chopper in the fourth quadrant, S1 and S3 are turned OFF by setting the amplitude of G1 equal to zero. Also, S2 is turned OFF by applying a constant value of zero at the gate terminal. Now, only S4 is operated through G2. The parameters set for Pulse Generator G2 are as follows: Amplitude—5, period—0.02 s, pulse width—60%, and phase delay—0 s. The voltage terminals of the load battery are reversed by setting the amplitude equal to –190 V. Now when S4 is ON, positive current flows through S4, D2, L, and E and the energy is stored in the load inductor. When S4 is OFF, inductor current is fedback to the source through D2 and D3. Output voltage and current waveforms obtained after simulating the model are shown in Fig. 7.19. It can be observed that the output voltage varies from –184 to –195 V and current varies from 0 to 3 A. Also, there is a sudden variation in the output voltage and current, and the power delivered to the load is negative.

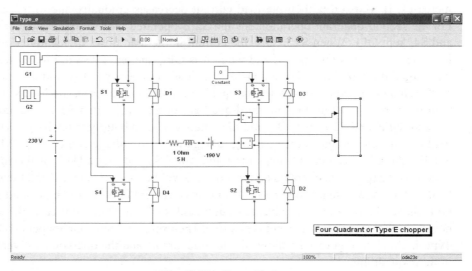

Fig. 7.18 Type-E chopper

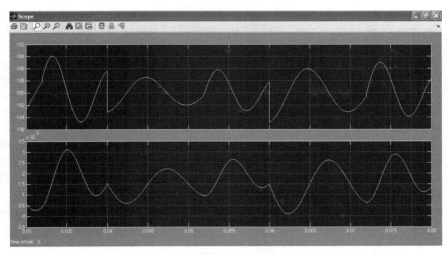

Fig. 7.19   Voltage and current waveforms of R-L load

7.8 CHOPPER COMMUTATION TECHNIQUES

Commutation is process of turning OFF a thyristor. A thyristor can be turned ON by applying a gate pulse, but it cannot be turned OFF by removing or applying a negative gate pulse once it starts conducting. Until now we have used ideal switches or MOSFETs for constructing a chopper circuit. These devices are used for low power applications and they can be turned OFF by removing the gate pulse, i.e., magnitude of the gate pulse is zero. If chopper is designed for high power applications, thyristors are generally used as a switch. While designing a thyristor chopper, a commutation circuit for turning OFF the main thyristor is also required. The thyristor is turned OFF when the current flowing through it falls below the holding current limit. Holding current of a thyristor is defined as the minimum current required for keeping the thyristor in conducting state if it is already conducting the current. This makes the chopper circuit containing thyristor switches more complex. There are three methods employed for commutation of a thyristor—current commutation, voltage commutation, and load commutation. These techniques vary in the way the commutation is achieved. All these methods are further discussed in this section.

7.8.1 Current Commutation Technique

In this technique, the externally generated current pulse is passed through the main conducting thyristor in the reverse direction for a short duration of time so that the p-n junction would get rebuilt and device goes to OFF state. When the thyristor current falls below the holding current limit it stops conducting and is turned OFF. The current pulse is generated by an initially charged capacitor. The commutation is initiated by triggering an auxiliary thyristor of the commutation circuit. The role of the commutation circuit is to turn OFF the main thyristor as it can be again turned ON by applying a gate pulse.

A model of a current-commutated chopper circuit containing a DC supply of 230 V, commutation circuit consisting of thyristor T2, diodes D1 and D2, inductance of 20 μH, capacitance of 50 μF, and a charging resistance of 10 MΩ is shown in Fig. 7.20. Pulse generators G1 and G2 are used to trigger T1 and T2. The load is of 35 Ω, 23 mH, and a freewheeling diode FD is connected across it. The charging resistance is so large that it can be considered as open circuit during the commutation interval. Initially, the commutating capacitor is charged up to 230 V through the charging resistance of 10 MΩ by turning ON switch S1. After this, the main thyristor T1 is turned ON by gate pulse G1. When T1 conducts, the load is connected to the DC supply of 230 V. The commutation process starts after turning ON the commutating thyristor T2. When T2 is turned ON, the capacitor discharges itself through T2 and the inductor. Charging resistance is removed from the circuit by switching OFF switch S1 as the commutation process starts. So when the capacitor discharges itself completely, the current in it still flows in the same direction due to the inductance present in the loop and the capacitor again gets charged and the voltage across it becomes –230 V. This capacitor voltage reverse biases T2 and it is turned OFF.

Now at the instant when T2 is turned OFF, the capacitor voltage has been build up to –230V and the inductor current is zero. Now the capacitor has only one path from which it can discharge itself, i.e., through inductor, D2 and T1. So now, the total current which flows through the main thyristor T1 is the load current I_L minus the capacitor current I_c, i.e., $I_L - I_c$. When the capacitor discharging current becomes equal to the load current, the current through T1 becomes zero and it is turned OFF. As the main thyristor is turned OFF due to the capacitor discharging current, this commutation technique is called as current commutation. This technique is widely used in traction for commutating thyristor choppers. The only consideration is that the capacitor should be charged with the correct polarity and the load current should be less than the peak capacitor current. The load voltage and current waveforms along with the capacitor voltage and inductor current waveforms are shown in Fig. 7.21. It can be observed that the output voltage varies from 0 to 230 V and the output current becomes constant.

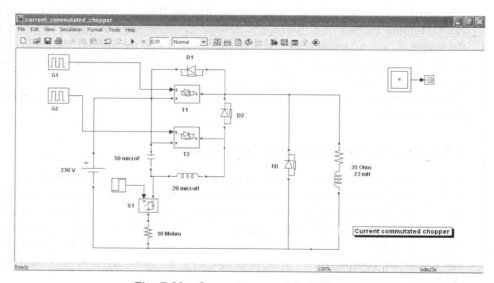

Fig. 7.20   Current-commutated chopper

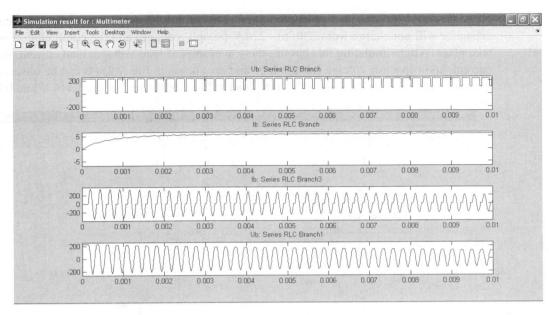

Fig. 7.21 Output voltage, output current, inductor current, and capacitor voltage waveforms

7.8.2 Voltage Commutation Technique

In this technique, a voltage pulse is applied to the main thyristor in the reverse direction for a short duration of time until it stops conducting. When a voltage pulse is suddenly applied to a conducting thyristor, the current flowing through it starts decaying, and when the current falls below the holding current limit, the thyristor stops conducting and is turned OFF. The voltage pulse is generated by charging a capacitor and then connecting it in parallel with the main thyristor. The commutation is initiated by triggering an auxiliary thyristor. Since a large reverse voltage is applied in reverse to the conducting thyristor, the commutation time required in voltage commutation is less when compared to the current commutation technique. It is normally used for power loads where the load fluctuation is quite low.

A voltage-commutated chopper circuit consisting of input DC supply 220 V, main conducting thyristor T1, load of 0.5 Ω, 2 mH, and commutating components $L = 20$ μH, $C = 50$ μF, diode D and auxiliary thyristor T2 is shown in Fig. 7.22. A freewheeling diode FD is connected across the load in order to provide an alternate path for the load current. Pulse generators G1 and G2 trigger thyristors T1 and T2 respectively. Initially, the capacitor is charged up to 220 V. Now the main chopper thyristor T1 is turned ON. When T1 conducts, the load gets supply from the input DC voltage source. Simultaneously, the capacitor gets charged in the opposite polarity through T1, L, and D (as the voltage at the anode of T1 is 220 V). So now the voltage across capacitor is –220 V. When T1 is to be turned OFF, T2 is gated by G2. As soon as T2 is turned ON, the negative 220 V of the capacitor is applied to the conducting thyristor T1 and T1 gets OFF. As T1 is commutated by the negative voltage across capacitor, this commutation method is termed voltage commutation. This method requires an initial charge on the commutation capacitor and it requires load at the output terminals of the chopper. If no load is connected, the capacitor will not charge from 220 to –220 V (supply voltage) and the commutation will not take place. After simulating this circuit, the waveforms of load voltage, load current, inductor current, and capacitor voltage are obtained and are shown in Fig. 7.23. It can be observed that the load current is almost constant and the load voltage fluctuates between 440 V and 0 V.

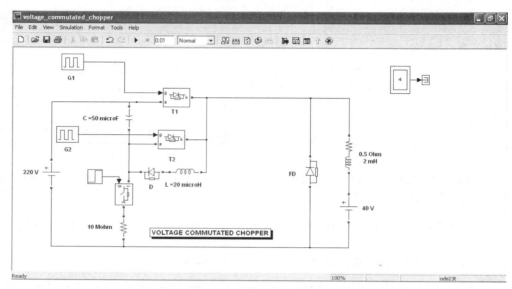

Fig. 7.22 Voltage-commutated chopper

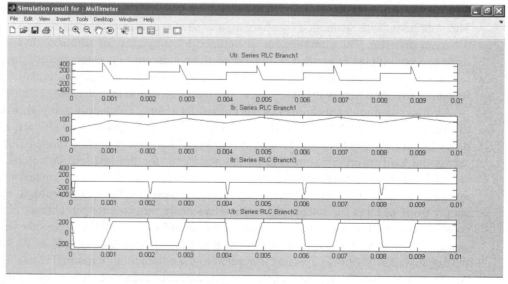

Fig. 7.23 Output voltage, output current, inductor current, and capacitor voltage waveforms

7.8.3 Load Commutation Technique

In the load commutation technique, the conducting thyristor is turned OFF due to either the nature of the load or the current transferred to the other device. In both these cases, the current flowing in the thyristor becomes almost zero. In this case, the commutation is provided by the load itself and as such, no separate commutation circuit is required.

A load-commutated chopper circuit containing a 240 V DC source at the input, thyristors T1, T2, T3, and T4, 20 µF capacitor, and load of 40 Ω and 20 mH is shown in Fig. 7.24. Freewheeling diode FD is connected across the load terminals to provide an alternate for the load inductor to discharge the load current. G1 triggers thyristor T1 and G2 triggers thyristor T2. If thyristor pair T1 and T2 is conducting, then T3, T4, and C are the commutating elements. Similarly, if T3 and T4 are conducting then T1, T2, and C act as commutating elements. Initially, the capacitor is charged up to the supply voltage, i.e., 240 V. When T1 and T2 are turned ON, the load current flows through the commutating capacitor C. Now the polarity of the voltage at C is reversed. In this process of supplying current to the load, C gets slightly overcharged which in turn forward biases diode FD. Now the load current is transferred from T1, C, and T2 to the freewheeling diode. Now as T3 and T4 are turned ON, the capacitor voltage reverse biases T1 and T2 and they are turned OFF. In a similar manner, when T3 and T4 are the main conducting thyristors, they are turned OFF by the capacitor voltage by triggering T1 and T2. The load voltage, capacitor voltage, and load current waveforms obtained after simulating the model of Fig. 7.24 are shown in Fig. 7.25. This commutating technique has the advantage that it can commutate any amount of load current, no commutating inductor is required and it works well on higher frequencies. However, this method requires sensing of capacitor current and higher PIV rating of the freewheeling diode.

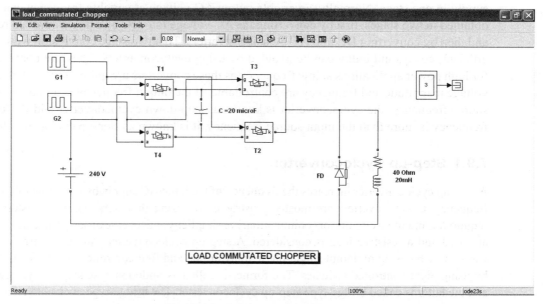

Fig. 7.24 Load-commutated chopper

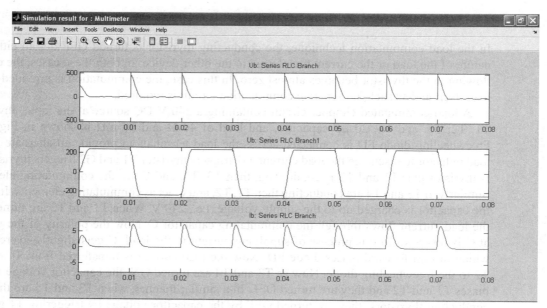

Fig. 7.25 Load voltage, capacitor voltage, and load current waveforms

7.9 CYCLOCONVERTERS

A cycloconverter is a device that converts one frequency AC supply to another frequency AC supply in one stage. Normally, a variable output AC voltage at variable frequency can also be obtained in two stages, first by converting the AC voltage into DC by using controlled rectifiers and then by converting DC into AC by using inverters. This two stage conversion which is less efficient, costly, and bulky can be avoided by using cycloconverters. They are normally used for high power applications at low frequency as these converters are quite slow. Both the output voltage magnitude and frequency are controllable. If the output frequency is less than the input source frequency, the cycloconverter is known as step-down cycloconverter and if the output frequency is more than the input source frequency, it is known as step-up cycloconverter.

7.9.1 Step-up Cycloconverter

A step-up cycloconverter increases the frequency of the input AC supply by a fraction of the supply frequency. Cycloconverters are mostly constructed by using thyristors. These thyristors further require commutation circuits for commutation. For simplicity and for ease of designing, ideal switches are used and a resistive load is considered. A step-up cycloconverter can be constructed in two ways—one by using midpoint single-phase transformer, and four controlled switches; and another by using eight controlled switches. The former is called as midpoint-type step-up cycloconverter and the latter is called bridge-type step-up cycloconverter. The bridge-type cycloconverter is more compact and economical as it does not require a center-tapped transformer.

A model of a bridge-type step-up cycloconverter consisting of input supply of 220 V, 50 Hz, eight ideal switches, four pulse generators, load resistance of 100 Ω, and a multimeter is shown in Fig. 7.26. This model is designed to convert 50 Hz input frequency to 100 Hz output frequency. The pulse generator G1 supplies gate pulse to P1 and P2, G2 supplies gate pulse to N1 and N2, G3 supplies gate pulse to P3 and P4, and G4 supplies gate pulse to N3 and N4. Switches P1 to P4 are of the positive group, i.e., output voltage is positive when they conduct, and switches N1 to N4 are of negative group, i.e., output voltage is negative when they conduct. The time period of the input voltage waveform is $T = 1/50 = 0.02$ s. So, for time duration 0 to 0.01 s, the input voltage is positive and for time duration 0.01 to 0.02 s, the input voltage is negative and it continues so on. At time, $t = 0$ s when the input supply is positive, P1 and P2 are turned ON. So, the load voltage is positive and it follows the input voltage. Now at $t = 0.005$ s, P1 and P2 are turned OFF and N1 and N2 are turned ON.

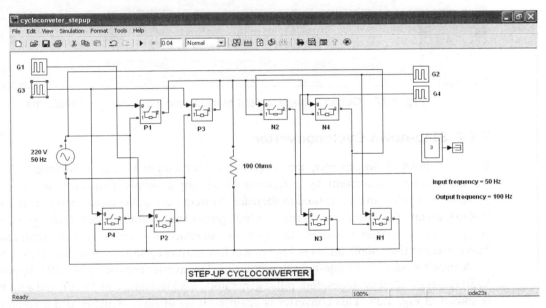

Fig. 7.26 Step-up bridge-type cycloconverter

Now the load voltage traces the input voltage but is negative. At $t = 0.01$ s, N1 and N2 are turned OFF and P3 and P4 are turned ON. Now the supply voltage has gone negative but the load voltage becomes positive and it traces the input voltage. At $t = 0.015$ s, P3 and P4 are turned OFF and N3 and N4 are turned ON. Now the load voltage again becomes negative and follows the input voltage waveform. So, one input cycle is completed in this manner and the process repeats by turning ON P1 and P2 at $t = 0.02$ s. The output voltage and current waveforms obtained after simulation are shown in Fig. 7.27. It can be observed from Fig. 7.27 that the frequency of the output voltage and current is 100 Hz, i.e., time period is 0.001 s. In this model, the parameters of Pulse Generators are set according to the described switching sequence. Also, the conversion of input frequency of 50 Hz to the output frequency of 100 Hz is achieved in one stage.

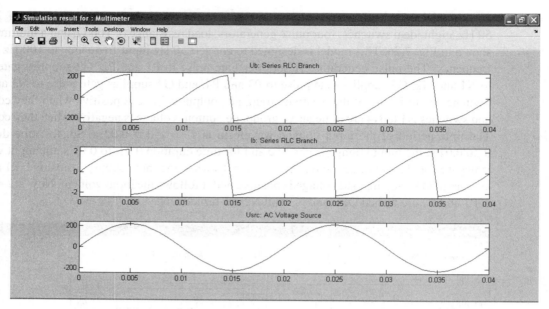

Fig. 7.27 Output voltage, output current, and input voltage waveforms

7.9.2 Step-down Cycloconverter

In a step-down cycloconverter, the frequency of the output voltage waveform is less than the input voltage waveform by a fraction of the input supply frequency. In other words, a step-down cycloconverter is used to decrease the frequency of the input supply. This converter is also built up of thyristors as it is used for high power applications. A step-down cycloconverter is again of two types based on the type of construction—bridge type or midpoint type. This converter finds applications in high power low frequency applications like electric traction.

A circuit model of a bridge-type step-down cycloconverter consisting of a DC supply of 220 V, 50 Hz, eight ideal switches, four pulse generators, load resistance of 100 Ω, and a multimeter is shown in Fig. 7.28. This converter is again built up of ideal switches and a resistive load is considered to further simplify the circuit model. This converter converts the input supply frequency of 50 Hz to 25 Hz, i.e., output frequency is half the input frequency. Switches P1 to P4 are of the positive group and N1 to N4 are of negative group.

Now for converting 50 Hz to 25 Hz at $t = 0$ s, P1 and P2 are turned ON. Now the output voltage is positive and it follows the input voltage waveform. At $t = 0.01$ s, P1 and P2 are turned OFF by removing the gate pulse and P3 and P4 are turned ON. Now again the output voltage is positive and it follows the input voltage waveform. Thus, the first half positive cycle of the output voltage wave is obtained. For negative half cycle at $t = 0.02$ s, N1 and N2 are turned ON. Now the output voltage is negative and it follows shape of the input voltage waveform. At $t = 0.03$ s, N1 and N2 are turned OFF and N3 and N4 are turned ON. The output voltage is again negative but follows the input voltage waveform. The output waveform of 25 Hz is obtained in this manner. This switching sequence is further continued by turning ON P1 and P2 at $t = 0.04$ s and so on for continuous operation. The output voltage and current waveforms thus obtained after simulation are shown in Fig. 7.29. The time period of the output waveform is 0.04 s, i.e., the frequency is 25 Hz, in conversion, i.e., 50 to 25 Hz is obtained in a single stage.

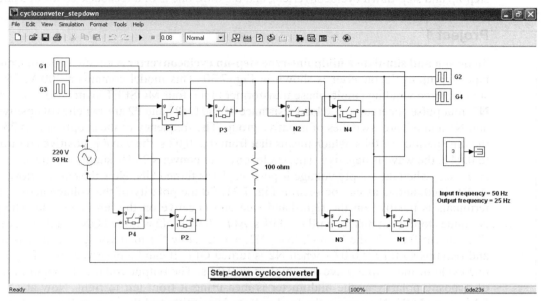

Fig. 7.28 Step-down bridge-type cycloconverter

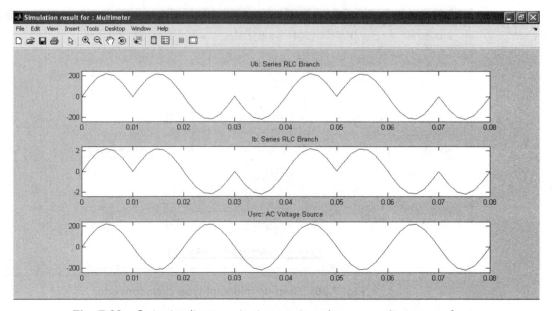

Fig. 7.29 Output voltage, output current, and source voltage waveforms

7.10 SIMULATION PROJECTS

Three simulation projects are discussed in this section. The first two projects are of midpoint-type step-up and step-down cycloconverters, and the third one is of a feedback-controlled chopper.

Project 1

To design and simulate a midpoint-type step-up cycloconverter A circuit model of a midpoint-type step-up cycloconverter is shown in Fig. 7.30. This model contains a 200 V, 25 Hz AC supply, three winding single-phase transformer (1:1), four MOSFET switches P1, P2, N1, and N2, four pulse generators, and load resistance of 100 Ω. P1 and P2 are switches of positive group and N1 and N2 are switches of negative group. The frequency of the supply side is 25 Hz. So the time period is 0.04 s, which means that from 0 to 0.02 s, the wave is positive and from 0.02 to 0.04 s, the wave is negative. This cycloconverter converts 25 Hz supply frequency to 50 Hz. At $t = 0$ s, when the supply voltage is positive, P1 is turned ON. Now the current flows in loop A, P1, load, and C as can be seen in Fig. 7.31. So the polarity of the voltage across the load terminals is positive on the right-hand side and negative on the left-hand side. The current continues to flow in this loop till $t = 0.01$ s. At $t = 0.01$ s, N2 is turned ON and P1 is turned ON. The current flows in the loop C, load, N2, and B. Now the load voltage polarity is reversed and remains so till $t = 0.02$ s when N2 is turned OFF. It can be observed from Fig. 7.30 that one cycle of the output wave is complete in 0.02 s. The output voltage shown in Fig. 7.30 is of opposite polarity as the multimeter is measuring it from left to right. Now at $t = 0.02$ s, P2 is turned ON. Now again the load voltage is positive and the current flows in loop B, P2, load, and C. Now at $t = 0.03$ s, P2 is turned OFF and N1 is turned ON. Now the current flows

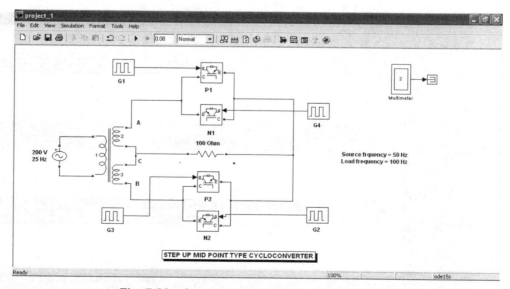

Fig. 7.30 Step-up midpoint-type cycloconverter

in loop C, load, N1, and A. The load voltage is of opposite polarity. Thus, in one cycle of input waveform, two cycles of output waveforms are obtained. So the frequency of the output voltage waveform is 50 Hz as can be seen from Fig. 7.31.

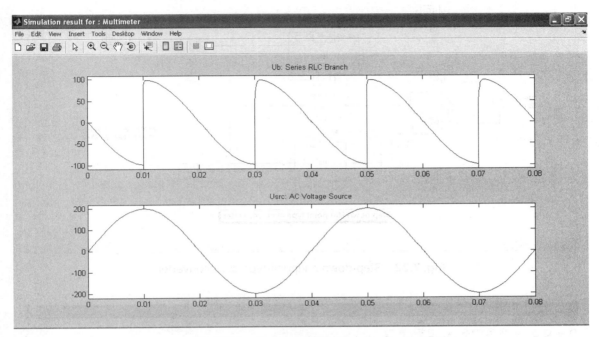

Fig. 7.31 Output and input voltage waveforms

Project 2

To design and simulate a midpoint-type step-down cycloconverter A step-down midpoint-type cycloconverter is shown in Fig. 7.32. This circuit is same as that of a step-up cycloconverter except that the firing sequence of switches M1, M2, M3, and M4 is different. This cycloconverter converts input frequency of 50 Hz, i.e., time period 0.02 s, to output frequency of 25 Hz, i.e., time period of 0.04 s. At $t = 0$ s, P1 is turned ON and the current starts flowing in loop A, P1, load, and C. The polarity of the load voltage is shown in Fig. 7.32. Now at $t = 0.01$ s, when the input voltage goes negative, P2 is turned ON and P1 is turned OFF. Now the current flows in loop B, P2, load, and C. The load voltage is again positive. At $t = 0.02$ s, P2 is turned OFF and N2 is turned ON. Now the load current flows from C, load, N2 to B and the load voltage is negative. At $t = 0.03$ s, N2 is turned OFF and N1 is turned ON. Now the current flows from C, load, N1 to A and the load voltage is again negative. Now at $t = 0.04$ s, N1 is turned OFF and again P1 is turned ON and the process repeats. So in two cycles of input voltage waveform, one cycle of output voltage waveform is obtained. As can be observed from Fig. 7.33, the frequency of the output voltage waveform is 25 Hz. So this cycloconverter converts 50 Hz frequency to 25 Hz in one stage.

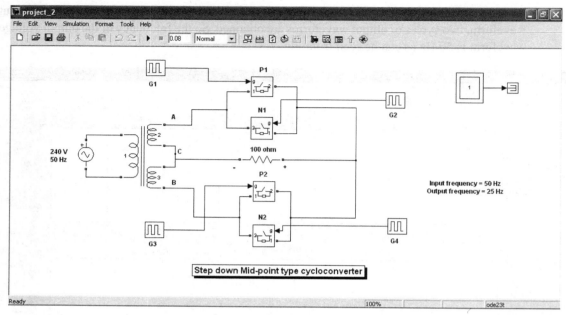

Fig. 7.32 Step-down midpoint-type cycloconverter

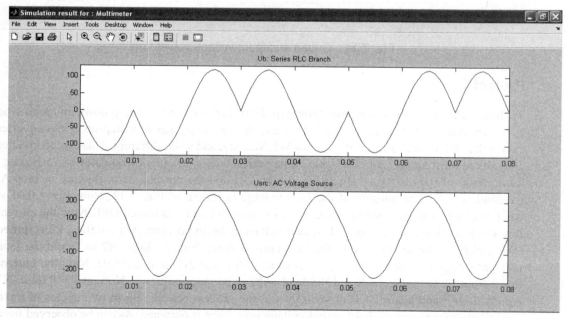

Fig. 7.33 Output voltage and input voltage waveforms

Project 3

To design and simulate a feedback control chopper A feedback control chopper is shown in Fig. 7.34. For this chopper, the input supply is of 20 V and is connected to the load by a switch S1. The output inductance is 0.05 mH and the load resistance is 5 Ω. A freewheeling diode is also connected across the inductance and load resistance. The desired load current for this chopper is 3.5 A. The feedback circuit consists of current measuring block, saturation block, two constant blocks, and a switch. The upper and lower limits for the saturation block are kept as 4 and 2. The threshold of the switch block is 3.5 and input1 of this switch is connected to 0 and input 3 is connected to 5. The condition for passing input1, i.e., 0, is that the load current is greater than 3.5, else input3, i.e., 5, will pass through the switch. The chopper switch S1 is initially closed.

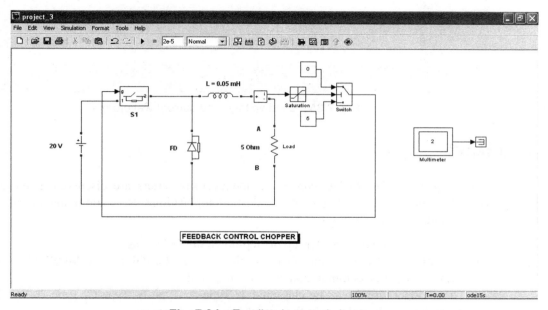

Fig. 7.34 Feedback control chopper

Now as switch S1 is closed, the supply is connected to the load. In this case, the approximate load current is 20/5 = 4 A. As this value is greater than the threshold of the switch, the output of the switch block is zero, i.e., switch S1 will open. Now since S1 is open, the load current will start decreasing. When the load current will reach the threshold value of 3.5 A, the switch will pass input3, i.e., 5. Now again switch S1 will be closed by input3 and the load will be connected to the supply. It is clear that a feedback control chopper is more complex than a duty cycle control chopper. The load voltage and current waveforms obtained are shown in Fig. 7.35. It can be observed that the ripples in the load current are quite low. The load current settles to a value of 3.5 A and the load voltage is at a value of 17.5 V. One can vary the load resistance and observe the variations in the load current with the load. This control technique is, therefore, quite useful if a low ripple output current is desirable from the chopper. This model is simulated for 20 μs as it takes more time for simulating this model because it contains an algebraic loop.

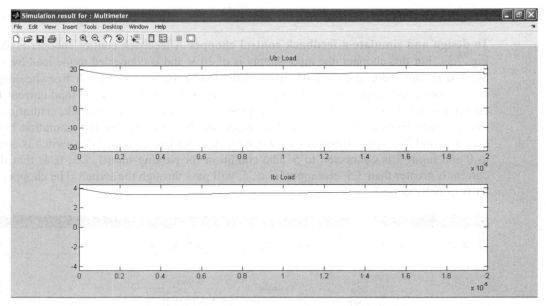

Fig. 7.35 Load voltage and current waveforms

SUMMARY

The choppers or DC to DC converters and cycloconverters are discussed in the chapter. Both these devices are widely used in industrial applications. The main points of this chapter can be summarized as follows:

- Choppers can be used to step-up or step-down the DC voltage.
- They are built by fully controlled switches like SCRs, IGBTs, and MOSFETs. For SCR switch, a separate commutation circuit is required.
- The chopper circuit can be classified as Type-A, Type-B, Type-C, Type-D, and Type-E depending on the polarity of the output voltage and current.
- For SCR choppers, three types of commutation techniques are available—voltage commutation, current commutation, and load commutation.
- Choppers are mostly used for controlling DC motors.
- Fixed frequency or variable frequency techniques can be used for controlling a chopper.
- A cycloconverter transfers power from the AC source to the load at controlled voltage and frequency which is an integral fraction of the source frequency.
- Cycloconverters perform direct conversion of the frequency through controlled switches. The source frequency can be step up or step down.
- Depending on the circuit components used, there are two types of cycloconverters— midpoint type and bridge type.
- Cycloconverters are slow operating devices and they are used in high power applications such as traction and industrial drives.

REVIEW QUESTIONS

1. Explain why a DC chopper can be considered equivalent to an AC transformer.
2. Explain the basic principle of operation of a chopper.
3. Mention the major advantages and disadvantages of choppers over other DC to DC conversion techniques.
4. Mention the various techniques that can be used for DC to DC conversion.
5. Explain the operation of a buck converter and its applications.
6. Explain the operation of a boost converter and mention the applications in which it can be used.
7. Mention the advantages and disadvantages of buck–boost converter.
8. Classify choppers according to their quadrant of operation.
9. Which type of chopper can be used for regenerative braking? Explain.
10. Mention the merits and demerits of feedback control technique of chopper control over duty cycle control technique.
11. Explain the functioning of a step-down cycloconverter with the help of suitable waveforms.
12. Mention the advantages of direct AC to AC conversion over rectifier–inverter based DC link, AC to DC to AC conversion technique.
13. Explain the operation and working of a step-up cycloconverter with suitable waveforms.
14. Mention the important industrial applications of cycloconverters.
15. What are the merits of midpoint-type cycloconverters over the bridge-type cycloconverters?

SIMULATION EXERCISES

1. A DC chopper has an input voltage of 15 V. It delivers 5 V output at power varying between 10 to 40 W. The switching frequency of the chopper is 10 KHz. Construct a circuit model to perform this function and determine value of load inductance necessary to keep the ripple within 5% limit.

2. Construct a buck converter which has an input supply of 20 V. This converter has to supply a 40 V to the inductive load. The load power varies between 40 to 100 W. Determine the load parameters for 100 W and plot the load voltage and current.

3. A buck regulator has an input voltage of 15 V and it is required to supply an average output voltage of 5 V. Take the switching frequency as 15 kHz, determine the duty cycle of the switch and load inductance and resistance to limit the load current to 0.5 A.

4. A DC chopper is to be operated at a duty cycle of 0.6. The load resistance is 10 Ω and inductance is 10 mH. Determine the voltage gain in case of a buck converter and boost converter.

5. Construct a boost converter with an input supply of 50 V. The filter capacitor and resistor values are 20 μF and 40 Ω. The switching frequency of the switch is 1 kHz. Vary the duty cycle of the chopper from 0.1 to 0.9 and examine the average output voltage obtained.

6. Develop the model for a Type-A chopper and plot the output voltage waveforms for duty cycle $D = 0.2$ and $D = 0.8$. Determine the average

and rms value of the output voltage obtained for both these duty cycles and the output power delivered in case of a resistive load of 10 Ω.

7. A chopper circuit has the following data: switching period T_s = 1,000 μs, R = 2 Ω, and L = 5 mH. If the input DC voltage is 20 V, determine the duty cycle of the chopper so that the minimum load current does not fall below 0.5 A. Plot the output voltage obtained.

8. Develop the model of a voltage-commutated chopper which delivers the power to a load of R = 0.1 Ω, L = 8 mH, and E = 40 V. The chopper frequency is 200 Hz and the peak value of the load current is to be limited to 40 A. Plot the output voltage and current after developing the model.

9. Construct a model of a step-down bridge-type cycloconverter. The cycloconverter is supplied by a 220 V, 50 Hz AC supply. The output frequency required is 12.5 Hz. Determine the switching patterns of the cycloconverter switches in order to obtain the desired output frequency and plot the output voltage.

10. A single-phase cycloconverter is to be used to obtain an output frequency of one-third the input supply frequency. The turn ratio of the primary winding to the upper secondary winding is 1:1 and to the lower secondary winding is 1:2. Plot the output voltage and current waveforms for a resistive load of 150 Ω.

11. A bridge-type step-up cycloconverter receives a single phase AC supply of 220 V, 50 Hz. It is desired to supply a load of 50 Ω with an AC supply of frequency 150 Hz. Construct a cycloconverter model and plot the output voltage and current obtained after simulation. Also determine the rms value of the output voltage obtained.

12. A single-phase midpoint cycloconverter is controlled by setting the delay angle at $\pi/3$. The output frequency desired is to be one-eighth of the input source frequency. The input AC source is of 120 V, 60 Hz. Plot the output voltage waveform and obtain its harmonic profile.

13. A 100 V DC source is to be used to charge a 150 V DC battery. Construct a boost converter operating at a switching frequency of 40 kHz. The peak load current required for charging the battery is 50 A.

14. A boost converter has the following parameters: Switching frequency = 100 kHz, input voltage = 10 V, output voltage = 48 V, duty cycle = 0.6, L = 10 μH, and C = 0.47 μF. The power delivered to the load is 50 W. If the load power is 10 W, determine average output voltage.

15. An electric train gets supply from a 1,500 V DC supply through a DC chopper. The minimum ON time and OFF time of the chopperswitch are 40 μs and 50 μs respectively. Develop a DC chopper if the lowest output voltage required is 20 V. Modify this model if the maximum switching frequency is restricted to 2 kHz.

16. Develop a three phase to three phase set-down cycloconverter fed from a 400 V, 50 Hz supply if delivering power to a star-connected load of 40 Ω at 25 Hz. Develop a Simulink model of this converter and plot the output voltage obtained.

SUGGESTED READING

Agrawal, J.P., *Power Electronic Systems: Theory and Design*, Chapter 6, Pearson Education, Delhi, 2006.

Barton, T.H., *Rectifiers, Cycloconverters, and AC Controllers*, Clarendon Press, Oxford, 1994.

Berkovich, Y. and A. Ioinovici, 'Dynamic model of PWM zero-voltage-transition DC-DC boost converter', *Proceedings of IEEE International Symposium on Circuits and Systems*, Orlando, FL, Vol. 5, pp. 254–257.

Bimbhra, P.S., *Power Electronics*, Chapter 7, Khanna Publishers, New Delhi, 2009.

Forsyth, A.J. and S.V. Mollov, 'Modelling and control of DC-DC converters', *Power Engineering Journal*, Vol. 12, No. 5, Sch. of Electron and Electr. Engg., Birmingham Univ., pp. 229–236, 1998.

Higham, N.J. and Higham, D.J., *MATAB Guide*, 2nd Ed., SIAM, Korea, 2005.

Ioannidis, G., A. Kandianis, and S.N. Manias, 'Novel control design for the buck converter', *IEE Proceedings: Electric Power Applications*, Vol. 145, No. 1, Dept. of Electr. and Comput. Engg., Nat. Tech. Univ. of Athens, pp. 39–47, January 1998.

Krein, P.T., *Elements of Power Electronics*, Chapter 4, Oxford University Press, New York, 2009.

Matsuo, H., F. Kurokawa, H. Etou, Y. Ishizuka, and C. Chen Changfeg, 'Design oriented analysis of the digitally controlled DC-DC converter', *Proc. IEEE Power Electronics Specialists Conference,* Galway, Ireland, pp. 401–407, 2000.

McMurray, W., *The Theory and Design of Cycloconverters*, MIT Press, Cambridge, MA, 1972.

Mitchell, D.M., *DC-DC Switching Regulator*, Chapters 2 and 4, McGraw-Hill, New York, 1988.

Okoro, O.I., *The Essential MATLAB and Simulink for Engineers and Scientists*, Juta and Company Ltd., Cape Town, South Africa, 2010.

Pressman, A.I., *Switching Power Supply Design*, McGraw-Hill, New York, 1991.

Rashid, M.H., *Power Electronics: Circuits, Devices, and Applications*, Chapter 5, Pearson Education, New Delhi, 2005.

Rombaut, C., G. Seguier, and R. Bausiere, *Power Electronic Converters*, Vol. 2, McGraw-Hill, New York, 1987.

Severns, R.P., and E.J. Bloom, *Modern DC to DC Switch Mode Power Converter Circuits*, Van Nostrand Reinhold, New York, 1985.

Ziogas, P.D., Y.G. Kang, and V.R. Stefanovic, 'Rectifier-inverter frequency changers with suppressed DC link components', *IEEE Transactions on Industry Applications*, Vol. IA-22, No. 6, Department of Electrical Engineering, Concordia University, pp. 1026–1036, November 1986.

8

POWER SYSTEM ENGINEERING

8.1 INTRODUCTION

Power system engineering deals with the generation, transmission, and distribution of electric power as well as the electrical devices connected to it including circuit breakers, relays, generators, motors, and transformers. A large part of this field is concerned with three-phase AC power which is a standard for large-scale power transmission and distribution. A significant part of power system engineering is concerned with the specialized power systems built for aircraft or electric railway networks. Michael Faraday's accidental discovery, in 1931, that a change in magnetic flux induces an electromotive force in a loop of a wire was the greatest discovery with respect to the field of power system engineering. It gave the principle of electromagnetic induction which helps in explaining the principle of working of electrical machines and transformers.

The world's first power station (AC) was developed at Godalming in England in 1881 followed by the first steam power station (DC) at New York City in 1882. By the end of 1890, thousands of power stations were built all over the world, dedicated mainly for the lightning loads. Many important developments in the field of power system engineering came from other emerging fields like computer technology, electronics, and telecommunication engineering. These developments allowed better control and operation of switchgears and generators in the remote power system.

Power, in general, is defined as the rate of doing work. According to mathematics, the electrical power is the product of two quantities—voltage and current. If these two quantities vary with respect to time, it is termed as AC power; and if they remain constant, it is termed as DC power. Most of the industrial applications like air conditioners, pumps, and induction heaters require AC power, whereas DC power is required mostly by digital equipments, batteries, and computers. In the case of AC power, the voltage levels can be transformed easily with the help of transformers. So, the AC power can be transmitted over longer distances with less loss of higher voltages. This is beneficial in conditions where the generation is distant from the load and it is desirable to step up the voltage level at the generation and step down the voltage level at the load. Also, in the case of AC power, the mismatch between the voltages can be managed easily. Power electronic converters, i.e., choppers, can be utilized to transform the voltage levels in case of DC power but they are more expensive in comparison to their traditional counterparts. Thus, the AC power is most widely used.

Power system engineering is a network of interconnected components which convert different forms of energy into electrical energy and then transmits it. Modern power system engineering consists of three main subsystems—the generation subsystem, the transmission subsystem, and the distribution subsystem. In the generation subsystem, the power plant produces electricity. The transmission subsystem transmits the electricity to different load centers. The distribution subsystem continues to distribute the power to the customers.

The generation subsystem contains an electrical power generation plant. The electrical power generation is a process whereby other forms of energies like potential, solar, wind, heat, and nuclear are transformed into electrical energy. As discussed earlier, the other forms of energies cannot be transmitted easily to distant locations, it is better to convert them into electrical form. The electrical energy can be utilized efficiently and can be readily and expeditiously converted into other forms of energies like heat energy (electric heater), light energy (electric bulb), and mechanical energy (electric motor). The process of electromechanical energy conversion is mostly used to convert energy from coal, water flow, uranium, petroleum, and natural gas into electrical energy. This energy conversion process takes advantage of synchronous AC generator, mechanically coupled to a steam, gas, or water flow, which converts its (sources) energy into kinetic energy. The synchronous generator then converts this kinetic energy of the turbine into electric energy. This turbine generator conversion process is very economical and is mostly used in the industry.

The electrical power is transmitted from the power station to the load, through transmission system. The transmission system can be further divided into bulk transmission system and sub-transmission system. The bulk transmission system transfers electrical power from interconnected generators to various area networks and major load centers and the sub-transmission system interconnects the bulk power system to the distribution system. The transmission circuit or line can be of two types—overhead transmission line and underground cable. For a given transmission voltage, the overhead conductors are less expensive than the underground cables. The underground cables are, therefore, used in urban areas, where overhead transmission is not possible. The transmission system in general is a highly integrated system containing substations, transformers, and transmission lines. The substation contains transformers, relays, and circuit breakers.

The transformers are used to step-up or step-down the voltage on the transmission line. A relay is a level detector which performs a switching action when the input voltage (or current) exceeds the specified limit. A circuit breaker is an automatically operated electrical switch used to protect the electric network from damage caused by overload or short circuit. The amount of power flow through a transmission line is limited by three main factors—thermal overload, voltage instability, and rotor angle instability. The thermal overload is due to the excessive current flow which causes overheating in the electric circuit. Voltage instability occurs when the power required to maintain the voltages at or above the acceptable levels exceeds the available power. The rotor angle instability is a dynamic problem which may occur due to faults, such as short circuit fault in the transmission system or due to undamped oscillatory response of the rotor in motion during starting.

The distribution system is required to transmit power from the transmission system to the consumer. The equipments associated with the distribution system include the substation transformers which are connected to the transmission system, the distribution lines from the transformer to the customer, and the protection and control equipments used between the transformer and the customer. The protection equipments include lightning protectors, circuit breakers, disconnectors, and fuses; while the control equipments include voltage regulators, capacitors, relays, and demand side management equipments.

8.2 ELECTRIC SUPPLY SYSTEM

A typical electric supply system model is shown in Fig. 8.1. It comprises the following three principal components:

1. Generating station
2. Transmission lines
3. Distribution lines

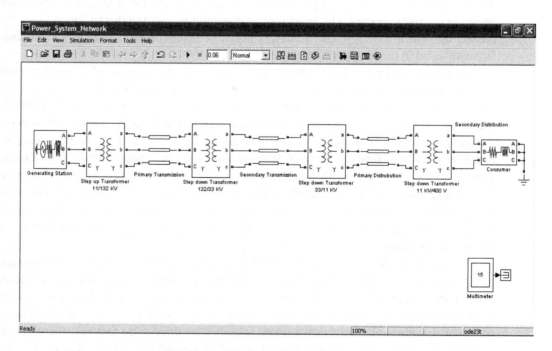

Fig. 8.1 Electric power supply system

The various components of the power system model shown in Fig. 8.1 are elaborated as follows:

Generating station At the generating station, the electrical generators convert mechanical power into three-phase AC power. The normal generation voltage is 11 kV (rms) phase to phase and the generator frequency is 50 Hz (Fig. 8.2). At the generating station, three-phase transformer steps up this voltage to a voltage level for suitable transmission (66 kV). The voltage is stepped up to 132 kV (Fig. 8.2). The primary transmission is generally carried out at 66, 132, 220, or 400 kV.

Primary transmission Electricity is transmitted at high voltages (132 kV or above) in order to reduce the energy lost in long distance transmission as shown in Fig. 8.2. Lower voltages such as 66 kV or 33 kV are usually considered sub-transmission voltages but are occasionally used on long lines with light loads. Voltages less than 33 kV are used for distribution and voltages above 230 kV are considered extra high and require different designs as compared to equipments used at lower voltage levels.

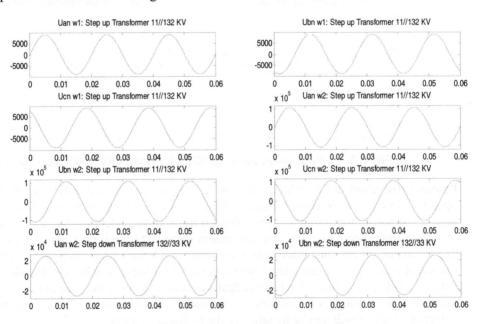

Fig. 8.2 Voltage waveforms of 11/132 kV and 132/33 kV transformers

Primary transmission is usually done through overhead transmission lines. Underground power transmission is costly and has greater operational limitations but is useful in urban areas or for sensitive locations. The high voltage overhead conductors are generally constructed by aluminum alloy made up of several strands and reinforced with steel strands. These conductors are un-insulated and design of these lines requires minimum clearances to be observed in order to maintain safety (line-to-line fault). Adverse weather conditions of high wind and low temperature can lead to power outages.

Secondary transmission The primary transmission ends at receiving station where the voltage is stepped down to 33 kV by step-down transformers as shown in Figs 8.2 and 8.3. The electric power is transmitted at 33 kV by the receiving station through three-phase three-wire overhead transmission lines. This system is called secondary transmission system.

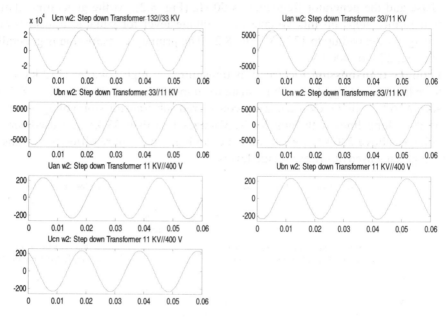

Fig. 8.3 Voltage waveforms of 33/11 kV and 11 kV/400 V transformer

Primary distribution The primary distribution begins from the substation where the secondary lines terminate. The voltage is reduced from 33 kV to 11 kV at the substation as shown in Fig. 8.3. A substation transfers power from transmission system to the distribution system of an area. Connecting the electricity consumers directly to the high voltage main transmission line is uneconomical, unless the consumers consume large amount of power, so the substation reduces the voltage to a suitable voltage level to fit for local distribution. Besides reducing the voltage level, a substation also isolates faults occurred in either the transmission or distribution network. The consumers consuming large amount of power (>50 kW) are normally supplied power at 11 kV, which can be handled by their own substation.

Secondary distribution The secondary distribution is the final stage of delivery of electricity to the end user (small consumers). This distribution network takes electricity from the primary distribution system at 11 kV and delivers it to the consumers at 400 V as shown in Fig. 8.3. The secondary distribution network includes pole-mounted transformers, low voltage (<1 kV) distribution lines, and sometimes electricity meters. The electricity is generally supplied at 400 V by a three-phase four-wire poles mounted overhead distribution line which can be seen in any locality. The single-phase residential load is connected between any phase and neutral (230 V) and three-phase motor load is connected to three-phase lines (400 V) directly. The residential power supplied in India is at 240 V, 50 Hz. The permitted variation in the voltage is 6% and the maximum load is allowed is approximately 40 A.

8.3 ELECTRICAL POWER QUALITY

It is generally expected that the supply voltages at each point in any power system (distribution as well as transmission) should be sinusoidal and should have constant and equal magnitudes, fixed frequency, equal phase displacement of 120° among the three phases, and above all, there should be no zero sequence voltage. However, because of the nonlinear loads connected to the distribution system, the supply voltages no longer are found to exhibit these characteristics. It may be pointed out that the good quality of power supply voltages implies voltage quality and supply reliability. The typical parameters influencing voltage quality are explained in the following subsections.

8.3.1 Waveform Distortion and Power System Harmonics

It is well known that a nonlinear load draws highly distorted current from the source, which consists of harmonics and fundamental active and reactive current components. If the source or the load is unbalanced, the source also contains negative sequence currents. The harmonic currents in combination with line impedance of the distribution network causes distortion in the supply voltage. Further, because of the non-ideal characteristics of the AC source, it also contributes to this distortion and thus aggravates the problem. The distorted supply voltages, thus, adversely affect the performance of the other equipments connected to the network. Table 8.1 summarizes some of the adverse effects of voltage harmonics on the consumer equipments and power system components (Akagi, 1996). The reactive currents deteriorate the power factor and causes extra losses in transmission, generation, and distribution systems. Also, the presence of these reactive current requires higher volt-ampere rating of all the power system components and higher voltage regulation.

Table 8.1   Effects of harmonics on equipments

Equipments	Effects
Capacitor banks	Overheating, insulation breakdown, and failure of internal fuses
Protection equipment	False tripping or no tripping when required
Measuring devices	Wrong measurements
Transformers and reactors	Overheating
Motors	Increased noise level, overheating, and additional mechanical vibrations
Telephones	Noise at respective frequencies
Lines	Overheating
Electronic devices	Wrong pulses on data transmission, over/under-voltage, and flickering of screens
Incandescent lamps	Reduced life time and flickering

Therefore, the reactive and harmonic currents must be fully compensated. Filters are often required to eliminate the distortion from the supply voltages and improve the power factor of the system. The classical solution has always been to install passive filters, but they have significant inherent problems, e.g., tuning to specific load and frequency, in conjunction with its bulky size. As such, new solutions need to be developed to overcome the detrimental effects of harmonics and reactive currents as mentioned earlier.

8.3.2 Voltage Unbalance

An unbalance or asymmetry in the magnitude and phase of the supply voltage is normally caused by single-phase faults and asymmetric three-phase loads. The electric motors constitute the major load on the power system and even a small voltage unbalance severely affects their performance and life (Quinn and Mohan, 1992). Increased motor current, noise, vibration and decreased rotor speed, and poor efficiency are some of the detrimental effects of voltage unbalance on the performance of the electric motors. It may be pointed out that a small negative sequence voltage can produce motor currents in excess magnitude as compared to those present under balanced conditions. These high unbalance currents will therefore cause vibration and higher heating in some windings. Excessive heating accelerates insulation failure and the vibration increases bearing wear and tear. Both of these effects are detrimental to the performance and the life of electric machine. The negative sequence voltage decreases the speed and torque of the motor. Thus, the overall efficiency of the motor is reduced because of unbalance voltages.

8.3.3 Voltage Regulation

Ideally, the output of most power supplies should be a constant voltage. Unfortunately, this is very difficult to achieve. There are two factors that can cause the output voltage to change. First, the AC line voltage is not constant. The so-called 115 V AC can vary from about 105 V AC to 125 V AC. This means that the peak AC voltage to which the rectifier responds can vary from about 148 to 177 V. Only the AC line voltage can be responsible for nearly a 20% change in the DC output voltage. The second factor that can change the DC output voltage is a change in the load resistance. In complex electronic equipments, the load can change as circuits are switched in and out. In a television receiver, the load on a particular power supply may depend on the brightness of the screen, the control settings, or even the channel selected. These variations in load resistance tend to change the applied DC voltage because the power supply has a fixed internal resistance. If the load resistance decreases, the internal resistance of the power supply drops more voltage. This causes a decrease in the voltage across the load.

Many circuits are designed to operate with a particular supply voltage. When there is a change in the supply voltage, the operation of that circuit may be adversely affected. Consequently, some types of equipments must have power supplies that produce the same output voltage regardless of changes in the load resistance or changes in the AC line voltage. This constant output voltage may be achieved by adding a circuit, called the voltage regulator, at the output of the filter. There are many different types of regulators in use today and a discussion of all of these is beyond the scope of this chapter.

8.3.4 Voltage Interruptions

The supply voltage in the mains dips and short interruptions are caused by a wide variety of phenomena. They can be caused by nearby events, such as a faulty load on an adjacent branch circuit causing a circuit breaker to trip, or perhaps by a large motor or heater on the same circuit being switched ON. They can also be caused by faraway events such as lightning storms or a downed power line. In case of a fault in the power distribution grid, an automatic circuit recloses, may open and close several times within a short period attempting to clear the fault, thus resulting in a sequence of short interruptions as seen by downstream loads. Voltage variations are typically caused by high power loads that have continuously varying power levels, rather than by abruptly switching ON and OFF the loads. In any case, the voltage changes can affect the operation of certain equipments or even damage the nearby electrical and electronic equipments. Therefore, immunity testing for these types of events should be performed to ensure a safe and reliable product operation.

8.4 POWER DEFINITIONS UNDER NON-SINUSOIDAL CONDITIONS

The definitions of power for sinusoidal AC systems are unique and unequivocal. However, under non-sinusoidal conditions, several sets of power definitions are still in use. For instance, the conventional concepts of reactive and apparent power lose their usefulness in non-sinusoidal cases. This problem has been existing for many years. Unfortunately, no agreement for a universally applicable power theory has been achieved.

In the late 1920s, two important approaches for the definition of power under non-sinusoidal conditions were introduced by Fryze and Budeanu. Fryze defined the powers in the time domain, whereas Budeanu worked in the frequency domain. Unfortunately, those power definitions are not complete and in some cases may lead to misinterpretations. No relevant contributions have emerged until the 1970s, since the power generating systems were satisfactorily represented as balanced and sinusoidal AC systems. However, the problem has become significant after the advancements in power electronics, as an increasing number of nonlinear equipments are being connected to the electric power system.

Many approaches for the power calculation can be found in the literature (Czarneck, 1995). At present, the state of the power definitions in the time domain seems to be more useful than those in the frequency domain, since it is possible to realize faster control algorithms for active line conditioners using the definitions in the time domain.

8.4.1 Power Definitions in Frequency Domain

The power definitions established by Budeanu in 1927 still remain an important statement for analysis in the frequency domain. If a single-phase AC circuit with generic load and source is at steady state, its voltage and current waveforms can be decomposed by Fourier series. Then, the rms value of each harmonic can be calculated and the following definitions of powers can be achieved.

Apparent power S

$$S = V * I \tag{8.1}$$

where V and I represent the rms value of the voltage and the current respectively, which are calculated as follows:

$$V = \sqrt{\left| \left(\frac{1}{T} \right) \int_0^T V^2(t)dt \right|}$$

or

$$V = \sqrt{\{\Sigma V_n^2\}}$$

$$I = \sqrt{\left| \left(\frac{1}{T} \right) \int_0^T i^2(t)dt \right|} = \sqrt{\Sigma I_n^2} \tag{8.2}$$

where V_n and I_n correspond to the rms value of the nth harmonic and T is the time period of the fundamental component.

Active power P

$$P = \sum_0^n P_n = \sum_0^n V_n I_n \cos \varphi_n \tag{8.3}$$

Reactive power Q

$$Q = \sum_0^n Q_n = \sum_0^n V_n I_n \sin \varphi_n \tag{8.4}$$

Harmonic power D

$$D^2 = S^2 - P^2 - Q^2 \tag{8.5}$$

The powers defined in these equations are well known and are widely used. However, only the active power P, as defined in Eqn (8.3), describes a clear physical meaning under non-sinusoidal conditions. The active power P represents the average value of the instantaneous active power. In the other words, it represents the average ratio of the energy transfer between two electrical subsystems.

In contrast, the reactive power Q from Eqn (8.4) and the harmonic power D from Eqn (8.5) are mathematical formulations that may lead to false interpretations, particularly when these concepts are extended to the analysis of three-phase circuits. All these equations treat electric circuits under non-sinusoidal conditions as a sum of several independent circuits excited at different frequencies. The calculated powers do not provide any tool for solving the related problems. Consequently, they do not provide any consistent basis for designing passive filters or active power line conditioners.

Although the apparent power (S) is considered as the fundamental rating of electrical equipment, this definition is perhaps the worst, since it shows up four different ways in a widely recognized technical dictionary. The apparent power given in Eqn (8.1) is one of them, which is called 'the rms volt-ampere'. It is more difficult is to find reasons for the applicability of the apparent power if a three-phase, four-wire, non-sinusoidal power supply–connected generic load is considered. These problems are reported in several works and many researchers have tried to solve them.

8.4.2 Power Definitions in Time Domain

In the early 1930s, Fryze proposed a set of power definitions in the time domain. An interesting characteristic of this approach is the absence of Fourier analysis which was very important at that time since no measurement equipment, such as spectrum analyzers, were available. The basic equations according to the Fryze's approach are as follows:

Active power P_w

$$P_w = \frac{1}{T} \int V(t)I(t)dt = V_w I = V I_w \tag{8.6}$$

The rms values of voltage and current are calculated as given in Eqn (8.2). Three units—P_w, V, and I—form the basis of the Fryze's approach. All other units can be calculated from these, i.e.,

Apparent power P_s

$$P_s = VI \tag{8.7}$$

Active power factor λ_w

$$\lambda_w = P_w/P_s = P_w/VI \tag{8.8}$$

Reactive power P_q

$$P_q = \sqrt{(P_s^2 - P_w^2)} = V_q I = V I_q \tag{8.9}$$

Reactive power factor λ_q

$$\lambda_q = \sqrt{(1 - \lambda_w^2)} \tag{8.10}$$

Active voltage V_w and active current I_w

$$V_w = \lambda_w V; \ I_w = \lambda_w I \tag{8.11}$$

Reactive voltage V_q and reactive current I_q

$$V_q = \lambda_q V; \ I_q = \lambda_q I \tag{8.12}$$

Fryze understood reactive power as comprising all the portions of voltage and current that does not contribute to the active power P_w. This concept of active and reactive powers is well accepted nowadays. For example, Czarnecki improved this approach, going in detail, by dividing the reactive power P_q into four subparts, according to their respective physical origins in the electric circuits.

It can be seen that there is no difference between the active and the apparent powers defined by Fryze and Budeanu. The active power calculated from Eqns (8.3) and (8.6) is the same. The apparent powers from Eqns (8.1) and (8.7) are also same.

Fryze verified that the active power factor λ_w reaches its maximum ($\lambda_w = 1$) if and only if the instantaneous current is proportional to the instantaneous voltage, otherwise $\lambda_w < 1$. However, under non-sinusoidal conditions, the fact of having currents proportional to voltages does not ensure an optimal power flow from an electromechanical energy conversion viewpoint. If the above-defined concepts are applied in the analysis of three-phase systems, they may lead to cases where the instantaneous active three-phase power contains an oscillating component even if the voltages and currents are proportional $(\lambda_w = 1)$. This oscillating electric power causes mechanical vibration in electric machines.

8.4.3 Electric Power in Three-phase Systems

The analysis of three-phase circuits is usually simplified as a sum of three single-phase power or the sum of the three separate powers. The physical meaning of the powers is assumed to be identical in both representations of the system. This is a crude simplification, especially in cases involving static converters.

The reactive power does not describe the same phenomenon in three-phase and single-phase circuits. For example, an ideal three-phase generator supplying a balanced capacitor bank has no mechanical torque if the losses are neglected. On the other hand, a single-phase generator supplying a capacitor has an oscillating mechanical torque. Therefore, it is false to think that the three-phase reactive power represents an oscillating energy between source and the load, if all the phases of the system are considered together. On the other hand, three-phase systems with neutral wire can present unbalances due to zero sequence components, which cause unknown problems in single-phase circuits.

8.4.4 Apparent Three-phase Power

Two definitions of apparent three-phase power are often used even under unbalanced, non-sinusoidal conditions:

1. 'Per phase' calculation

$$S_{3\varphi} = \Sigma S_k = \Sigma V_k I_k; \, k = (a, b, c); \tag{8.13}$$

2. 'Aggregate rms value' calculation

$$S_z = \sqrt{(\Sigma V_k^2)} \times \sqrt{(\Sigma I_k^2)}; \, k = (a, b, c); \tag{8.14}$$

where V_a, V_b, V_c and I_a, I_b, I_c are the rms values of the phase voltages and the line currents respectively. It is possible to demonstrate that for a balanced, sinusoidal case the apparent powers from Eqns (8.13) and (8.14) are equivalent, but under non-sinusoidal and/or unbalanced conditions, the result always holds that, $\Sigma S_k \geq S_{3\varphi}$ are mathematical definitions without a physical meaning, although some authors choose to assign the sense 'maximum reachable active power at unity power factor' to the apparent power.

It is not possible to establish a consistent set of definitions using the apparent power as a basic equation because the subsequent definitions of reactive and harmonic powers also seem to be mathematical definitions without any physical meaning. Therefore, the instantaneous active power is taken as the fundamental equation for the power definitions in three-phase system.

8.4.5 Instantaneous Active Three-phase Power

All power definitions hitherto have had a precondition that the system is in steady state. For the design of active power line conditioners, it is imperative to establish power definitions that are valid also during the transient periods. For a three-phase system, with or without a neutral conductor, the instantaneous active three-phase power describes the total instantaneous energy flow per time unit between two subsystems, and is given by

$$P_{3\varphi}(t) = V_a(t) I_a(t) + V_b(t) I_b(t) + V_c(t) I_c(t) \equiv P_{3\varphi} = V_a I_a + V_b I_b + V_c I_c$$

where V_a, V_b, and V_c are the instantaneous phase voltages and I_a, I_b, and I_c are the instantaneous line currents.

8.4.6 Instantaneous Reactive Three-phase Power

The instantaneous reactive three-phase power comprises all portions of the phase powers that do not contribute to the instantaneous active three-phase power.

The above-mentioned active and reactive powers have the same fundamental principle as the Fryze's definitions, but here the instantaneous values of a three-phase system are considered. The idea that the reactive power is related to an oscillating energy flux is abandoned. This oscillating energy flux between two subsystems is now treated as an instantaneous active three-phase power that has an average value equal to zero.

Since the instantaneous values of voltages and current are used here, the instantaneous active and reactive powers defined earlier are also valid during transient's period. Further, no restrictions were imposed on the definitions and these can be used also under non-sinusoidal conditions.

In short, some important conclusions regarding power definitions can be summarized as follows:

1. The powers defined in frequency domain do not form a basis for efficient algorithm in the control of active line conditioners, because of the time delay incurred in calculating the rms values of voltages and currents, which is necessary in this approach.
2. The compensation algorithms established through minimization methods are relatively simple to implement. However, they are inapplicable in three-phase four-wire systems and cannot guarantee constant active power to the source.
3. The generalized Fryze's current method results in currents being proportional to the voltages, and gives the smallest rms value for the compensated currents. However, it is not an 'instantaneous' algorithm (it is not an algebraic set of equations).

8.5 POWER SYSTEM HARMONICS

The objective of the electric utility is to deliver sinusoidal voltage at a fairly constant magnitude throughout their system. This objective is complicated by the fact that there are loads on the system that produce harmonic currents. These currents result in distorted voltages and currents that can adversely impact the system performance in different ways. As the number of harmonic producing loads has increased over the years, it has become increasingly necessary to address their influence when making any additions or changes to an installation. To fully appreciate the impact of these phenomena, there are two important concepts to bear in mind with regard to power system harmonics. The first is the nature of harmonic-current producing loads (nonlinear loads) and the second is the way in which harmonic currents flow and how the resulting harmonic voltages develop.

8.5.1 Linear and Nonlinear Loads

A linear element in a power system is a component in which the current is proportional to the voltage. In general, this means that the current wave shape will be the same as the voltage wave shape (see Fig. 8.4). Typical examples of linear loads include motors, heaters, and incandescent lamps.

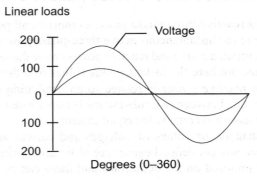

Fig. 8.4 Voltage and current waveforms for linear loads

On the other hand, the current wave shape on a nonlinear load is not the same as the voltage wave shape (see Fig. 8.5). Typical examples of nonlinear loads include rectifiers (power supplies, UPS units, and discharge lighting), adjustable speed motor drives, ferromagnetic devices, DC motor drives, and arcing equipment.

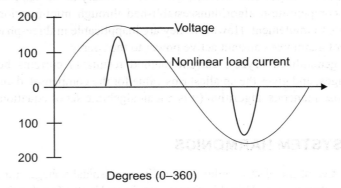

Fig. 8.5 Voltage and current waveforms for nonlinear loads

The current drawn by nonlinear loads is not sinusoidal, but periodic, i.e., the current wave looks the same from cycle to cycle. Periodic waveforms can be described mathematically as a series of sinusoidal waveforms that have been summed together (see Fig. 8.6).

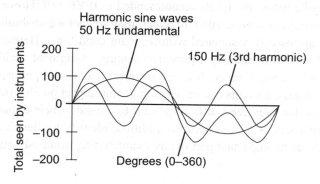

Fig. 8.6 Waveform with symmetrical harmonic components

The sinusoidal components are integer multiples of the fundamental where the fundamental, in India, is 50 Hz. The only way to measure a voltage or current that contains harmonics is to use a true rms reading meter. If an averaging meter is used, which is the most common type, the error can be significant.

Each term in the series is referred to as a harmonic of the fundamental. The third harmonic would have a frequency of three times 50 Hz or 150 Hz. Symmetrical waves contain only odd harmonics and unsymmetrical waves contain both even and odd harmonics.

A symmetrical wave is one in which the positive portion of the wave is identical to the negative portion of the wave. An un-symmetrical wave contains a DC component (or offset) or the load is such that the positive portion of the wave is different from the negative portion. An example of un-symmetrical wave would be a half-wave rectifier.

Most power system elements are symmetrical. They produce only odd harmonics and have no DC offset. There are exceptions, of course, and normally-symmetrical devices may produce even harmonics due to component mismatches or failures. Arc furnaces are another common source of even harmonics but they are notorious for producing both even and odd harmonics at different stages of the process.

8.5.2 Harmonic Current Flow

When a nonlinear load draws current, that current passes through all the impedance that is between the load and the system source. As a result of the current flow, harmonic voltages are produced by impedance in the system for each harmonic. The sum of these voltages when added to the nominal voltage produces voltage distortion. The magnitude of the voltage distortion depends on the source impedance and the harmonic voltages produced. If the source impedance is low, then the voltage distortion will be low. If a significant portion of the load becomes nonlinear (harmonic currents increase) and/or when a resonant condition prevails (system impedance increases), the voltage can increase dramatically. Power systems are able to absorb a considerable amount of current distortion without problems and the distortion produced by a facility may be below the levels recommended in IEEE 519. However, the collective effect of many industrial customers, taken together, may impact a distribution system. When problems arise, they are usually associated with resonant conditions. Harmonic currents can produce a number of problems, namely equipment heating, equipment malfunction, equipment failure communications, interference, fuse and breaker misoperation, process problems, and conductor heating. Harmonic currents can have a significant impact on electrical distribution systems and the facilities that they feed. It is important to consider their impact when contemplating the additions or changes to a system. In addition, identifying the size and location of nonlinear loads should be an important part of any maintenance, troubleshooting, and repair programs.

8.6 SIMULATION OF POWER SYSTEM FOR DIFFERENT FAULT CONDITIONS

In this section, a power system model is simulated under different fault conditions and their effects are analyzed. This model is simulated for under- or over-voltage, voltage unbalance, voltage sag/swell, and harmonic distortion. The following program is used to set the values of various parameters of the power system model shown in Fig. 8.7.

```
%%%%%%%%%%%%%%%%%%%%%%%%%%%%%%%%%%%%%%%%%%%%%%%%%%%%%%%%%%%%%%%%%%%%
% This program sets the parameters for three phase power system %
%         model under different fault conditions            %
%%%%%%%%%%%%%%%%%%%%%%%%%%%%%%%%%%%%%%%%%%%%%%%%%%%%%%%%%%%%%%%%%%%%

clear all;
clc;


%%%%%%%%%%%%%%%%%%%%%%%%%%%%%%%%%%%%%%%%%%%%%%%%%%%%%%%%%%%%%%%%%%%%
%              Three-Phase power System Parameters          %
%%%%%%%%%%%%%%%%%%%%%%%%%%%%%%%%%%%%%%%%%%%%%%%%%%%%%%%%%%%%%%%%%%%%

f = 50;  %frequency of the power system in Hz
T = 1/f; % Time period in seconds
R1 = 1; % Resistance of the transmission line in Ohms
L1 = 10e-6; % Inductance of the transmission Line in H
R2 = 50; % load resistance in Ohms
Ton = 0.02; % On Time of the power supply in seconds


%%%%%%%%%%%%%%%%%%%%%%%%%%%%%%%%%%%%%%%%%%%%%%%%%%%%%%%%%%%%%%%%%%%%
%                    Three voltage source                  %
%%%%%%%%%%%%%%%%%%%%%%%%%%%%%%%%%%%%%%%%%%%%%%%%%%%%%%%%%%%%%%%%%%%%
```

```
Pia = 0; %Phase angle of the voltage Source Va in Degrees
Pib = 120; %Phase angle of the voltage Source Vb in Degrees
Pic = 240; %Phase angle of the voltage Source Vc in Degrees
V = 230;   %Peak voltage per phase

%%%%%%%%%%%%%%%%%%%%%%%%%%%%%%%%%%%%%%%%%%%%%%%%%%%%%%%%%%%%%%%%%
%              Under Voltage, continuous Reduction parameters           %
%%%%%%%%%%%%%%%%%%%%%%%%%%%%%%%%%%%%%%%%%%%%%%%%%%%%%%%%%%%%%%%%%

k1 = 20.0; % under voltage, continuous reduction of voltage in %
V1 = (k1/100) * V; %Voltage to be reduced
Va = V - V1; % Actual reduced peak voltage

%%%%%%%%%%%%%%%%%%%%%%%%%%%%%%%%%%%%%%%%%%%%%%%%%%%%%%%%%%%%%%%%%
%                  Voltage Unbalance Parameters                        %
%%%%%%%%%%%%%%%%%%%%%%%%%%%%%%%%%%%%%%%%%%%%%%%%%%%%%%%%%%%%%%%%%

UF = 0.5; % Voltage Unbalance factor
Vneg = Va * UF; % Negative sequence voltage

%%%%%%%%%%%%%%%%%%%%%%%%%%%%%%%%%%%%%%%%%%%%%%%%%%%%%%%%%%%%%%%%%
%                  Voltage Sag/Swell parameters                        %
%%%%%%%%%%%%%%%%%%%%%%%%%%%%%%%%%%%%%%%%%%%%%%%%%%%%%%%%%%%%%%%%%

N = 3.0;% duration of sag/swell for number of fundamental cycles
T1 = 0.05; % Time in sec at which sag will be initiated
k = 50;     % reduction in magnitude in % for + values sag
            % will be there & for -ive values swell will be there
Vu = Va * (k/100); % 50% Reduction in phase voltage
T3 = T * N; % end of sag after N cycles
T2 = T1 + T3;

%%%%%%%%%%%%%%%%%%%%%%%%%%%%%%%%%%%%%%%%%%%%%%%%%%%%%%%%%%%%%%%%%
%                  Parameters for Harmonic Distortion                  %
%%%%%%%%%%%%%%%%%%%%%%%%%%%%%%%%%%%%%%%%%%%%%%%%%%%%%%%%%%%%%%%%%

h5 = 0.05; % Harmonic Factor of 5th harmonic
h7 = 0.03; % Harmonic Factor of 7th harmonic
f5 = f * 5; % 5th harmonic frequency
f7 = f * 7; % 7th harmonic frequency
Vh5 = V * h5; % 5th harmonic, 5% 0f Fundamental component
Vh7 = V * h7; % 7th harmonic, 3% of Fundamental component

%%%%%%%%%%%%%%%%%%%%%%%%%%%%%%%%%%%%%%%%%%%%%%%%%%%%%%%%%%%%%%%%%
%                       End of the program                             %
%%%%%%%%%%%%%%%%%%%%%%%%%%%%%%%%%%%%%%%%%%%%%%%%%%%%%%%%%%%%%%%%%
```

In the model shown in Fig. 8.7, the parameters are as follows: block 'Va': peak amplitude—Va, phase (deg)—Pia, frequency (Hz)—f, sample time—0, measurements—voltage; block 'Vb': peak amplitude—Va, phase (deg)—Pib, frequency (Hz)—f, sample time—0, measurements—voltage; block 'Vc': peak amplitude—Va, phase (deg)—Pia, frequency (Hz)—f, sample time—0, measurements—voltage; block 'Vaneg': peak amplitude—Vneg, phase (deg)—Pia, frequency (Hz)—f, sample time—0, measurements—voltage; block 'Vbneg': peak amplitude—Vneg, phase (deg)—Pic, frequency (Hz)—f, sample time—0, measurements—voltage; block 'Vcneg': peak amplitude—Vneg, phase (deg)—Pib, frequency (Hz)—f, sample time—0,

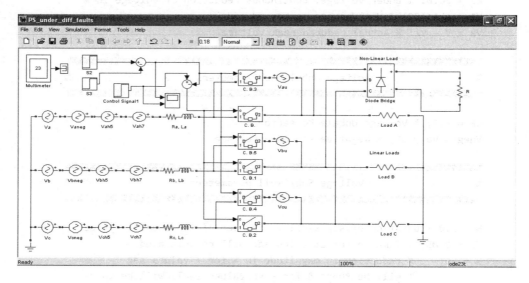

Fig. 8.7 Power system model for different fault conditions

measurements—voltage; block 'Vah5': peak amplitude—Vh5, phase (deg)—Pia, frequency (Hz)—f5, sample time—0, measurements—voltage; block 'Vbh5': peak amplitude—Vh5, phase (deg)—Pib, frequency (Hz)—f5, sample time—0, measurements—voltage; block 'Vch5': peak amplitude—Vh5, phase (deg)—Pic, frequency (Hz)—f5, sample time—0, measurements—voltage; block 'Vah7': peak amplitude—Vh7, phase (deg)—Pia, frequency (Hz)—f7, sample time—0, measurements—voltage; block 'Vbh7': peak amplitude—Vh7, phase (deg)—Pib, frequency (Hz)—f7, sample time—0, measurements—voltage; block 'Vch7': peak amplitude—Vc, phase (deg)—Pic, frequency (Hz)—f7, sample time—0, measurements—voltage; block 'Ra, La': resistance (ohms)—R1, inductance (H)—L1, measurements—branch current; block 'Rb, Lb': resistance (ohms)—R1, inductance (H)—L1, measurements—branch current; block 'Rc, Lc': resistance (ohms)—R1, Inductance (H)—L1, measurements—branch current; block 'Control Signal1': step time—Ton, initial value—0, final value—5, sample time—0; block 'S2': step time—T1, initial value—0, final value—5, sample time—0; block 'S3': step time—T2, initial value—0, final value—5, sample time—0; block 'Vau': peak amplitude—Vu, phase (deg)—Pia, frequency (Hz)—f, sample time—0, measurements—voltage; block 'Vbu': peak amplitude—Vu, phase (deg)—Pib, frequency (Hz)—f, sample time—0, measurements—voltage; block 'Vcu': peak amplitude—Vu, phase (deg)—Pic, frequency (Hz)—f, sample time—0, measurements—voltage; block 'Load A', 'Load B', and 'Load C': resistance (ohms)—R2, measurements—branch voltage; and block 'R': resistance (ohms)—R2, measurements—branch voltage and current. Numerical values of these parameters are stored in the MATLAB workspace, when we run the above given program.

In the model shown in Fig. 8.7 'Va', 'Vb', and 'Vc' are the voltages of the three phases as shown in Fig. 8.8. 'Vaneg', 'Vbneg', and 'Vcneg' are the three-phase negative sequence voltages for creating voltage unbalance in the power system as shown in Fig. 8.8. 'Vah5', 'Vbh5', and 'Vch5' are the three-phase voltages of the 5th harmonic of the fundamental frequency of the power system as shown in Fig. 8.8. 'Vah7', 'Vbh7', and 'Vch7' are the three-phase voltages of the 7th harmonic of the fundamental frequency of the power system as shown in Fig. 8.9.

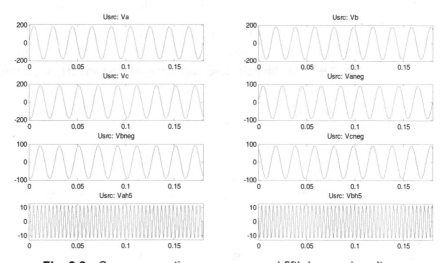

Fig. 8.8 Source, negative sequence, and fifth harmonic voltages

'Ra' and 'La' is the resistance and inductance of the first phase, 'Rb' and 'Lb' of the second phase, and 'Rc' and 'Lc' of the third phase of the transmission line. Figures 8.9 and 8.10 show the current of each phase of the power system. Circuit breakers 'C. B.', 'C. B. 1', and 'C. B. 2' connect the three-phase supply to the load when the control signal is high. Circuit breakers 'C. B. 3', 'C. B. 5', and 'C. B. 4' are used for voltage sag/swell. The three-phase linear load, i.e., 'Load A', 'Load B', and 'Load C', and a nonlinear load 'Diode Bridge' is also connected to the power system. This 'Diode Bridge' rectifier feeds a resistive load 'R'. Voltages 'Vau', 'Vbu', and 'Vcu' are shown in Fig. 8.10 along with the load voltages for voltage sag. Figure 8.11 shows the load voltages in case of voltage swell (50%).

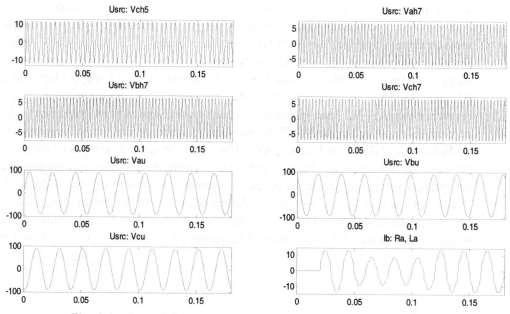

Fig. 8.9   The 7th harmonic, sag voltages, and line (R_a, L_a) current

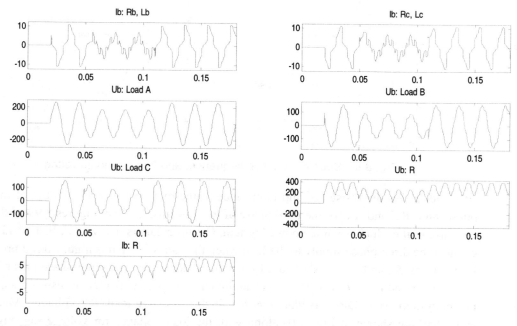

Fig. 8.10   Power line (R_b, L_b, R_c, L_c) currents, load voltage, and rectifier output voltage

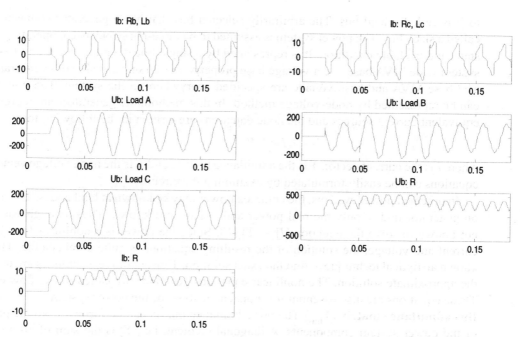

Fig. 8.11 Power line (R_b, L_b, R_c, L_c) currents, load voltage, and rectifier output voltage in case of voltage swell

8.7 LOAD FLOW STUDIES

The load flow (also known as power flow) techniques provide a basic calculation procedure used to determine the characteristics of power system under steady-state condition. This study reveals the electrical performance and power flow (complex) for specified conditions when the system is operating in a steady state. This study also provides information regarding the line and power transformer loads throughout the system. The power system can be evaluated by the load flow study as the voltages at different points of the power system can also be calculated. Due to the nonlinear nature of the power system, numerical techniques are employed to obtain a solution within an acceptable tolerance limit. These studies are performed using computer software like MATLAB that can simulate actual steady-state power system operating conditions, enabling the evaluation of bus voltage profiles, real and reactive power flow, and losses. A properly designed power system aids in determining the required initial capital investment and future operating costs. Load flow studies are generally used to determine the component or circuit loading, bus voltage profiles, real and reactive power flow, power system losses, and proper transformer tap settings.

The solution to the load flow problem begins with identifying the known and unknown variables in the system. These variables depend on the type of bus. A bus with at least one generator connected to it is called as a generator bus and a bus without any generator connected to it is called a load bus. The arbitrarily selected bus that has a generator connected to it is called a slack bus. The power system is assumed to be three-phase balanced, operating in steady state, and in stable condition. It is represented by a single line diagram on per unit basis with a system wide MVA base and a voltage base properly chosen on each side of every transformer. The base MVA and base voltage are specified everywhere in the system. The power system can be represented by node-voltage method. In this method, the generators are represented by equivalent current sources and the node equations are written in the following form:

$$I = Y V$$

where I is the current vector, Y is the admittance matrix, and V is the node voltage vector. These equations can be easily formulated by examining the circuit.

In practical power systems, the complex power may be known at load nodes, and sometimes on generator nodes, only the real power and voltage are known. Thus, enough variables are not known to solve the equation $[I] = [Y] [V]$. Since the power is a nonlinear function of the current and voltage, the solution of the resulting equations is quite cumbersome. There is no known analytical technique to find the exact solution. Thus, iterative techniques are used to find the approximate solution. The nonlinear equations generated are called power flow equations. These equations are also essential for transient analysis of the power system.

Bus admittance matrix (Y_{bus}) The bus admittance matrix can be formed from the parameters of the power system components. A diagonal element, i.e., Y_{ii} is the sum of all admittances connected to ith bus, and an off-diagonal element, i.e., Y_{ij} is negative of the total admittance, are directly connected between ith and jth buses. The following step-by-step procedure is quite useful for formulating the bus admittance matrix:

1. All nodes of the system are numbered from '0' to 'n'. Node '0' is the reference or ground node.
2. All the generators (voltage sources) are replaced by equivalent current sources in parallel with an admittance.
3. All the lines, transformers, and loads are replaced by their equivalent admittances wherever possible. If the load is known (in MVA) and the operating voltage is unknown, it is impossible to change the load to admittance. In this case, take $y = 1/z$, where y and z are complex numbers.
4. The bus admittance matrix Y is now formed by inspecting the system as follows:
   y_{ii} = sum of admittances connected to the ith node, and
   $y_{ij} = y_{ji} = -$(sum of admittances connected from ith node to jth node)
5. Now, the current vector (I) is found from the sources connected to the nodes '0' to 'n'. If there is no source connected, the injected current would be '0'.
6. The resulting equations are called node voltage equations and are given in matrix form by:

$$[I_{bus}] = [Y_{bus}] [V_{bus}]$$

This equation can also be given as $[V_{bus}] = [Y_{bus}]^{-1} [I_{bus}]$. It is emphasized that the matrix $[Y_{bus}]^{-1}$ is not the same as the matrix Z which results from solving a circuit using mesh equation. We define $[Z_{bus}] = [Y_{bus}^{-1}]$ and $[Y_{bus}]$ is same as $[Y]$. Thus, $[Y_{bus}] = [Y]$; however, $Z_{bus} \neq [Z]$.

In general, solving a large set of linear equations is never done by taking matrix inverse. Determining the inverse of a large matrix requires greater effort than is required to find a solution and the also the resulting solutions are less accurate. Usually, Gauss elimination method is used to find a direct solution. Iterative solutions are also very effective for larger systems. Also, in a real power system containing hundreds of nodes, each node is rarely connected to more than two or three nodes and therefore most of the elements in the admittance matrix are zero. This type of matrix is called sparse matrix and special techniques exist to solve systems with sparse matrices.

8.7.1 Gauss–Seidel Method

The Gauss–Seidel (G–S), Newton–Raphson (N–R), and quasi-Newton–Raphson methods are commonly used for solving nonlinear equations. The G–S method is also known as the method of successive displacements.

Consider a nonlinear system which can be represented by the equation, $fcn\ (x) = 0$. This equation can be broken into two parts and can be written as $x = g(x)$. Now, let us assume $x^{(0)}$ as an initial 'guess' of the solution. The solution can be further refined using the following:

$$x^{(1)} = g(x^{(0)})$$

And, it can be refined further as

$$x^{(2)} = g(x^{(1)})$$
$$x^{(3)} = g(x^{(2)})$$

and, so on.

After the nth iteration, we have $x^{(n)} = g(x^{(n-1)})$. If this process is convergent, the successive solutions approach a value that can be declared as the solution. Thus, if at some step $(k + 1)$, we get $|x^{(k+1)} - x^{(k)}| \leq \varepsilon$, where ε is the desired accuracy of the solution, we claim that the solution has been found to the accuracy specified.

Consider an equation, $f(x) = x^3 - 6x^2 + 9x - 5 = 0$. This equation can be represented as $x = -1/9\ (x^3 - 6x^2 - 5) = g(x)$.

Now, we will solve this equation by different methods with the help of MATLAB. The following MATLAB program depicts the process graphically. The output of the program is shown in Fig. 8.12.

```
%%%%%%%%%%%%%%%%%%%%%%%%%%%%%%%%%%%%%%%%%%%%%%%%%%%%%%%%%%%%%%%%%%%%%%%%%%%%
% The following program plots the equation f(x) = x^3 - 6x^2 + 9x - 5 = 0 %
%%%%%%%%%%%%%%%%%%%%%%%%%%%%%%%%%%%%%%%%%%%%%%%%%%%%%%%%%%%%%%%%%%%%%%%%%%%%

clc;
clear all;

x = 0:0.01:4.5;
g = -1 / 9 * x.^3 + 6 / 9 * x.^2 + 5 / 9;
```

```
k = 1;
dz = 10;
z = 2;
r(k) = z;
s(k) = z;

while dz > 0.001
    k = k + 2;
    r(k-1) = z;
    p = -1 / 9 * z^3 + 6 / 9 * z^2 + 5 / 9;
    s(k-1) = p;
    dz = abs(z-p);
    z = p;
    r(k) = z;
    s(k) = z;
end

plot(x, g,'-',r, s, '-'), grid
xlabel('x')
text(0.8, 4.2, 'g(x) =-1 / 9 * x.^3 + 6 / 9 * x.^2 + 5 / 9')
text(3.2, 3.0, 'x')
```

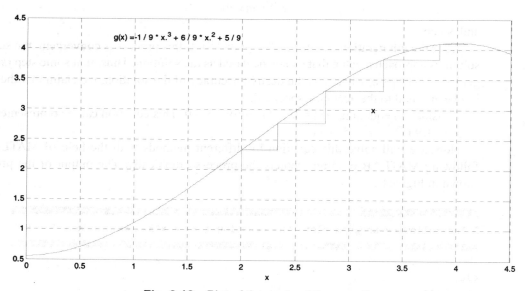

Fig. 8.12   Plot of the roots of the equation

The following MATLAB program finds the roots of the function $f(x)$. Figure 8.13 shows the plot of the function $f(x)$.

```
%%%%%%%%%%%%%%%%%%%%%%%%%%%%%%%%%%%%%%%%%%%%%%%%%%%%%%%%%%%%%%%%%%%%%%%%%%
%    The following program finds the roots and plots the function    %
%            f(x) = x^3 - 6x^2 + 9x - 5 = 0                          %
%%%%%%%%%%%%%%%%%%%%%%%%%%%%%%%%%%%%%%%%%%%%%%%%%%%%%%%%%%%%%%%%%%%%%%%%%%

clc;
clear all;

roots([1; -6; 9; -5])
x = 0 : 0.01 : 4.5;
zz = zeros(size(x));
f = x.^3 - 6 * x.^2 + 9.* x - 5;
plot(x, f, x, zz), grid

ans =

    4.1038
    0.9481 + 0.5652i
    0.9481 - 0.5652i
```

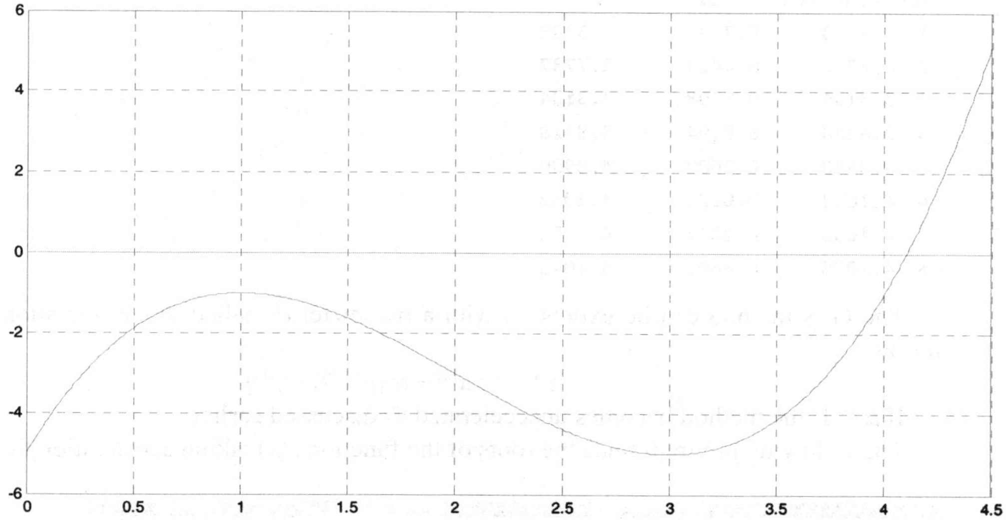

Fig. 8.13   Plot of the function $f(x)$

The following program finds the roots of the function $f(x)$ by iteration.

```
%%%%%%%%%%%%%%%%%%%%%%%%%%%%%%%%%%%%%%%%%%%%%%%%%%%%%%%%%%%%%%%%%%%%%%%%%%
%         The following program finds the roots of f(x) iteratively    %
%%%%%%%%%%%%%%%%%%%%%%%%%%%%%%%%%%%%%%%%%%%%%%%%%%%%%%%%%%%%%%%%%%%%%%%%%%
```

```
clc;
clear all;

dx = 1;
x = 2;
iteration = 0;
disp('Iteration    g(x)    dx     x')

while abs(dx) >= 0.001 & iteration < 100

    iteration = iteration + 1;
    f = -1 / 9 * x.^3 + 6 / 9 * x^2 + 5 / 9;
    dx = f-x;
    x = x+ dx;

    fprintf('%f', iteration), disp([f, dx, x])
end
```

The output of this program is as follows:

```
Iteration f(x)      dx         x
  1. 2.3333      0.3333     2.3333
  2. 2.7737      0.4403     2.7737
  3. 3.3134      0.5398     3.3134
  4. 3.8328      0.5194     3.8328
  5. 4.0930      0.2602     4.0930
  6. 4.1053      0.0123     4.1053
  7. 4.1036     -0.0017     4.1036
  8. 4.1038      0.0002     4.1038
```

The G–S method can be expressed with a parameter α, called the acceleration factor, as follows:

$$x^{(k+1)} = x^{(k)} + \alpha \, [g(x^{(k)}) - x^{(k)}]$$

If $\alpha = 1$, the method becomes unaccelerated as discussed earlier.

The following program finds the roots of the function $f(x)$ taking acceleration factor as 1.3.

```
%%%%%%%%%%%%%%%%%%%%%%%%%%%%%%%%%%%%%%%%%%%%%%%%%%%%%%%%%%%%%%%%%%%%%%%%%
%     The following program finds the roots of the function f(x)     %
%          by iteration taking acceleration factor = 1.3             %
%%%%%%%%%%%%%%%%%%%%%%%%%%%%%%%%%%%%%%%%%%%%%%%%%%%%%%%%%%%%%%%%%%%%%%%%%

clc;
clear all;

dx = 1;
```

```
x = 2;
iteration = 0;
af = 1.3;

disp('Iteration f(x)    dx         x')

while abs(dx) >= 0.001 & iteration < 100
    iteration = iteration + 1;
    f = -1 / 9 * x^3 + 6 / 9 * x^2 + 5 / 9;
    dx = f - x;
    x = x + af * dx;

    fprintf('%f', iteration), disp([f, dx, x])
end
```

The output of this program is as follows:

Iteration	f(x)	dx	x
1.	2.3333	0.3333	2.4333
2.	2.9021	0.4687	3.0427
3.	3.5976	0.5549	3.7641
4.	4.0755	0.3114	4.1689
5.	4.0916	−0.0773	4.0684
6.	4.1080	0.0396	4.1198
7.	4.1013	−0.0185	4.0958
8.	4.1049	0.0091	4.1076
9.	4.1033	−0.0044	4.1019
10.	4.1041	0.0021	4.1047
11.	4.1037	−0.0010	4.1034
12.	4.1039	0.0005	4.1040

8.7.2 Newton–Raphson Method

Newton–Raphson method is a technique for finding successively better approximations to the zeros (or roots) of a function. Consider a function $f(x)$ of one variable x and its derivative $f'(x)$. Start with an initial guess x_0. If the function is reasonably well behaved, a better approximation x_1 is

$$x_1 = x_0 - f(x_0)/f'(x_0)$$

Here, x_1 is the intersection point of the tangent line to the graph of $f(x)$, with the x-axis. This process is repeated until a sufficiently accurate value is reached. After n iterations, we get

$$x_{n+1} = x_n - \{f(x_n)/f'(x_n)\}$$

In this method, we start with an initial guess x_0 which is reasonably close to the true root. After this, the function is approximated by its tangent line and one computes the x-intercept of this tangent line. This x-intercept will be a better approximation to the roots of the function than the original guess and the method can be further iterated. The following algorithm can be followed for solving a problem by N–R method.

1. Take an initial guess x_0 and evaluate $f'(x)$, i.e., find the first derivative of the function $f(x)$ with respect to x.
2. Calculate the next estimate of the root, i.e., $x_{i+1} = \{f(x_i)/f'(x_i)\}$ and find the absolute relative approximate error. This error is given by

$$|\varepsilon_a| = |(x_{i+1} - x_i)/x_{i+1}| \times 100$$

3. Find if the absolute relative approximate error is greater than the pre-specified relative error tolerance.
4. If so, go to Step 2, else stop the algorithm.
5. Also check that the number of iterations should not exceed the maximum number of iterations.

Consider a function $f(x) = 2x^5 + x^3 + 4x^2 - 3x - 2$. The following MATLAB program finds the roots of this function by Newton–Raphson method.

```
%%%%%%%%%%%%%%%%%%%%%%%%%%%%%%%%%%%%%%%%%%%%%%%%%%%%%%%%%%%%%%%%%%%
% This program finds the roots of the given function f(x) by %
%                  Newton-Raphson method                       %
%            f(x)  = 2x^5 + x^3 + 4x^2 -3x- 2                   %
%            and f'(x) = 10x^4 + 3x^2 + 8x - 3                  %
%%%%%%%%%%%%%%%%%%%%%%%%%%%%%%%%%%%%%%%%%%%%%%%%%%%%%%%%%%%%%%%%%%%

clc;
clear all;

x0 =-2.0;  % Initial guess
x = x0;
n = 15; % Number of iterations

    fprintf('Itera     f(x)              dfdx           x(itera+1) \n');

    for itera = 1:n
        f = 2 * x^5 + x^3 + 4 * x^2 - 3 * x - 2;
        dfdx = 10 * x^4 + 3* x^2 + 8 * x - 3;
        x = x - f / dfdx;

        fprintf('%3d  %12.3e  %12.3e  %18.15f \n', itera - 1, f, dfdx, x);
    end
```

The output of this program is as follows:

Itera	f(x)	dfdx	x(itera+1)
0.	−5.200e+001	1.530e+002	−1.660130718954248
1.	−1.579e+001	6.794e+001	−1.427725081801916
2.	−4.338e+000	3.324e+001	−1.297232802986880
3.	−9.072e−001	1.999e+001	−1.251847638038164
4.	−8.652e−002	1.625e+001	−1.246521599153282
5.	−1.101e−003	1.583e+001	−1.246452048000949
6.	−1.861e−007	1.583e+001	−1.246452036241837
7.	−5.329e−015	1.583e+001	−1.246452036241837
8.	8.882e−016	1.583e+001	−1.246452036241837
9.	8.882e−016	1.583e+001	−1.246452036241837
10.	8.882e−016	1.583e+001	−1.246452036241837
11.	8.882e−016	1.583e+001	−1.246452036241837
12.	8.882e−016	1.583e+001	−1.246452036241837
13.	8.882e−016	1.583e+001	−1.246452036241837
14.	8.882e−016	1.583e+001	−1.246452036241837

The root value is $x = -1.246452036241837$ and is obtained after 14 iterations.

The following MATLAB program finds the roots of the function $f(x) = x^3 - 4x^2 + 2x - 10$ by using Newton–Raphson method up to ten iterations.

```
%%%%%%%%%%%%%%%%%%%%%%%%%%%%%%%%%%%%%%%%%%%%%%%%%%%%%%%%%%%%%%%%%%%
%      This program finds the roots of the given function f(x) by     %
%                    Newton-Raphson method                           %
%                 f(x) = x^3 - 4x^2 + 2x - 10                         %
%                 f'(x) = 3x^2 - 8x + 2                               %
%%%%%%%%%%%%%%%%%%%%%%%%%%%%%%%%%%%%%%%%%%%%%%%%%%%%%%%%%%%%%%%%%%%

clc;
clear all;

x0 = 3.0; %Initial guess
n = 10; % Number of iterations
x = x0;

fprintf('Itera      f(x)        dfdx        x(itera+1)\n');

for itera = 1:n
    f = x^3 - 4 * x^2 + 2 * x - 10;
    dfdx = 3 * x^2 - 8 * x + 2;
    x = x - f / dfdx;

    fprintf('%3d  %12.3e  %12.3e  %18.15f \n', itera - 1, f, dfdx, x);
end
```

The output of this program is as follows:

```
Itera f(x)            dfdx          x(itera+1)
0  −1.300e+001    5.000e+000    5.600000000000000
1   5.138e+001    5.128e+001    4.598127925117005
2   1.184e+001    2.864e+001    4.184686182960697
3   1.604e+000    2.106e+001    4.108535765150575
4   4.916e−002    1.977e+001    4.106049287201422
5   5.146e−005    1.973e+001    4.106046679146596
6   5.658e−011    1.973e+001    4.106046679143728
7  −5.329e−015    1.973e+001    4.106046679143728
8  −5.329e−015    1.973e+001    4.106046679143728
9  −5.329e−015    1.973e+001    4.106046679143728
```

The root value is $x = 4.106046679143728$ and it is obtained after 9 iterations.

8.8 POWER SYSTEM STABILITY

In any physical system, stability is defined as the capability of the system to return to its original equilibrium position on the occurrence of a disturbance or to another equilibrium state which is generally in proximity of the initial equilibrium state. If the system does not remain in the initial equilibrium state as time (t) tends to infinity, the system is said to be unstable. In power systems, instability means a condition denoting loss of synchronism or falling out of step, and stability is its ability to return to normal operation after experiencing some form of disturbances. In general, stability is that attribute of the power system or part of the system which enables it to develop restoring forces between the elements which are equal to or greater than the disturbing forces so as to restore the state of equilibrium between the elements. The stability limit is the maximum power flow possible through some particular point in the system when the entire system or part of the system is operating with stability. Two types of stability limits are in general use—small-disturbance (steady-state) limit and large-disturbance (transient) limit. In a given power system, the large-disturbance stability limit cannot exceed the small-disturbance stability limit, although in some cases the former can closely approach the latter. This may not always be apparent from the simulation results, where large-disturbance stability simulation results covering a short period of time which may suggest a higher limit than the small-disturbance limit as determined from a linearized analysis. Herein lies the importance of small-disturbance limit. A power system cannot be operated above its small-disturbance stability limit although, depending on the operating criteria, it may be permissible to operate the system above the large-disturbance limit. This means, whenever there is a disturbance of sufficiently large magnitude, there will be a system disruption and shutdown. Depending on the probability of occurrence of such disturbances, this may be an acceptable operating mode. Then the small-disturbance limit assumes special importance. This also points out to the justification of combining the two terms, and simply calling it as stability limit.

The study of stability of the system under conditions of gradual or relatively slow change in load is termed as steady-state stability analysis. The study of steady-state stability is mainly concerned with the determination of the upper limit for loading machines before losing synchronism if the load is increased gradually. Once the magnitude of power flow exceeds the steady-state limit in case of interconnected systems, synchronism between the ends may be lost. Suppose, at any time instant t_0 the power system is operating in a synchronous equilibrium and frequency equilibrium, i.e., the system frequency is constant and the rotor angles of the different machines with respect to a rotating synchronous reference frame are fixed. The occurrence of a disturbance tends to momentarily alter the synchronous equilibrium and the frequency equilibrium. The disturbances may be small or big and the system may become unstable in either event depending on the operating condition at t_0. In standard literature, the study of the stability of the system for small disturbances is termed as static or dynamic stability analysis. The ability of a power system to remain in synchronism after the initial swing, until it settles down to the new steady-state equilibrium condition, is the dynamic stability of the system. In a dynamically stable system, the oscillations do not acquire greater amplitudes and die out quickly. The mathematical model for such studies is a set of linear time-invariant differential equations. Computer simulation is the only effective means of studying dynamic stability problems. When the disturbances are large and the non-linearities inherent in the power system can no longer be ignored, the study of the stability under such circumstances is known as transient stability analysis. The maximum flow of power possible through a point in the system without losing the stability with sudden and large changes in the network conditions, such as brought about by faults or by sudden large loads, is termed as transient stability of the system. Transient stability of power transmission system is its inherent ability to recover normal operation following sudden or severe disturbances. The mathematical model of such studies is a set of nonlinear differential equations coupled with nonlinear algebraic equations. The dimensionality of the mathematical model both of dynamic and transient stabilities can be very large, for even moderately sized systems.

8.9 LOAD FREQUENCY CONTROL

The modern-day power systems are divided into various areas. For example, in India, there are various regional grids. Each grid supplies power to a particular area and is generally interconnected to its adjoining areas. The transmission lines that connect an area to its adjoining areas are called tie-lines. These lines are used to share power between the two regions. Load frequency control, as the name signifies, is used to regulate the power flow between different areas while holding the frequency constant. The supply frequency may drop if there is an increase in the load. The supply frequency is required to be constant, i.e., $\delta f = 0$. The power flow through different tie-lines is scheduled as follows: area '1' may export a pre-specified amount of power to area '2' while importing another pre-specified amount of power from area '3'. Moreover, it is expected that to fulfill this obligation, area '1' absorbs its own load change, i.e., increase its generation to supply extra load, in its area or decrease generation when the load demand in its area has reduced. While doing this, area '1' must, however, maintain its obligation to areas '2' and '3' as far as importing and exporting of power is concerned. Thus, the main objectives of load frequency control are to hold the supply frequency constant, i.e., $\delta f = 0$, against any load change and to maintain the tie-line power flow to its pre-specified value.

The close regulation of system frequency is essential in respect of the need for synchronous operations of power stations. When load on the turbine increases, the speed of the turbine decreases. Nominal speed can be achieved by changing the setting of the governor. If the system consists of a single machine-supplying load, then speed and frequency changes will be in accordance with the characteristics of the governor. In case of system consisting of two machines running in parallel, the load will be shared by the machines according to their characteristics. The frequency control may be done by automatic or by manual regulators. Since manual regulation is not feasible in a large interconnected system, load frequency control (LFC) equipment is installed for each generator. The foremost task of LFC is to keep the frequency constant against the randomly varying active power loads, which are also referred to as unknown external disturbance. Another task of the LFC is to regulate the tie-line power exchange error. A typical large-scale power system is composed of several areas of generating units. In order to enhance the fault tolerance of the entire power system, these generating units are connected via tie-lines. The usage of tie-line power brings in a new error into the control problem, i.e., tie-line power exchange error. When a sudden active power load change occurs to an area, the area will obtain energy via tie-lines from other areas. But eventually, the area that is subject to the load change should balance it without external support. Otherwise, there would be economic conflicts between the areas. Hence, each area requires a separate load frequency controller to regulate the tie-line power exchange error so that all the areas in an interconnected power system can set their set-points differently. Another problem is that the interconnection of the power systems results in huge increases in both the order of the system and in the number of the tuning controller parameters. As a result, when modeling such complex high-order power systems, the model and parameter approximations cannot be avoided (Sauer and Pai, 2002). Therefore, the requirement of the LFC is to be robust against the uncertainties of the system model and the variations of system parameters in reality. In summary, the LFC has two major assignments—to maintain the standard value of frequency and to keep the tie-line power exchange under schedule in the presences of any load changes (Kothari et al., 1998; Kundur, 1994). In addition, the LFC has to be robust against unknown external disturbances and system model and parameter uncertainties. The high-order interconnected power system could also increase the complexity of the controller design of the LFC.

8.9.1 Existing Load Frequency Control Solutions

In industry, proportional-integral (PI) controllers have been widely used for decades as the load frequency controllers. A PI controller design on a three-area interconnected power plant is presented by El-Abiad and Stagg (1962), where the controller parameters of the PI controller are tuned using trial-and-error approach.

The LFC design based on an entire power system model is considered as a centralized method. It is introduced with a simplified multiple-area power plant in order to implement optimization techniques on the entire model (Kothari et al., 1998). However, the simplification is based on the assumption that all the subsystems of the entire power system are identical, while in reality they are not. The assumption makes the simulation model in the paper quite different from the real system. Another problem for the centralized methods is that even if the method works well on a low-order test system, it would face an exponentially increasing computation problem with the increase of the system size. Since the tie-line interfaces give rise to weakly coupled terms between areas, the large-scale power system can be decentralized into small subsystems through treating tie-line signals as disturbances. Numerous control techniques have been applied to the decentralized power systems. Fuzzy logic control is a method based on fuzzy set theory, in which the fuzzy logic variables can be of any value between 0 and 1 instead of just true and false. When the variables are selected, the decision will be made through specific fuzzy logic functions. Research results obtained from applying the fuzzy logic control technique to the decentralized LFC problem have been proposed by Ibraheem and Kothari (2005).

Genetic algorithm (GA) is one of the most popular computer intelligence algorithms. This algorithm been verified to be effective to solve complex optimization problems where PI-type controllers tuned via GA and linear matrix inequalities (GALMI) are presented on a decentralized three-area nine-unit power system (Rerkpreedapong, 2003). It is found that the structure of the GALMI-tuned PI controller is much simpler than that of the $H2/H\infty$ controller, although the performances of the two methods are equivalent (Ohba et al., 2007). Most of the reported solutions of the LFC problem have been tested for their robustness against large-step load change. However, very few of the published researches deal with parameter uncertainties. Sauer and Pai (2002) the authors set up a 15% floating rate for the parameters in one area and successfully controlled the system with an optimally tuned PID controller. A lot of approximations and simplifications were made during the modeling process of the power systems, on which the controller is designed. The simplified system model has deviated far from the real system. A control technique with a notable robustness against not only parameter uncertainties but also model uncertainties and external load change will be preferred by the power industry.

8.10 SIMULATION PROJECTS

Project 1

MATLAB Program to Solve three Nonlinear Equations by Newton–Raphson Method

```
%%%%%%%%%%%%%%%%%%%%%%%%%%%%%%%%%%%%%%%%%%%%%%%%%%%%%%%%%%%%%%%%%%%%%%%%%%%
%This Program solves three Nonlinear equations by Newton-Raphson method %
%%%%%%%%%%%%%%%%%%%%%%%%%%%%%%%%%%%%%%%%%%%%%%%%%%%%%%%%%%%%%%%%%%%%%%%%%%%

%%%%%%%%%%%%%%%%%%%%%%%%%%%%%%%%%%%%%%%%%%%%%%%%
%    Let us take three nonlinear equations as:   %
% %%%%%%%%%%%%%%%%%%%%%%%%%%%%%%%%%%%%%%%%%%%%%%%
```

```
%%%%%%%%%%%%%%%%%%%%%%%%%%%%%%%%%%%%%%%%%%%%%%%%%%%%%%%%%
%   f1 = x1^2 – x2^2 + x3^2 – 10 = 0 (First equation)    %
%   f2 = x1 * x2 + x2^2 – 4x3 - 4 = 0 (Second equation)  %
%   f3 = x1– x1* x3 + x2 * x3 – 5 = 0  (Third equation)  %
%%%%%%%%%%%%%%%%%%%%%%%%%%%%%%%%%%%%%%%%%%%%%%%%%%%%%%%%%

%%%%%%%%%%%%%%%%%%%%%%%%%%%%%%%%%%%%%%%%%%%%%%%%%%%%
%  Let the initial conditions be x1 = x2 = x3 = 1  %
%%%%%%%%%%%%%%%%%%%%%%%%%%%%%%%%%%%%%%%%%%%%%%%%%%%

%%%%%%%%%%%%%%%%%%%%%%%%%%%%%%%%%%%%%%%%%%%%%%%%%%%%%%%
% The Jacobian matrix of the above equations is:  %
%%%%%%%%%%%%%%%%%%%%%%%%%%%%%%%%%%%%%%%%%%%%%%%%%%%%%%%

%%%%%%%%%%%%%%%%%%%%%%%%%%%%%%%%%%%%%
% J = [2x1   –2x2      2x3     %
%       x2   x1+2x2    –4       %
%       1–x3   x3     –x1+x2];  %
%%%%%%%%%%%%%%%%%%%%%%%%%%%%%%%%%%%%

clc;
clear all;

x = [1;1;1]; % Initial conditions

ini=1;
iter=0;
while ini>1e–6

    f =[x(1)^2–x(2)^2+x(3)^2–10;x(1)*x(2)+x(2)^2–4*x(3)–4;
    x(1)–x(1)*x(3)+x(2)*x(3)–5];

    J=[2*x(1) –2*x(2) 2*x(3);x(2) x(1)+2*x(2) –4;1–x(3) x(3) –x(1)+x(2)];

    inix=–inv(J)*f;
    x=x+inix;
    ini=max(abs(f));
    iter=iter+1;

end

'The Solution converges in iterations',iter,pause
'Final values of variable x are',x
```

The output of this program is as follows:

ans =

The solution converges in iterations

iter =
     7

After pressing a key from the keyboard we get

ans =

Final values of variable x are

x =
   2.5870
   3.2369
   3.7128

Project 2

MATLAB Program to Develop Bus Admittance Matrix Y_{bus}

```
%%%%%%%%%%%%%%%%%%%%%%%%%%%%%%%%%%%%%%%%%%%%%%%%%%%%%%%%%%%%%%%%%%%%%%%%%
%       This program generates bus admittance matrix Y_BUS            %
%%%%%%%%%%%%%%%%%%%%%%%%%%%%%%%%%%%%%%%%%%%%%%%%%%%%%%%%%%%%%%%%%%%%%%%%%

function [yb,yc]= Y_BUS

clc;
clear all;

%%%%%%%%%%%%%%%%%%%%%%%%%%%%%%%%%%%%%%%%%%%%%%%%%%
% line impedances of the bus in p.u. are %
%%%%%%%%%%%%%%%%%%%%%%%%%%%%%%%%%%%%%%%%%%%%%%%%%%

    ZL  =[0 0.02+0.2i 0 0 0.03+0.35i
        0.02+0.01i 0 0.04+0.3i 0 0.04+0.25i
        0 0.04+0.2i 0 0.05+0.25i 0.08+0.4i
        0 0 0.05+0.25i 0 0.09+0.5i
        0.05+0.25i 0.05+0.25i 0.08+0.4i 0.1+0.5i 0];

%%%%%%%%%%%%%%%%%%%%%%%%%%%%%%%%%%%%%%%%%%
 % line charging admittances %
%%%%%%%%%%%%%%%%%%%%%%%%%%%%%%%%%%%%%%%%%%

    yc =j*[0 0.02 0 0 0.03
        0.03 0 0.025 0 0.021
        0 0.025 0 0.02 0.01
        0  0 0.02 0 0.075
        0.02 0.02 0.01 0.065 0];
```

```
%%%%%%%%%%%%%%%%%%%%%%%%%%%%%%%%%%%%%%%%%
%       Bus admittance matrix Y_BUS       %
%%%%%%%%%%%%%%%%%%%%%%%%%%%%%%%%%%%%%%%%%

    for m=1:5
        for n=1:5
            if ZL(m,n) == 0
                yb(m,n)=0;
            else
                yb(m,n)=-1/ZL(m,n);
            end
        end
    end
    for m=1:5
        y_sum=0;
        c_sum=0;
        for n=1:5
          y_sum=y_sum+yb(m,n);
          c_sum=c_sum+yc(m,n);
        end
        yb(m,m)=c_sum-y_sum;
end
```

The bus admittance matrix generated is as follows:

```
Y_BUS =
 0.7382 − 7.7368i  −0.4950 + 4.9505i      0            0         −0.2431 + 2.8363i
−40.0000 +20.0000i 41.0607 −27.0993i  −0.4367 + 3.2751i  0        −0.6240 + 3.9002i
    0      −0.9615 + 4.8077i  2.2115 −11.0027i  −0.7692 + 3.8462i  −0.4808 + 2.4038i
    0           0        −0.7692 + 3.8462i  1.1179 − 5.6884i  −0.3487 + 1.9372i
−0.7692 + 3.8462i  −0.7692 + 3.8462i  −0.4808 + 2.4038i  −0.3846 + 1.9231i  2.4038
−11.9042i
```

Project 3

MATLAB Program for Power Flow Analysis by Gauss–Seidel Method

```
%%%%%%%%%%%%%%%%%%%%%%%%%%%%%%%%%%%%%%%%%%%%%%%%%%%%%%%%%%%%%%%%%
% The following program is for Gauss-Seidel power flow analysis  %
%%%%%%%%%%%%%%%%%%%%%%%%%%%%%%%%%%%%%%%%%%%%%%%%%%%%%%%%%%%%%%%%%

clc;
clear all;

d2r=2*pi/360;
w=2*50*pi;
```

```
%%%%%%%%%%%%%%%%%%%%%%%
% The Y_Bus matrix is  %
%%%%%%%%%%%%%%%%%%%%%%%

[Y_BUS,yc]=Y_BUS;
r=real(Y_BUS);g=imag(Y_BUS);

%%%%%%%%%%%%%%%%%%%%%%%%%%%%%%%%%%%%%%%%%%%%%%%%
%  Bus parameters and initial conditions  %
%%%%%%%%%%%%%%%%%%%%%%%%%%%%%%%%%%%%%%%%%%%%%%%%

p=[0;-0.86;-0.37;-0.26;0.18];
q=[0;-0.32;-0.11;-0.08;-0.30];
nv=[1.15;1;1;1;1.01];
theta=[0;0;0;0;0];
v=[nv(1);nv(2);nv(3);nv(4);nv(5)];

alpha =input('Enter G-S acceleration constant: ');
e=1;iter=0;

while e>2e-6

    for i=2:4
      tmp1=(p(i)-j*q(i))/conj(v(i));
      tmp2=0;
      for k=1:5
         if (i==k)
             tmp2=tmp2+0;
         else
             tmp2=tmp2+Y_BUS(i,k)*v(k);
         end
      end
      vt=(tmp1-tmp2)/Y_BUS(i,i);
      v(i)=v(i)+alpha*(vt-v(i));
    end

      q5=0;
      for i=1:5
         q5=q5+Y_BUS(5,i)*v(i);
      end
      q5=-imag(conj(v(5))*q5);
      tmp1=(p(5)-j*q5)/conj(v(5));
      tmp2=0;
      for k=1:4
         tmp2=tmp2+Y_BUS(5,k)*v(k);
```

```
        end
    vt=(tmp1–tmp2)/Y_BUS(5,5);
    v(5)=abs(v(5))*vt/abs(vt);
```

```
                        %%%%%%%%%%%%%%%%%%%%%%%%%%%%
                        % Active and reactive power %
                        %%%%%%%%%%%%%%%%%%%%%%%%%%%%
```

```
    for i=1:5
        sm=0;
        for k=1:5
            sm=sm+Y_BUS(i,k)*v(k);
        end
        s(i)=conj(v(i))*sm;
    end
```

```
                    %%%%%%%%%%%%%%%%%%%%%%%%%%%%%%%%%%%%
                    % Active and Reactive power mismatch %
                    %%%%%%%%%%%%%%%%%%%%%%%%%%%%%%%%%%%%
```

```
    e_p=p–real(s)';
    e_q=q+imag(s)';

    e_pq=[e_p(2:5);
         e_q(2:4)];
    e=max(abs(e_pq));
    iter=iter+1;
    if iter==1
        pause
    end
end
```

```
'Number of iteration required for convergence are:',iter,pause
'Magnitudes of the final voltages are:',abs(v)',pause

'Voltage angles ins degrees are:',angle(v)'/d2r,pause
'Active powers in each bus in MW are:',(real(s)+[0 0 0 0 0.18])*100,pause
'Reactive powers in each bus in MVAr are:',(-imag(s)+[0 0 0 0 0.11])*100
```

This program will ask the following input before executing:

```
Enter G–S acceleration constant: 1.3
```

The output of this program for acceleration constant of 1.3 is

```
ans =
```

Number of iterations required for convergence are as follows:

```
iter =

    12

ans =
```

Magnitudes of the final voltages are as follows:

```
ans =

    1.1500    1.1214    1.0560    1.0350    1.0100

ans =
```

Voltage angles in degrees are as follows:

```
ans =

         0    0.3291   -2.5725   -3.8197    0.7751

ans =
```

Active powers in each bus in MW are as follows:

```
ans =

   -2.5790   -86.0002   -37.0000   -26.0000    36.0000

ans =
```

Reactive powers in each bus in MVAr are as follows:

```
ans =

   56.1074   -32.0001   -11.0000    -8.0000  -121.3827
```

SUMMARY

The main points of this chapter can be summarized as follows:

- The power system consists of generating station, transmission system, sub-transmission system, primary distribution system, and secondary distribution system.
- The underground transmission and distribution lines are more reliable but are more costly as compared to overhead transmission and distribution lines.
- Many approaches for the power calculation can be found in the literature. At present, the state of the power definitions in the time domain seems to be more useful than those in the frequency domain, since it is possible to realize faster control algorithms for active line conditioners using the definitions in time domain.
- Any unbalance set of three-phase voltages can be resolved into three-balanced sets, i.e., positive sequence, negative sequence, and zero sequence components.

- Single line-to-ground fault is the most commonly occurring fault.
- Harmonic currents can produce a number of problems: equipment heating, equipment malfunction, equipment failure communications, interference, fuse and breaker misoperation, process problems, and conductor heating.
- Numerical techniques are used to obtain a solution within an acceptable tolerance limit due to the nonlinear nature of the power system. Load flow studies are performed using computer software that can simulate actual steady-state power system operating conditions, enabling the evaluation of bus voltage profiles, real and reactive power flows, and losses.
- In general, stability is that attribute of the power system or part of the system which enables it to develop restoring forces between the elements which are equal to or greater than the disturbing forces so as to restore the state of equilibrium between the elements; and stability limit is the maximum power flow possible through some particular point in the system when the entire system or part of the system are operating with stability.
- Transient stability of power transmission system is its inherent ability to recover normal operation following sudden or severe disturbances.
- According to the equal area criteria, the kinetic energy gained by the rotor during swing must be equal to the kinetic energy returned as it swings.
- For maintaining good voltage profile and voltage stability, control of reactive power of the power system in necessary.
- The foremost task of LFC is to keep the frequency constant against the randomly varying active power loads.
- The requirement of the LFC is to be robust against the uncertainties of the system model and the variations of system parameters in reality.

REVIEW QUESTIONS

1. Mention the main components of electrical supply systems.
2. List the merits and demerits of AC and DC transmission and distribution systems.
3. State the economic considerations which effect the selection of the size of the conductor of a transmission line.
4. What do you mean by power system harmonics?
5. How are harmonics generated in a power system?
6. Define steady-state stability and transient-state stability of a power system.
7. Mention the techniques for improving the transient-state stability of a power system.
8. Define load frequency control.
9. What are the effects of power system harmonics?
10. How can power system harmonics be reduced?
11. Mention the power system components which generate or absorb power.
12. What do you mean by power quality?
13. Explain voltage sag, voltage swell, under-voltage, and over-voltage.
14. Mention the various types of compensators used in power system.
15. Explain the various methods of voltage control in transmission system.

16. How is bus admittance matrix formed?
17. What is the difference between $[Y_{bus}]$ and $[Z_{bus}]$?
18. What is a Jacobian matrix?
19. Explain G–S method for solution of nonlinear equations.
20. Discuss N–R method for solving nonlinear equations.
21. Distinguish between steady-state, transient, and dynamic stabilities.
22. What is power angle equation?
23. Explain the concept of equal area criterion.

SIMULATION EXERCISES

1. Write a MATLAB program to solve the following differential equation:
   $dy/dt = 2t^2 - y^3$, for $0 \le t \le 5$. At $t = 0$, $y = 1$.
2. Develop a Simulink model for a three-phase 240 V rms, 50 Hz power system to depict the following conditions:
   (a) Under-voltage of 50% after 0.1 s of the simulation time
   (b) Voltage sag of 20% for 3 cycles
   (c) Voltage unbalance of 40%
   (d) Third and Fifth harmonics of the fundamental frequency
3. Write a MATLAB program to evaluate the roots of the function $g(x) = x^4 - x^3 + 7x^2 + 8x - 11$ by using N–R method.
4. Develop a MATLAB program to determine the roots of a function $h(x) = 7x^5 + x^4 + 4x^2 - 2x - 5$ by N–R method and plot the function $h(x)$.
5. Write a MATLAB program to evaluate the roots of the function $f(x) = x^3 - x^2 + 11x - 1$ by G–S method and plot the function $f(x)$. Consider the acceleration factor of 1.2 and solve the problem again.
6. A 60 Hz synchronous generator having inertia constant $H = 5$ MJ/MVA and a direct axis transient reactance $Xd' = 0.3$ p.u. is connected to an infinite bus through a purely reactive circuit as shown in Fig. 8.13. The reactances are shown on the diagram in a common system base. The generator is delivering reactive power $P_e = 0.8$

p.u. and $Q = 0.074$ p.u. to the infinite bus at a voltage of 1 p.u. A three-phase fault occurs at the middle of one line and is cleared by isolating the faulted circuit simultaneously at both ends as shown in Fig. 8.14 . The fault is cleared in 0.3 s. Obtain the numerical solution for 1.0 s using the modified Euler method with a step size of _t=0.01 second in MATLAB. Use the swing curve to determine the system stability and the critical clearing time. Repeat the simulation using Simulink and obtain the swing plots.

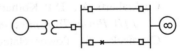

Fig. 8.14 Diagram of the system

7. Develop a MATLAB program for power flow analysis of power system by N–R method. The bus admittance matrix Y_{bus} of the system is known.
8. Develop a MATLAB program which can construct bus impedance matrix $[Z_{bus}]$ for a given system.
9. Construct a MATLAB function lfsnr which can compute the bus voltages of the power system by N–R iteration taking the following data as input:
   (a) Power flow data of the bus
   (b) Number of buses in the system
   (c) Number of lines in the system

(d) Number of voltage-controlled buses

(e) Line data table

10. Write a MATLAB program to solve the following nonlinear equations by N–R method:

$$f_1 = x_1 x_2 - x_3 x_2 + x_1 x_3 - 1 = 0$$

$$f_2 = x_1^3 - 2x_2 x_4 + x_1 x_3 + 2x_3 x_2 + 2 = 0$$

$$f_3 = x_1^2 - 2x_2^3 + x_3 x_4 - 10 = 0$$

$$f_4 = x_1 x_2 + x_2^3 - 4x_3 x_4 - 4 = 0$$

11. A three-phase synchronous generator is connected through a line whose reactance is 0.13 p.u. to an infinite bus whose voltage is 1.0 p.u. and is delivering 0.9 p.u. real power at 0.75 power factor (lagging) to the bus. Develop a MATLAB program to compute the following: (a) the magnitude of the power input which can be suddenly increased without the generator losing synchronism and (b) the sudden rise in input power without the generator losing synchronism. Take the initial input power to be zero.

SUGGESTED READING

Arillaga, J., C.P. Arnold, and B.J. Harker, *Computer Modeling of Electrical Power Systems*, John Wiley, New York, 1986.

Akagi, H., 'New trends in active filters for power conditioning', *EPE-95, European Conference on Power Electronics and Applications*, Vol. 32, Issue 6, Sevilla, Espain Nov/Dec. 1996, pp. 1312–1322.

Bergen, A.R. and V. Vittal, *Power System Analysis*, 2nd Ed., Pearson Education Asia, New Delhi, 2001.

Barret, J.P., P. Bornard, and B. Meyer, *Power System Simulation*, 1st Ed., Chapman & Hall, Great Britain, 1997.

Chakrabarti, A., D.P. Kothari, and A.K. Mukhopadhyay, *Performance Operation and Control of EHV Power Transmission Systems*, 1st Ed., Wheeler Publication, India, 1995.

Czarneck, L.S., 'Power related phenomena in three phase unbalanced systems', *IEEE Transactions on Power Delivery*, Vol. 10, No. 03, Dept. of Electr. and Comput. Engg., Louisiana State Univ., Baton Rouge, LA, pp. 1168–1176, July 1995.

El-Abiad, A.H. and G.W. Stagg, 'Automatic evaluation of power system performance: Effects of line and transformer outages', *AIEE Transactions*, Vol. 81, Issue 3, pp. 712–716, 1962.

Greenwood, A., *Electrical Transients in Power Systems*, Wiley Interscience, New York, 1971.

Gross, C.A., *Power System Analysis*, 2nd Ed., John Wiley, New York, 1983.

Glover, J.D. and M.S. Sharma, *Power System Analysis and Design*, 3rd Ed., Thomson Asia Pvt. Ltd., Bangalore, 2003.

Grainger, J.J. and W.D. Stevenson Jr., *Power System Analysis*, Tata McGraw Hill, New Delhi 2003.

Ibraheem, Kumar, P., and D. Kothari, 'Recent philosophies of automatic generation control strategies in power systems', *IEEE Transactions on Power Systems,* Vol. 20, No. 1, Dept. of Electr. Engg., Fac. of Engg. and Technol., New Delhi, pp. 346–357, February 2005.

Kusic, G.L., *Computer Aided Power System Analysis*, Prentice Hall of India, New Delhi, 1989.

Kothari, M., N. Sinha, and M. Rafi, 'Automatic generation control of an interconnected power system under deregulated environment,' *Power Quality*, Vol. 18, Dept. of Electr. Engg., Indian Inst. of Technol., New Delhi, pp. 95–102, June 1998.

Kundur, P., *Power System Stability and Control*, McGraw-Hill, New York, 1994.

Kothari, D.P. and I.J. Nagrath, *Modern Power Analysis*, 3rd Ed., Tata McGraw Hill, New Delhi, 2003.

Mohalanabis, A.K., D.P. Kothari, and S.I. Ahson, *Computer Aided Power System Analysis and Control*, 1st Ed., Tata McGraw- Hill, New Delhi, 1988.

Mariani, E. and S.S. Murthy, *Control of Modern Integrated Power Systems*, 1st Ed., Springer-Verlag, London, 1997.

Nagrath, I.J. and D.P. Kothari, *Power System Engineering*, 1st Ed., Tata McGraw Hill, New Delhi, 1994.

Ohba, S., H. Ohnishi, and S. Iwamoto, 'An advanced LFC design considering parameter uncertainties in power systems', *Proceedings of IEEE Conference on Power Symposium,* Tokyo Electr. Power Co., Inc., Tokyo, pp. 630–635, September 2007.

Qunin, C.A. and N. Mohan, 'Active filtering of harmonic currents in three phase four wire systems with three phase and single phase nonlinear load', IEEE-APEC'92, *Applied Power Electronics Conference*, Dept. of Electr. Engg., Minnosota Univ., Minneapolis, MN pp. 829–836, 1992.

Rerkpreedapong, D., A. Hasanovic, and A. Feliachi, 'Robust load frequency control using genetic algorithms and linear matrix inequalities', *IEEE Transactions on Power Systems*, Vol. 18, No. 2, Dept. of Electr., Eng. Kasetsart Univ., Bangkok, pp. 855–861, May 2003.

Saadat, H., *Power System Analysis*, Tata Mcgraw Hill, India, 2002.

Singh, L.P., *Advanced Power System Analysis and Dynamics*, 3rd Ed., Wiley Eastern, India, 1992.

Sauer, P.W. and M.A. Pai, *Power System Dynamics and Stability*, 1st Reprint, Pearson Education, Asia, 2002.

Stevenson Jr., W.D., *Elements of Power System Analysis*, 4th Ed., McGraw-Hill, Singapore, 1982.

Taylor, C.W., *Power System Voltage Stability*, 1st Ed. (EPSR), McGraw-Hill Inc., US, 1994.

Weedy, B.M., *Electric Power Systems*, 3rd Ed., John Wiley, London, 1979.

9

CONTROL SYSTEM ENGINEERING AND ELECTRICAL MACHINES

9.1 INTRODUCTION TO CONTROL SYSTEM ENGINEERING

Control system engineering focuses on the design and modeling of various dynamic physical systems and their controllers so that these systems behave in the desired manner. The control engineers are concerned with understanding and controlling the segments of their environment called systems to provide a useful economic product for the society. This field of engineering is based on the foundations of feedback theory and linear system analysis, and it integrates the concepts of network theory and communication engineering. Thus, it is not limited to any particular engineering discipline but is equally applicable to aeronautical, chemical, mechanical, environmental, civil, and electrical engineering. Furthermore, as the understanding of the dynamics of business, social, and political systems increases, the ability to control these systems will also increase.

The control theory has two major divisions—classical control theory and modern control theory. The classical control theory is limited to the design of single-input and single-output systems. The analysis of these systems is carried out by ordinary differential equations in time domain. In the s-domain or frequency domain, it is carried out with the help of Laplace transform. A controller designed using classical theory requires fine-tuning due to approximate design methods used. Owing to their simple design and easy physical implementation, these controllers are preferred in industrial applications. Proportional (P), proportional integrator (PI), proportional differentiator (PD), and proportional-integrator–differentiator (PID) controllers are the example of some classical controllers. Modern control theory deals with multi-input and multi-output systems and their analysis is carried out in the frequency domain. This overcomes the limitations of classical control theory in more sophisticated design problems like space missile control.

A control system is an interconnection of components forming a system configuration that provides a desired system response. The basis for analysis of a system is the foundation provided by classical control theory, which assumes a cause–effect relationship for the components of a system. Depending on the presence of feedback, control system may be classified as open-loop control system and closed-loop control system or feedback control system. In the open-loop control system, the output of the system or process has no effect on the control action.

These systems operate on time basis. For instance, a room heater without any temperature sensing device is an open-loop control system. The temperature (control variable) of the room is risen by the heat generated by the heating element. The room temperature depends on the time during which the supply to the heater remains ON. On the other hand, a closed-loop control system utilizes an additional measure of the actual output to compare it with the desired output response. The measure of the output is called the feedback signal. The provision of feedback automatically corrects the changes in the output due to disturbances. Thus, the closed-loop control system is often called as automatic control system. In the closed-loop control system, the feedback loop makes the system response relatively insensitive to external disturbances and internal variations in system parameters. Thus, it is possible to use relatively inaccurate and inexpensive components to obtain the accurate control of a given system which, on the other hand, is impossible in case of open-loop systems. In the open-loop control systems, stability is not a major problem and hence they are easier to build whereas in the closed-loop control systems, stability is a major concern as they may tend to overcome errors that can cause oscillations. So, the open-loop control systems are advisable for systems in which the inputs are known in advance and no disturbances of significant amount are present, and the closed-loop control systems are advisable when unpredictable disturbances or variations in the system components are observed.

9.2 TIME RESPONSE ANALYSIS

The time response of a dynamic system provides information about the response of the system to certain inputs. The stability of the system and the performance of the controller can be determined by analyzing its time response. The time response of a control system includes its transient response and the steady-state response. The transient response, also known as dynamic response of a system, is defined as that part of the time response that dies out as time approaches infinity, i.e., very large. The steady-state response is simply that part of the total response that remains after the transient has died out. The step signal, impulse signal, ramp signal, and sinusoidal signal are the commonly used test signals for time response analysis. As these signals are simple functions of time mathematically and experimentally, analysis of control systems can be carried out by these signals easily. The typical input signal to be used for analyzing the system characteristics can be determined from the inputs that the system will be subjected to most frequently under normal operation. For example, if the inputs to the control system are gradually changing functions of time, then ramp signal will be a better choice. Once a control system is designed on the basis of test signals, its performance for the actual inputs is satisfactory.

9.2.1 First-order System

A first-order system is shown in Fig. 9.1. The transfer function, i.e., output/input in frequency domain, of this system is given by $\text{TF} = G(s)/\{1 + G(s)H(s)\} = 1/\{1 + 2s\}$, as $G(s) = 1/2s$ and $H(s) = 1$. The time constant of this system is 2. Physically, this system can represent an R-L or R-C circuit. The response of this first-order system is analyzed for unit step, impulse, and ramp function assuming the initial conditions to be zero. Mathematically, the response of this system to these signals can be analyzed as follows:

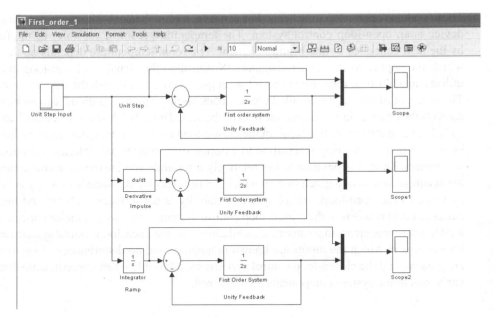

Fig. 9.1 Model of a first-order system with unit step, impulse, and ramp test signals

For Unit Step Input

TF = $C(s)/R(s) = 1/\{1 + 2s\}$. In case of unit step signal, $R(s) = 1/s$, thus, $C(s) = 1/\{s(1 + 2s)\}$ or $c(t) = 1 - e^{-t/2}$. Now for $t = 2$ s, $c(2) = 0.632$. So, smaller the time constant of the system, the faster is its response to a unit step input. At $t = \infty$, the output of the system is equal to 1. The plot of the response of the system to unit step signal is shown in Fig. 9.2.

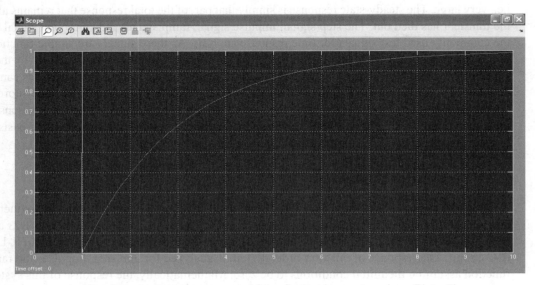

Fig. 9.2 Unit step response of the first-order system (see Plate 7)

For Impulse Input

In case of impulse signal, $R(s) = 1$, thus, $C(s) = 1/\{1 + 2s\}$ or $c(t) = 0.5e^{-0.5t}$. So, as $t = \infty$, the output of the system decays down to zero. The larger the time constant of the system, the greater will be the time required to bring the system output to zero. The plot of the response of the system to impulse signal is shown in Fig. 9.3.

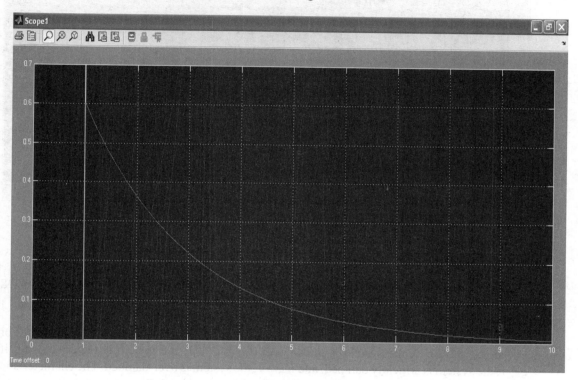

Fig. 9.3 Impulse response of the first-order system (see Plate 7)

For Ramp Input

In case of ramp input signal, $R(s) = 1/s^2$, thus, $C(s) = 1/\{s^2(2s + 1)\}$ or $c(t) = t - 2 + 2e^{-0.5t}$. As $t = \infty$, the output of the system also becomes infinity and error signal, i.e., $r(t) - c(t)$, becomes equal to 2, i.e., equal to the time constant of the system. So, greater the time constant of the system, the greater is the error. The plot of the response of the system to ramp signal is shown in Fig. 9.4.

By comparing the response of the three test signals it can be observed that the response of the impulse signal is differential, and response of ramp signal is integral of the response of the unit step signal. Also, it can be observed from Fig. 9.1 that impulse signal is obtained by differentiating the unit step signal and ramp signal is obtained by integrating the unit step signal in time domain.

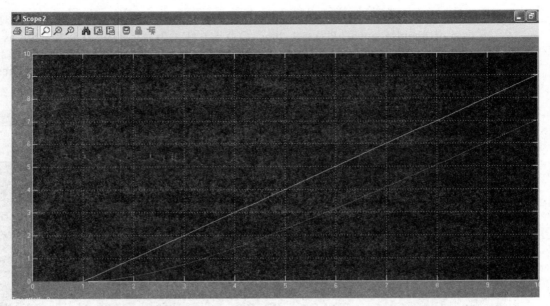

Fig. 9.4 Ramp response of the first-order system (see Plate 8)

9.2.2 Second-order System

Consider a second-order system, shown in Fig. 9.5, which can be expressed by the following transfer function:

$$C(s)/R(s) = \omega_n^2/\{s^2 + 2\xi \omega_n s + \omega_n^2\}$$

where ω_n is the undamped natural frequency and ξ is the damping factor of the system. For a system, if $\xi\omega = 0$, the system is undamped; if $0 < \xi < 1$, the system is under-damped; if $\xi = 1$, the system is critically damped; and if $\xi > 1$, the system is over-damped. The characteristic equation of the system under consideration is $s^2 + 2\xi \omega_n s + \omega_n^2 = 0$. The response of this system $c(t)$ and the error signal $e(t)$ for unit step input (see Fig. 9.6), i.e., $R(s) = 1/s$, is given by

$$c(t) = 1 - \{(e^{-\xi\omega_n t}/\sqrt{(1 - \xi^2)}) \sin(\omega_n\sqrt{(1 - \xi^2)}t) + \tan^{-1}(\sqrt{(1 - \xi^2)}/\xi)\}$$

$$e(t) = c(t) - r(t) = e^{-\xi\omega_n t} (\cos \omega_d t + (\xi/\sqrt{(1 - \xi^2)}) \sin \omega_d t)$$

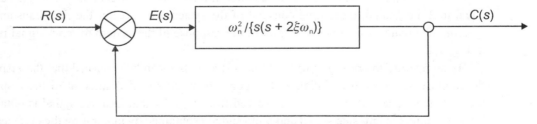

Fig. 9.5 Block diagram of a second-order system

Time Response Specification

The real-time control systems are generally designed for damping coefficient less than unity. Thus, the overshoots and undershoots decay exponentially and peak overshoot is the first overshoot of the system response. The time response specifications are defined as follows:

Delay time (t_d) This is the time required for the system response to reach 50% of the final value for the first time.

Rise time (t_r) This is the time required for the system response to rise from 0 to 100% of the final value in case of under-damped systems ($\xi < 1$) and from 10% to 90% of the final value in case of over-damped system ($\xi > 1$). For a second-order under-damped system, the rise time for unit step input is $t_r = (\pi - \theta)/\omega_d$, where $\theta = \tan^{-1}\{\sqrt{(1 - \xi^2)}/\xi$.

Peak time (t_p) This is the time required by the system response to reach the first peak of the overshoot. For a second-order under-damped system, the peak time for unit step input is $t_p = \pi/\omega_d$.

Peak overshoot (M_p) It is defined as the maximum peak overshoot of the system response measured from unity. If the final steady-state value of the system response is not unity, then maximum peak overshoot is given by $\{c(\text{peak time}) - c(\infty)\}/c(\infty)$. For a second-order under-damped system, the peak overshoot for unit step input is $M_p = e^{-\pi\xi}/\sqrt{(1 - \xi^2)}$.

Settling time This the time required for the system response to settle down and stay within a particular specified tolerance band of 2% or 5% of the final value. For a second-order under-damped system, the settling time for unit step input is $t_s = 4/(\xi\omega_n)$ for 2% criterion and $t_s = 3/(\xi\omega_n)$ for 5% criterion.

Steady-state error (e_{ss}) It is the difference between the actual output and the desired output as time tends to infinity, i.e., very large. Thus, $e_{ss} = \text{limit}_{t \to \infty}[r(t) - c(t)]$.

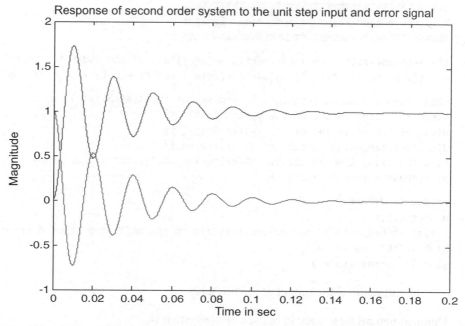

Fig. 9.6 Step response and error signal of the second-order system (see Plate 8)

The following MATLAB program plots the response and the error signal of a second-order system ($\omega_n = 314$ and $\xi = 0.1$) for unit step input signal. The response and the error signal of the system are shown in Fig. 9.6. This program also computes the damped natural frequency, rise time, peak time, settling time, and maximum peak overshoot of the system response and displays them on the command window as shown after the end of the following program.

```
%%%%%%%%%%%%%%%%%%%%%%%%%%%%%%%%%%%%%%%%%%%%%%%%%%%%%%%%%%%%%%%%%%%%%%%%
%  The following program plots the response of second-order system     %
%                     for unit step input                              %
%%%%%%%%%%%%%%%%%%%%%%%%%%%%%%%%%%%%%%%%%%%%%%%%%%%%%%%%%%%%%%%%%%%%%%%%

clc;
clear all;

t = 0:10e-6:0.2;
wn = 314;         % Undamped natural frequency in rad/sec
zeta = 0.1;       % Damping factor
theta = atan(sqrt(1 - zeta^2)/zeta); % Phase angle in rad.

wd = wn * sqrt(1 - (zeta)^2); % Damped natural frequency
tr = (pi - theta) / wd;       % Rise time
tp = pi / wd;                 % Peak time
ts1 = 4 / (zeta * wn);        % Settling time for 2% criterion
ts2 = 3 / (zeta * wn);        % Settling time for 5% criterion
Mp = exp((-pi * zeta) / sqrt(1 - zeta^2)); % Peak overshoot

%%%%%%%%%%%%%%%%%%%%%%%%%%%%%%%%%%%%%%%%%%%%%%%
% System output response and error signal %
%%%%%%%%%%%%%%%%%%%%%%%%%%%%%%%%%%%%%%%%%%%%%%%

c = 1-((exp(-zeta * wn * t)./sqrt(1 - (zeta)^2)) .* sin (wd * t + theta));
e = ((exp(-zeta*wn*t)).*(cos(wd*t)+((zeta / sqrt(1 - zeta^2))).* sin(wd*t)));

disp('Damped natural frequency in rad/sec of the system is:'), wd
disp('Rise time in sec of the system is:'), tr
disp('Peak time in sec of the system is:'), tp
disp('Settling time in sec for 2% criterion is:'), ts1
disp('Settling time in sec for 5% criterion is:'), ts2
disp('Peak overshoot is:'), Mp

orient landscape
plot(t,c,t,e)
title('\bfResponse of second-order system to the unit step input & error signal')
xlabel('\bfTime in sec')
ylabel('\bfMagnitude')
```

The following is the output of the program:

Damped natural frequency in rad/sec of the system is:

```
wd =
  312.4261
Rise time in sec of the system is:
tr =
    0.0053
Peak time in sec of the system is:
tp =
    0.0101
Settling time in sec for 2% criterion is:
ts1 =
    0.1274
Settling time in sec for 5% criterion is:
ts2 =
    0.0955
Peak overshoot is:
Mp =
    0.7292
```

The following MATLAB program plots the impulse, step, and ramp response of second-order system given by

$$\frac{C(s)}{R(s)} = \frac{1}{(s^2 + 0.2s + 1)}$$

The impulse, step, and ramp response of this system are shown in Figs 9.7, 9.8, and 9.9 respectively. It can be observed that the system output for ramp input increases till 3 s and then starts decreasing. The system response for impulse and step input is under-damped.

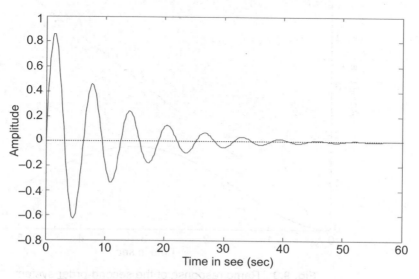

Fig. 9.7 Impulse response of the second-order system

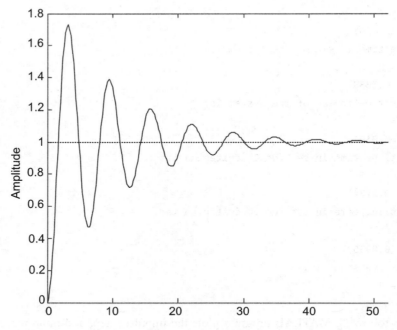

Fig. 9.8 Step response of the second-order system

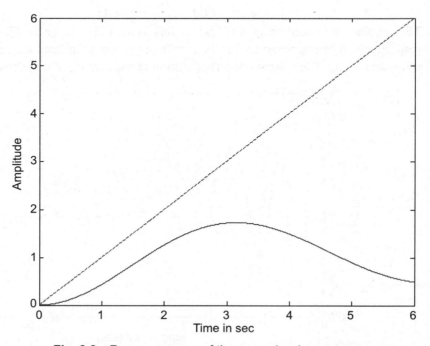

Fig. 9.9 Ramp response of the second-order system

```
%%%%%%%%%%%%%%%%%%%%%%%%%%%%%%%%%%%%%%%%%%%%%%%%%%%%%%%%%%%%%%%%%%%%%
%      This MATLAB program plots the impulse, step and ramp       %
%              response of a second-order system                  %
%%%%%%%%%%%%%%%%%%%%%%%%%%%%%%%%%%%%%%%%%%%%%%%%%%%%%%%%%%%%%%%%%%%%%

clc;
clear all;
clf;

num = [0 0 1];
den = [1 0.2 1];

figure(1)
impulse(num,den);
title('\bfImpulse response of the second-order system')
xlabel('\bfTime in sec')
ylabel('\bfAmplitude')

figure(2)
step(num,den);
title('\bfStep response of the second-order system')
xlabel('\bfTime in sec')
ylabel('\bfAmplitude')

t = 0:0.001:6;
c = step(num,den,t);

figure(3)
plot(t,c,'-',t,t,'--')
title('\bfRamp response of the second-order system')
xlabel('\bfTime in sec')
ylabel('\bfAmplitude')
```

Consider a second-order under-damped system, a critically damped system, and an undamped system as shown in Fig. 9.10. These systems are given as follows:

1. Under-damped system ($0 < \xi < 1$)
   For this system, open-loop gain $G(s) = 98{,}596/(s^2 + 62.85s + 98{,}596)$. Thus, for this system, $\omega_n = \sqrt{98{,}596} = 314$ rad/s and $2\xi\omega_n s = 62.85$ or $\xi = 0.1$.
2. Critically damped system ($\xi = 1$)
   For this system, open-loop gain $G(s) = 98{,}596/(s^2 + 6{,}285s + 98{,}596)$. Thus, for this system, $\omega_n = \sqrt{98{,}596} = 314$ rad/s and $2\xi\omega_n s = 6{,}285$ or $\xi = 10$.
3. Undamped system ($\xi = 0$)
   For this system, open-loop gain $G(s) = 98{,}596/(s^2 + 98{,}596)$. Thus, for this system, $\omega_n = \sqrt{98{,}596} = 314$ rad/s and $2\xi\omega_n s = 0$ or $\xi = 0$.

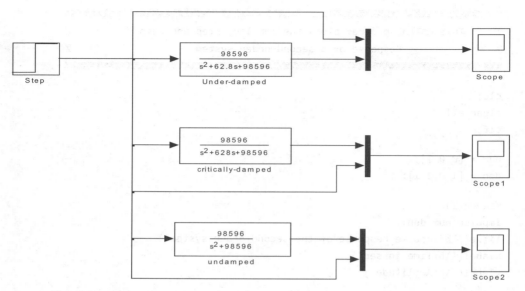

Fig. 9.10   Under-damped, critically damped, and undamped second-order systems

The step response of these systems are shown in Fig. 9.11 (under-damped), Fig. 9.12 (critically damped), and Fig. 9.13 (undamped). It can be observed that for critically damped system, the response rises linearly and then becomes constant, i.e., 1, while for undamped system, the response is oscillatory.

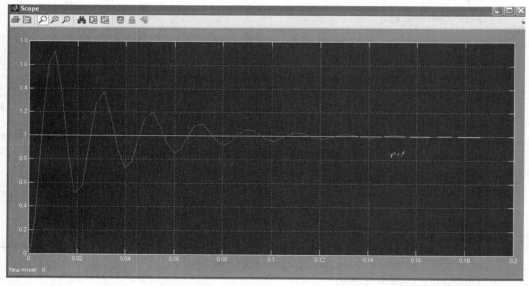

Fig. 9.11   Unit step response of under-damped second-order system (see Plate 9)

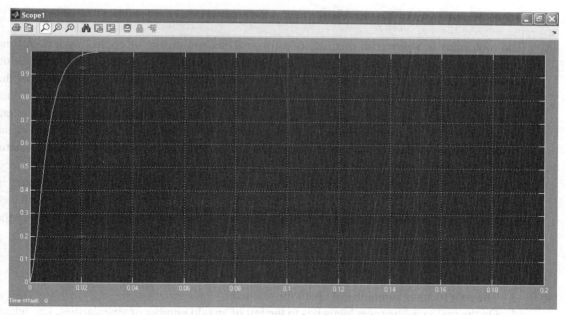

Fig. 9.12 Unit step response of critically damped second-order system (see Plate 9)

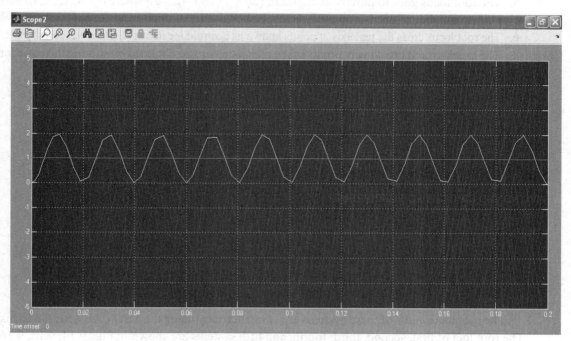

Fig. 9.13 Unit step response of undamped second-order system (see Plate 10)

9.2.3 Root Locus Technique

The root locus is a powerful method used for analysis and design of control system in time domain and to study its stability. Being a graphical technique, it provides quick information about the dynamics of the system. This is a graphical technique used for examining how the roots of a system vary with the gain of the system. In addition, it also provides information about the stability and relative stability of the system. The stability and transient response of a system depends upon the location of the system poles and can be easily determined by the root locus. In some cases, the desired system performance can be obtained by varying the system static gain K. Although a variation in K also causes the system transient response parameters to vary. This method is very useful in designing a linear control system as it indicates the manner in which the open-loop poles and zeros should be modified so as to satisfy the desired performance parameters of the system response. The following rules can be followed for constructing the root locus:

1. The branches in a root locus are equal to the number of open-loop poles.
2. The locus starts from the open-loop poles and ends at the open-loop zeros.
3. The root locus is symmetrical about the real axis.
4. On the real axis, the roots locus lies to the left of an odd number of singularities, i.e., where the sum of poles and zeros lying on the real axis is odd beginning from the left.
5. The root locus goes to infinity at angles of $(2q + 1)\ 180°/(p - z)$, where $q = 0, 1, \ldots, (p - z)$, p is the number of open-loop poles, and z is the number of open-loop zeros.
6. The root locus cross the real axis at points known as centroid located at $(\Sigma \text{ poles} - \Sigma \text{ zeros})/p - z$.
7. The break-in and breakaway points of the root locus can be obtained by solving the equation $dK/ds = 0$. These points can be complex also.

The following MATLAB program plots the root locus for the following systems:

1. First system

   $G(s)\ H(s) = (0.0345s^2 + 0.08675s + 0.0987)/(s^4 + 3s^3 + 2.8s^2 + 0.914s + 0.0908)$

2. Second system

   $G(s)\ H(s) = (s^2 + 2s + 10)/s^2(s + 2)$

3. Third system

   $G(s)\ H(s) = (s + 1.34)/s^2(s + 12)$

4. Fourth system

   $G(s)\ H(s) = 1/s\ (s^3 + 20s^2 + 25s + 50)$

5. Fifth system

   $G(s)\ H(s) = (s + 1)/(s^3 + 3s^2 + 12s - 16)$

The root locus plots of these systems are constructed by using the MATLAB function locus. The root loci of first, second, third, fourth, and fifth systems are shown in Figs 9.14, 9.15, 9.16, 9.17, and 9.18 respectively (see colour plates for references).

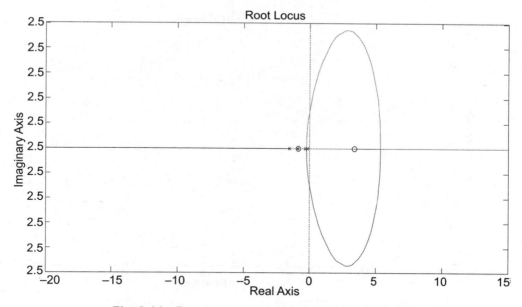

Fig. 9.14   Root locus of the first system (see Plate 10)

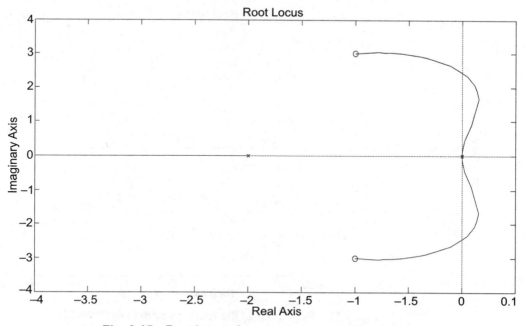

Fig. 9.15   Root locus of the second system (see Plate 11)

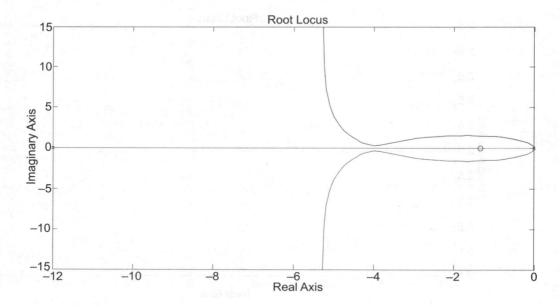

Fig. 9.16 Root locus of the third system (see Plate 11)

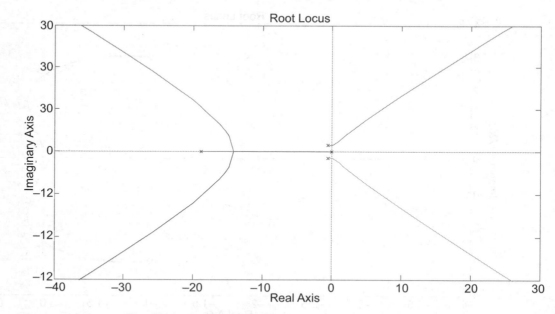

Fig. 9.17 Root locus of the fourth system (see Plate 12)

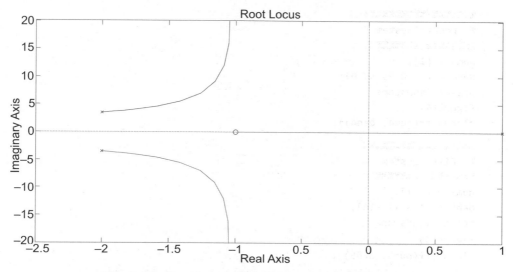

Fig. 9.18   Root locus of the fifth system (see Plate 12)

```
%%%%%%%%%%%%%%%%%%%%%%%%%%%%%%%%%%%%%%%%%%%%%%%%%%%%%%%%%%%%%%%%%%%%%%%%%
%  This MATLAB program plots the 'root locus plot' of various systems  %
%%%%%%%%%%%%%%%%%%%%%%%%%%%%%%%%%%%%%%%%%%%%%%%%%%%%%%%%%%%%%%%%%%%%%%%%%

%%%%%%%%%%%%%%%%%%%%%%
%  First system  %
%%%%%%%%%%%%%%%%%%%%%%
num1 = [0 0 -0.0345 0.0867 0.0987];
den1 = [1 3 2.8 0.914 0.0908];
orient landscape
figure(1)
rlocus(tf(num1,den1));

%%%%%%%%%%%%%%%%%%%%%%%%
%  Second System  %
%%%%%%%%%%%%%%%%%%%%%%%%
num2 = [1 2 10];
den2 = [1 2 0 0];
orient landscape
figure(2)
rlocus(tf(num2, den2));

%%%%%%%%%%%%%%%%%%%%%%%
%  Third System  %
%%%%%%%%%%%%%%%%%%%%%%%
num3 = [1 1.34];
den3 = [1 12 0 0];
orient landscape
figure(3)
rlocus(tf(num3, den3));
```

```
%%%%%%%%%%%%%%%%%%%
%  Fourth System %
%%%%%%%%%%%%%%%%%%%%%
num4 = [1];
den4 = [1 20 25 50 0];
orient landscape
figure(4)
rlocus(tf(num4, den4));

%%%%%%%%%%%%%%%%%%%%
%  Fifth System  %
%%%%%%%%%%%%%%%%%%%%
num5 = [1 1];
den5 = [1 3 12 -16];
orient landscape
figure(5)
rlocus(tf(num5, den5));
```

9.3 FREQUENCY RESPONSE ANALYSIS

The frequency response of a system is its open-loop response to sinusoidal signals of various frequencies. The response of a linear-time invariant system for sinusoidal input signal is sinusoidal but with a change in frequency and phase of the signal. The difference between the input and output sinusoidal signals is defined as the frequency response of the system. To plot the frequency response of a system, it is necessary to create a frequency vector of varying frequencies, i.e., from 0 to ∞, and to compute the system responses at these frequencies. The open-loop frequency response of a system can be used to predict the behavior of the closed-loop system. The frequency response method is sometimes less intuitive but is useful in modeling the transfer function from the physical data of a real-time system. The frequency response analysis has the following advantages: (1) its measurements are done under steady-state conditions of the system and are therefore simpler to analyze in comparison to transient responses; (2) it is performed on open-loop systems and therefore instability problems do not arrive; and (3) its results conveniently provide system order, resonant frequencies, error constants, and gain. It also provides useful insights about the stability and performance characteristics of the control system. However, it is not possible to generate low frequency signals, so frequency analysis normally starts from 10 to 100 Hz frequencies depending on the system dynamics.

9.3.1 Bode Plot

A Bode plot is a combination of two plots—one plot of logarithm of frequency (x-axis) versus magnitude of the transfer function (y-axis) in dB, and second plot of logarithm of frequency (x-axis) and the phase angle of the transfer function (y-axis). The logarithm of frequency is taken on x-axis so that both high and low frequency positions can be easily represented on the same graph. These plots can be used to determine the stability and relative stability of the systems. They are also very useful in the design of control systems. We should be familiar with the Bode plots of some basic factors so as to utilize them while constructing a composite Bode

Plate 9

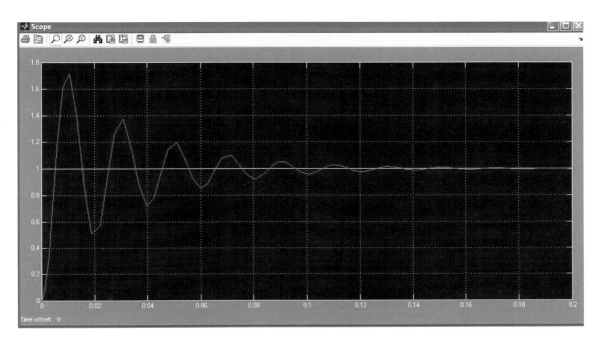

Unit step response of under-damped second-order system (*Chapter 9, page 314*)

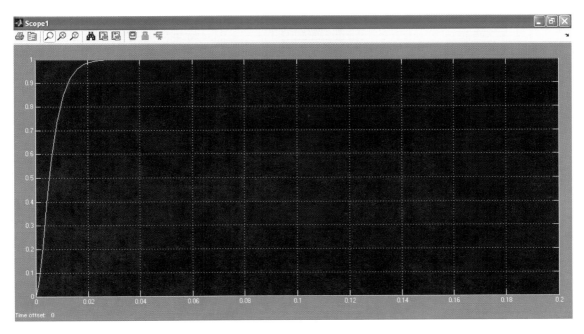

Unit step response of critically damped second-order system (*Chapter 9, page 315*)

Plate 10

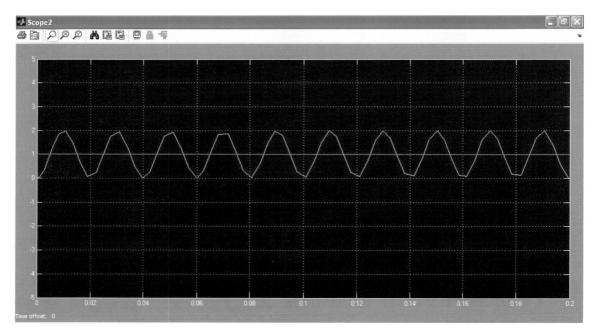

Unit step response of undamped second-order system (*Chapter 9, page 315*)

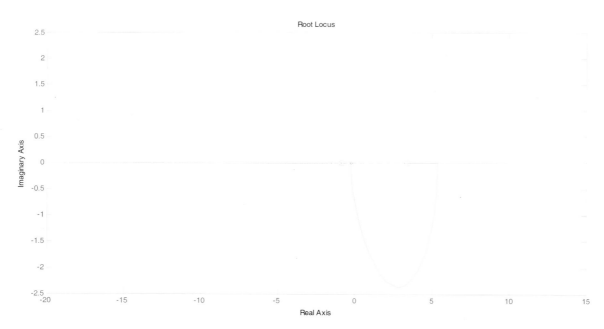

Root locus of the first system (*Chapter 9, page 317*)

Plate 11

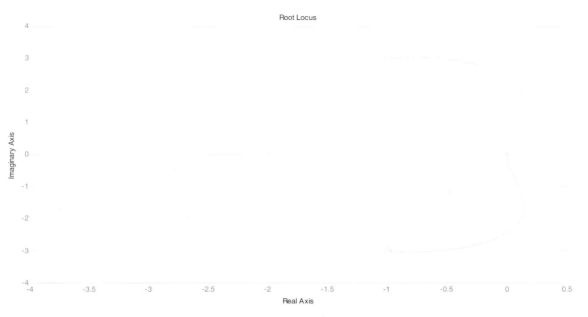

Root locus of the second system (*Chapter 9, page 317*)

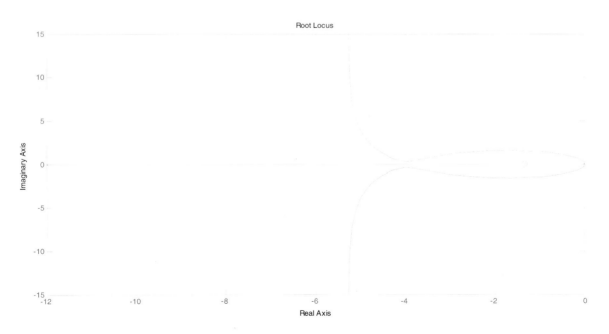

Root locus of the third system (*Chapter 9, page 318*)

Plate 12

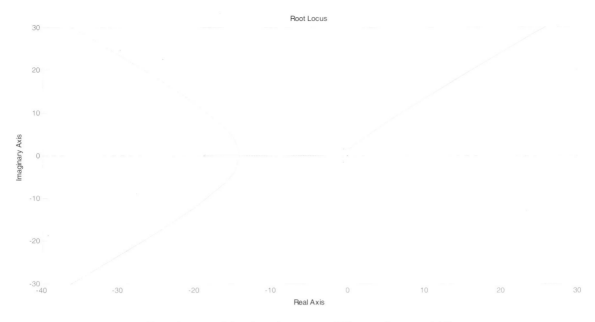

Root locus of the fourth system (*Chapter 9, page 318*)

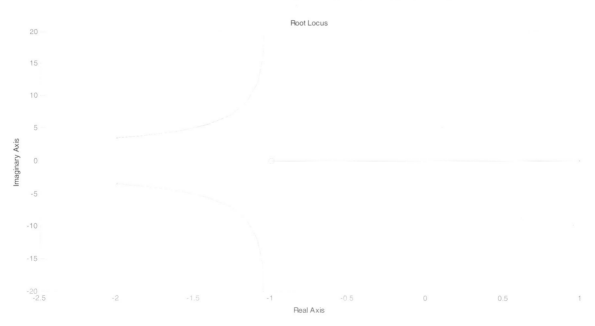

Root locus of the fifth system (*Chapter 9, page 319*)

plot of any general form by sketching the curves for each factor and adding individual curves graphically as adding the logarithms of the gains corresponds to multiplying them together. The following MATLAB program plots the Bode plot for some basic factors. The Bode plots for gain $K = 100$, pole at origin $(1/j\omega)$, three poles at origin $(1/j\omega)^3$, zero at origin $(j\omega)$, three zeros at origin $(j\omega)^3$, first-order pole on the real axis $\{1/(1 + i\omega)\}$, first-order zero on the real axis $(1 + j\omega)$, quadratic poles $\{1/(1 + 20j\omega + (j\omega)^2)\}$, and quadratic zeros $\{1 + 20j\omega + (j\omega)^2\}$ are shown in Figs 9.19, 9.20, 9.21, 9.22, 9.23, 9.24, 9.25, 9.26 and 9.27 respectively.

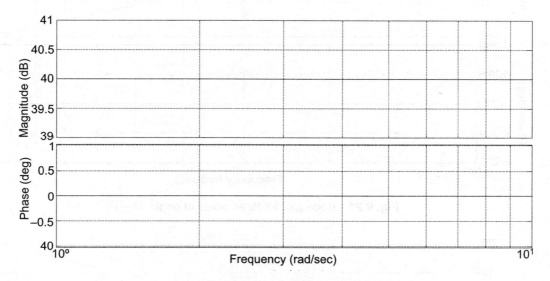

Fig. 9.19 Bode plot for gain $K = 100$

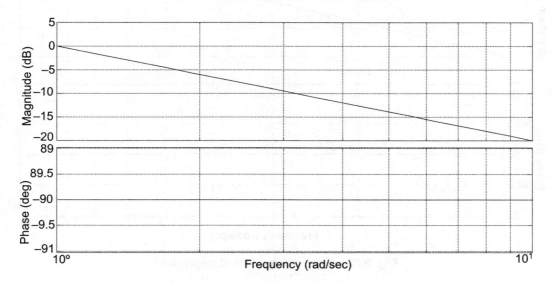

Fig. 9.20 Bode plot for pole at origin $(1/j\omega)$

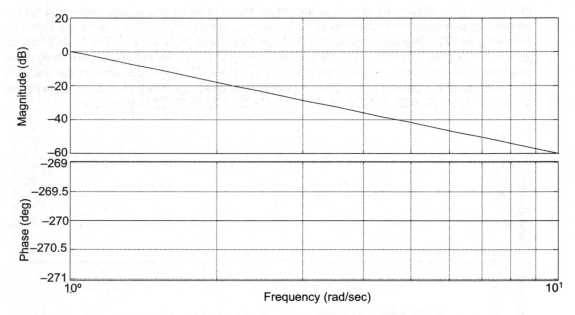

Fig. 9.21 Bode plot for three poles at origin $1/(j\omega)^3$

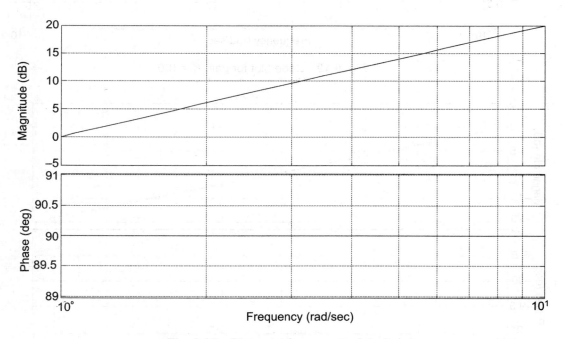

Fig. 9.22 Bode plot for zero at origin $(j\omega)$

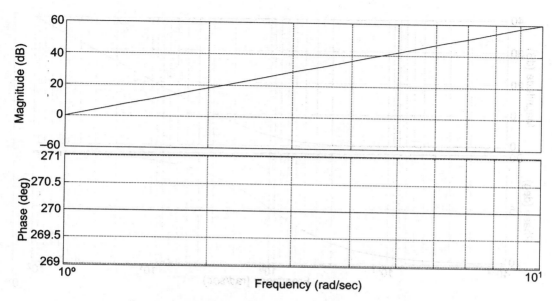

Fig. 9.23   Bode plot for three zeros at origin $(j\omega)^3$

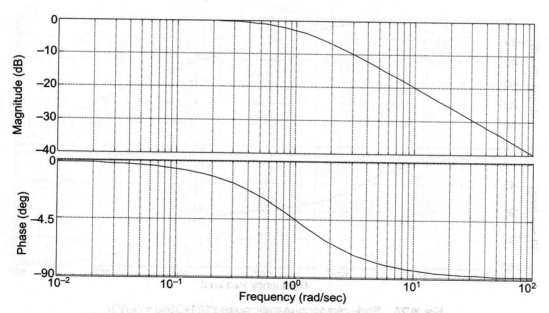

Fig. 9.24   Bode plot for first-order pole on the real axis $\{1/(1 + j\omega)\}$

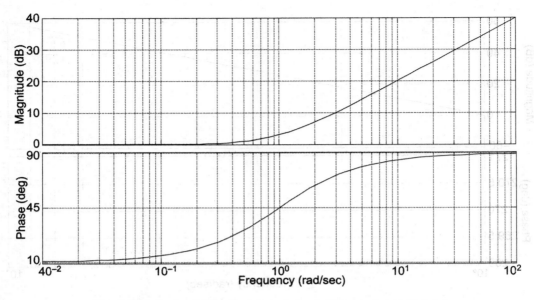

Fig. 9.25 Bode plot for first-order zero on the real axis $(1 + j\omega)$

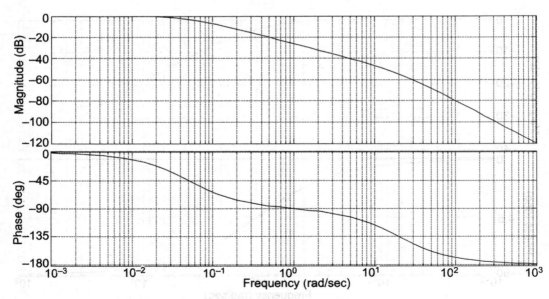

Fig. 9.26 Bode plot for quadratic poles $\{1/(1+20j\omega + (j\omega)^2)\}$

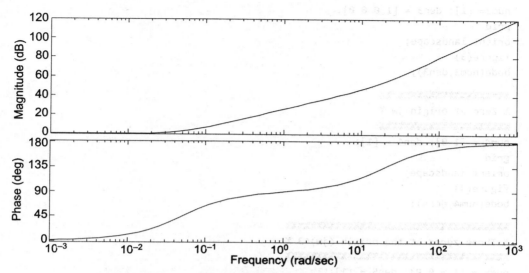

Fig. 9.27   Bode plot for quadratic poles {1+20$j\omega$ + ($j\omega$)²}

```
%%%%%%%%%%%%%%%%%%%%%%%%%%%%%%%%%%%%%%%%%%%%%%%%%%%
% This MATLAB program plots the Bode plot of some   %
%                  basic factors                    %
%%%%%%%%%%%%%%%%%%%%%%%%%%%%%%%%%%%%%%%%%%%%%%%%%%%%
clc;
clear all;
clf;

%%%%%%%%%%
% Gain K %
%%%%%%%%%%
num1 = [100]; den1 = [1];
grid
orient landscape;
figure(1)
bode(num1,den1);

%%%%%%%%%%%%%%%%%%%%%%%%%%
% Pole at origin 1/jw %
%%%%%%%%%%%%%%%%%%%%%%%%%%
num2 = [1]; den2 = [1 0];
grid
orient landscape;
figure(2)
bode(num2,den2);

%%%%%%%%%%%%%%%%%%%%%%%%%%%%%%%%%%%%%%%
% Three poles at origin 1/(jw)^3 %
%%%%%%%%%%%%%%%%%%%%%%%%%%%%%%%%%%%%%%%
```

```
num3= [1]; den3 = [1 0 0 0];
grid
orient landscape;
figure(3)
bode(num3,den3);

%%%%%%%%%%%%%%%%%%%%%%
% Zero at origin jw %
%%%%%%%%%%%%%%%%%%%%%%
num4 = [1 0]; den4 = [1];
grid
orient landscape;
figure(4)
bode(num4,den4);

%%%%%%%%%%%%%%%%%%%%%%%%%%%%%%%%%%%%%%
% Three zeros at the origin (jw)^3 %
%%%%%%%%%%%%%%%%%%%%%%%%%%%%%%%%%%%%%%
num5 = [1 0 0 0]; den5 = [1];
grid
orient landscape;
figure(5)
bode(num5,den5);

%%%%%%%%%%%%%%%%%%%%%%%%%%%%%%%%%%%%%%%%%%
% First-order pole on real axis 1/(1+jw)  %
%%%%%%%%%%%%%%%%%%%%%%%%%%%%%%%%%%%%%%%%%%
num6 = [1]; den6 = [1 1];
grid
orient landscape;
figure(6)
bode(num6,den6);

%%%%%%%%%%%%%%%%%%%%%%%%%%%%%%%%%%%%%%%%%%
% First-order zero on real axis (1 + jw) %
%%%%%%%%%%%%%%%%%%%%%%%%%%%%%%%%%%%%%%%%%%
num7 = [1 1]; den7 = [1];
grid
orient landscape;
figure(7)
bode(num7,den7);

%%%%%%%%%%%%%%%%%%%%%%%%%%%%%%%%%%%%%%%%%%%%%%%
% Quadratic poles 1/{1 + 2*10*(jw/wn)+(jw/wn)^2} %
%%%%%%%%%%%%%%%%%%%%%%%%%%%%%%%%%%%%%%%%%%%%%%%
num8 = [1]; den8 = [1 20 1];
grid
orient landscape;
```

```
figure(8)
bode(num8,den8);

%%%%%%%%%%%%%%%%%%%%%%%%%%%%%%%%%%%%%%%%%%%%%%
% Quadratic zeros {1+2*10(jw/wn)+(jw/wn^2} %
%%%%%%%%%%%%%%%%%%%%%%%%%%%%%%%%%%%%%%%%%%%%%%
num9 = [1 20 1]; den9 = [1];
grid
orient landscape;
figure(9)
bode(num9,den9);
```

9.3.2 Nyquist Plot

A Nyquist plot is used for analyzing the stability of a feedback control system. In this plot, the gain and phase of system's frequency response are plotted in polar coordinates. This plot can be obtained simply by plotting the locus of the imaginary $\{G(j\omega)\}$ versus real $\{G(j\omega)\}$ for frequencies varying from $-\infty$ to $+\infty$. It readily provides the information about the degree of stability and the adjustments required to make the system stable. According to Nyquist stability criteria, if the Nyquist plot of a system encircles the $(-1 + j0)$ point in the counterclockwise direction as many times as the number of right-hand side poles of $G(s)\,H(s)$ (open-loop transfer function) in the s plane, the system is said to be stable. It can be expressed mathematically as follows:

$$Z = N + P$$

where

Z is the number of zeros of $1 + G(j\omega)\,H(j\omega)$ on the right-hand side of the s-plane
P is the number of poles of $G(j\omega)\,H(j\omega)$ in the right-hand side of the s-plane
N is the number of encirclements of $(-1 + j0)$ point by the Nyquist plot in counterclockwise direction

If $Z = 0$, $N = -P$, i.e., number of encirclements are equal to the number of poles. The negative sign indicates counterclockwise direction of encirclements. The system is unstable if there is any clockwise encirclement of the $-1 + j0$ point.

The gain margin and phase margin are used to measure the relative stability of a system. The gain margin is the amount the gain is less than unity when the Nyquist plot crosses $-180°$ axis, i.e., negative x-axis or phase crossover, and phase margin is the angle the phase is less than $180°$ when the gain is unity. Mathematically,

$$GM = -20 \log\{G(j\omega_c)\}$$

and

$$PM = \theta$$

where ω_c is the frequency at which the phase angle of the open-loop transfer function is $-180°$ and θ is the angle made by the line joining the point at which the plot cuts the unity gain circle and the origin, from the $-180°$ axis.

9.3.3 Nichols Plot

Nichols chart is another technique for analyzing frequency responses. It is similar to a Nyquist plot but shows the gain on the logarithmic scale, i.e., decibels versus phase on a linear scale in degrees with an axis origin at point (0 dB, −180°). This chart is symmetrical about the −180° axis. The gain and phase margins can be easily determined by a Nichols chart. The gain margin is the negative value of the gain axes intersect and the phase margin is equal to the distance between the axis origin and the phase axes intersect. The resonant peak magnitude and frequency, and the bandwidth of the closed-loop system can also be easily determined from the plot of the open-loop locus.

The following MATLAB program plots the Bode plot, the Nyquist plot, and the Nichols plot for the following two systems:

1. First system: $G(s)\,H(s) = (s^2 + 2s + 3)/(3s^2 + 2s + 1)$
2. Second system: $G(s)\,H(s) = (2s + 6)/(s(s − 1))$

The poles and zeros of the first system are complex and for the second system, the poles are at $s = 0$, $s = 1$ and zero is at $s = −3$. The Bode plot, the Nyquist plot, and the Nichols plot for the first system are shown in Figs 9.28, 9.29 and 9.30 respectively. For second system, Bode plot, Nyquist plot, and the Nichols plot are shown in Figs 9.31, 9.32, and 9.33 respectively. It can be observed from Nyquist plot (Figs 9.29 and 9.32) that the first system is stable while the second one is unstable. Also, the response of the first system does not intersect −180° axis (Fig. 9.30) but intersects the gain crossover point (0 dB). The phase margin is the horizontal distance between the gain crossover point and the origin (0 dB, −180°) in degrees. On the other hand, the response of the second system intersects the phase crossover point (−180°) and the gain crossover point (0 dB) as well. The gain margin is the distance between the phase crossover point and the origin. A negative gain or phase margin indicates that the system is unstable.

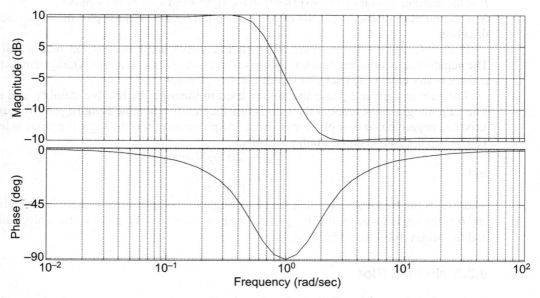

Fig. 9.28 Bode plot of the first system

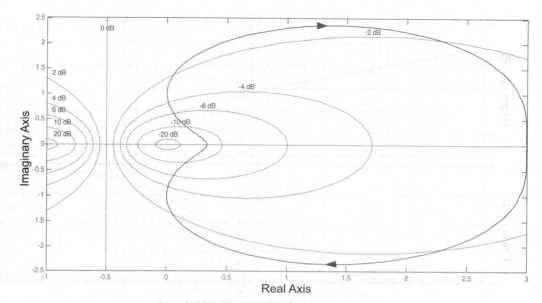

Fig. 9.29 Nyquist plot of the first system

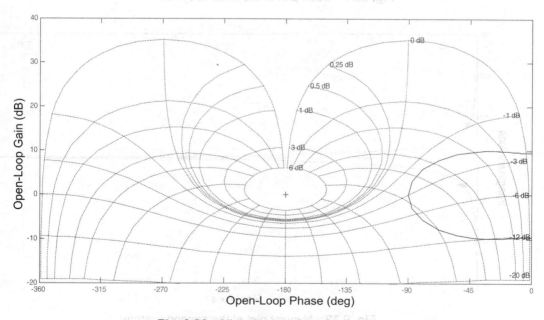

Fig. 9.30 Nichols plot of the first system

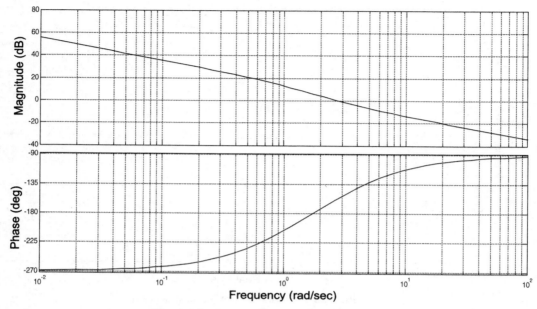

Fig. 9.31 Bode plot of the second system

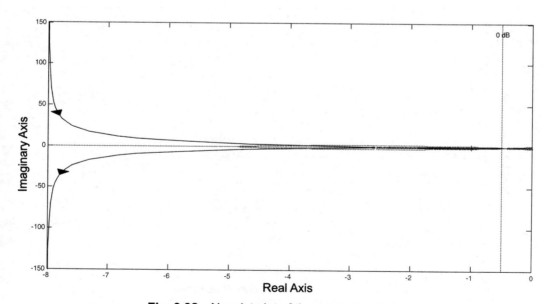

Fig. 9.32 Nyquist plot of the second system

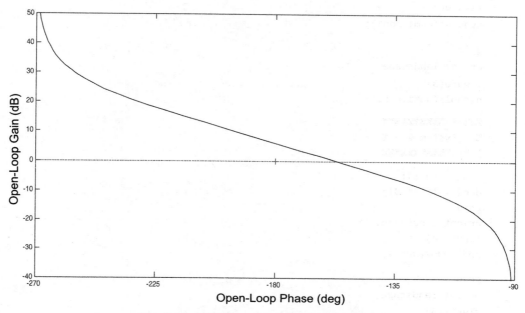

Fig. 9.33 Nichols plot of the second system

```
%%%%%%%%%%%%%%%%%%%%%%%%%%%%%%%%%%%%%%%%%%%%%%%%%%%%%%%%%%%%%%%%%%%%%%%%%%%
% This program plots the Bode plot, Nyquist plot and Nichols plot        %
%                      for the given systems                             %
%%%%%%%%%%%%%%%%%%%%%%%%%%%%%%%%%%%%%%%%%%%%%%%%%%%%%%%%%%%%%%%%%%%%%%%%%%%

clc;
clear all;
clf;

%%%%%%%%%%%%%%%%%
%  System 1  %
%%%%%%%%%%%%%%%%%

num1 = [1 2 3];
den1 = [3 2 1];

grid
orient landscape;
figure(1)
bode(num1,den1);

grid
orient landscape
```

```
figure(2)
nyquist(num1,den1);

grid
orient landscape
figure(3)
nichols(num1,den1);

%%%%%%%%%%%%%%%%%
%   System 2   %
%%%%%%%%%%%%%%%%%

num2 = [2 6];
den2 = [1 -1 0];
grid
orient landscape;
figure(4)
bode(num2,den2);

grid
orient landscape
figure(5)
nyquist(num2,den2);

grid
orient landscape
figure(6)
nichols(num2,den2);
```

9.4 INTRODUCTION TO ELECTRICAL MACHINES

An electrical machine is an electromechanical energy conversion device that converts electrical energy into mechanical energy or vice versa, e.g., motor and generator. The transformers are also included in the family of electric machines as they utilize electromagnetic phenomena. All rotating electric machines have an equivalent linear electric machine where the stator moves along a straight line instead of rotating. Electric machines can be broadly classified as DC machines and AC machines. The DC machines are the electromechanical energy converters which work from a DC power source and generate mechanical power (motor action) or convert mechanical power into DC power (generator action), whereas the AC machines take power from a AC source and generate mechanical power (motor action) or convert mechanical power into AC power (generator action). Owing to their simple control structure, the DC machines were mostly used in closed-loop controlled drives in the past. With the recent development of efficient power electronic converters, AC machines are now used for closed-loop controlled drives as the cost of their maintenance is quite low because of the absence of a commutator. In the following sections, various types of electric machines are analyzed.

9.5 DC MACHINES

A DC machine is an electromechanical energy conversion device that works on the principle of electromagnetic induction. A DC machine mainly consists of an armature, field winding, commutator, and brush gear. Armature is the rotating part of a DC machine which rotates in the magnetic field. It consists of a core and windings, on the core, and is separated by a small distance from the field poles in order to avoid any rubbing inside the machine. The field windings are used for generating uniform magnetic field in which the armature rotates. If the permanent magnets are used for generating the magnetic field in the machine, it is called permanent magnet DC machine. A commutator is used to convert alternating voltage into direct voltage. It also provides electrical connection between the rotating armature windings and the brush gear. The brush gear collects current from the commutator and supplies it to the external load. It consists of brushes, brush holders, brush studs, brush rocker, and current collecting bus bars. In a DC machine, the effect of the magnetic field produced by the armature current on the flux of the main poles of the DC machine is called as armature reaction. The flux at the trailing pole tip strengthens and at the leading pole tips weakens due to the armature reaction in case of DC generators and the effect is just opposite in the case of a DC motor. The EMF generated in a DC generator is given by

$$\text{EMF} = (\phi \, PN/60) \times (Z/A)$$

where

P is the number of poles

ϕ is the flux per pole (Wb),

Z is the total number of armature conductors

N is the armature speed in rpm

A is the number of parallel paths in armature

In case of lap winding, $A = P$ and for wave winding, $A = 2$. This EMF is equal to the back EMF in case of a DC motor, i.e., $V = E_b + I_a R_a$ or $I_a = (V - E_b)/R_a$, where I_a is the armature current, R_a is the armature resistance, V is the terminal voltage, and E_b is the back EMF of the motor.

The DC machines based on the method of excitation can be classified as separately excited DC machines and self-excited DC machines. In separately excited DC machines, the field magnets are energized from an independent external source whereas in self-excited DC machines, the field magnets are energized by the current produced by the supply (in case of motor) or generator themselves. Self-excited DC machines are further divided, in accordance with how the field winding is connected in the machine, as DC series machine, DC shunt machine, and DC compound machine.

9.5.1 Separately Excited DC Machines

A separately excited DC generator is shown in Fig. 9.34. The field winding of this generator is energized by a DC battery. The field current is I_f, armature current is I_a, and the load current is

I. Let E_g be the EMF generated by the generator and V be the load voltage. The load current is same as the armature current. Thus,

$$I = I_a$$

and

$$V = E_g - IR_a$$

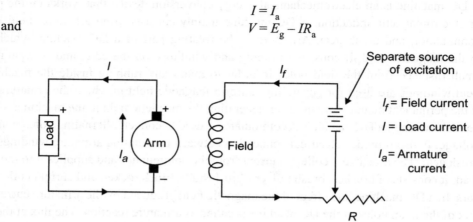

Fig. 9.34 Separately excited DC generator

In case of separately excited DC motors, $V = E_a + IR_a$. The speed of a DC motor can be given by $\omega = (V - IR_a)/K\phi$, where K is the machine constant and ϕ is the flux per pole. So, the speed of the DC motors can be varied by varying the armature resistance (which varies I_a) or by varying the field resistance (which varies the flux per pole). The armature resistance control method consists of a variable resistance connected in series with the armature as shown in Fig. 9.35. The speed of the motor can be reduced to any desired value depending on the armature resistance. The field current remains unaffected as it is connected to a separate voltage source. The 'subsystem' for varying the armature series resistance of a separately excited DC motor is shown in Fig. 9.36. As the circuit breakers connected in parallel are turned ON, the resistances connected in parallel to them gets short circuited. Initially, all the CBs are OFF and are turned ON one by one as can be seen from Fig. 9.36. The supply given to the armature of the DC motor is 180 V and to the field winding is 100 V. The initial speed of the motor is taken as 1 rad/s and is fed back to the motor torque (input terminal T_L) by multiplying it by a gain of 0.405. When the 'ideal switch' is turned ON at 0.01 s, the 180 V DC supply gets connected with the armature of the motor through the series resistances. Initially, all the CBs are OFF so the voltage supplied to the armature is low (20 V) as can be seen in Fig. 9.37. After 6 s when all the CBs are turned ON, the voltage at the armature terminal become 180 V. The speed of the motor in rad/s, armature current in A and electromagnetic torque developed in Nm are shown in Fig. 9.38. The graphs of armature current (I_a) versus speed (ω), and speed (ω) versus electromagnetic torque (T_e) are shown Figs 9.39 and 9.40 respectively. If the mechanical torque input, i.e., T_L, to the motor is negative it acts as a generator.

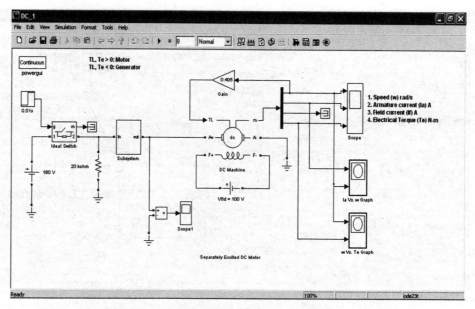

Fig. 9.35 Separately excited DC motor armature control

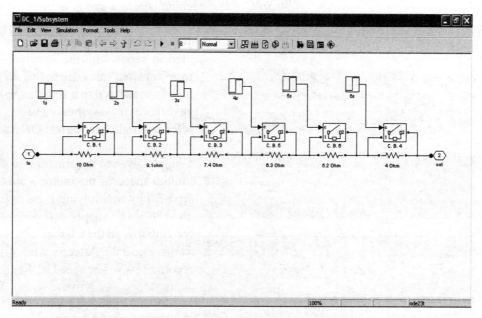

Fig. 9.36 Subsystem for varying the armature resistance

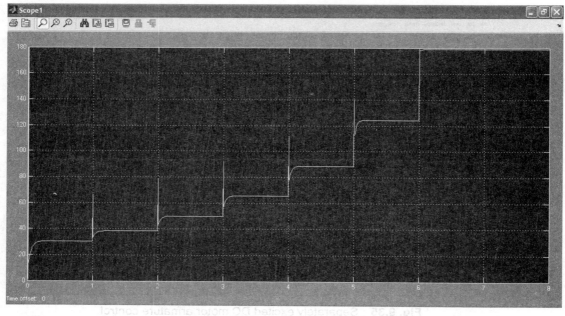

Fig. 9.37 Armature voltage

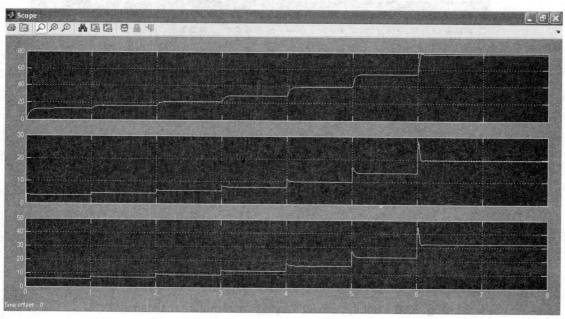

Fig. 9.38 Motor speed (rad/s), armature current (A), and electromagnetic torque (Nm) of the motor

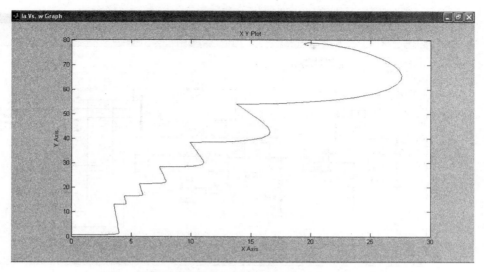

Fig. 9.39   Graph of armature current vs speed

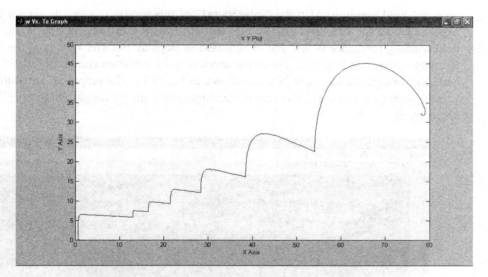

Fig. 9.40   Graph of speed vs electromagnetic torque

In field control method of speed control, a variable resistance is connected in series with the field winding of the motor as shown in Fig. 9.41. As the field resistance increases, the field current decreases with a consequent reduction in flux which is inversely proportional to the motor speed. Thus, as the field resistance increases, the speed of the motor also increases. This method of speed control does not depend on the motor load and therefore permits the remote control of the motor speed. The different parameters of the model shown in Fig. 9.41 are same as that of Fig. 9.35, except that the series resistances are now connected in series

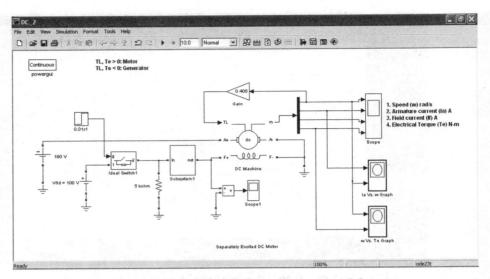

Fig. 9.41 Field control of separately excited DC motor

with the field winding. Also the 'subsystem1' of this model is same as shown in Fig. 9.36, expect that all the CBs are initially ON and are turned OFF one by one. Thus, initially the field winding resistance is low and is increased in steps of 1 s. The voltage at the field winding terminal is shown in Fig. 9.42. The motor speed in rad/s, armature current in A, field current in A, and electromagnetic torque in Nm are shown in Fig. 9.43. The graphs of armature current (I_a) versus speed (ω), and speed (ω) versus electromagnetic torque are shown in Figs 9.44 and 9.45 respectively.

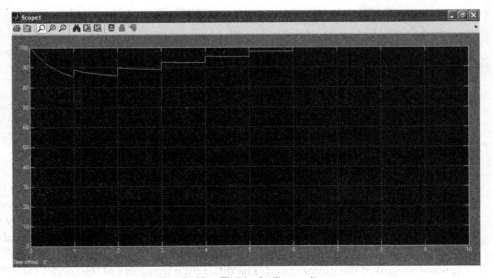

Fig. 9.42 Field winding voltage

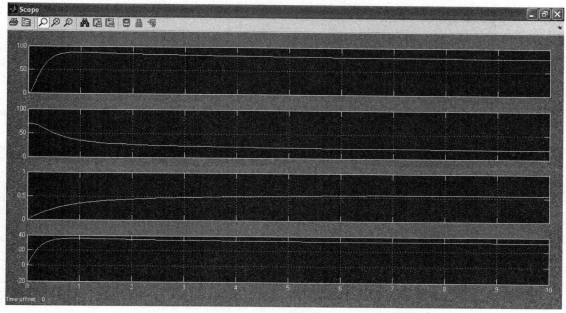

Fig. 9.43   Speed (rad/s), armature current (A), field current (A),
and electromagnetic torque (N m)

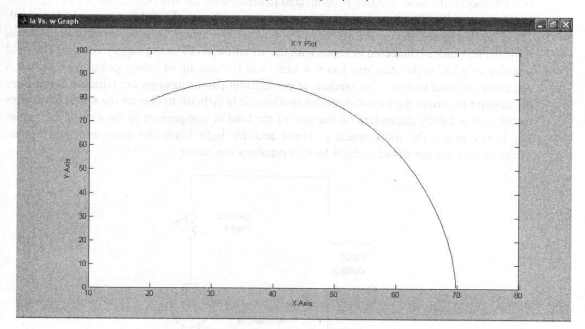

Fig. 9.44   Armature current vs speed

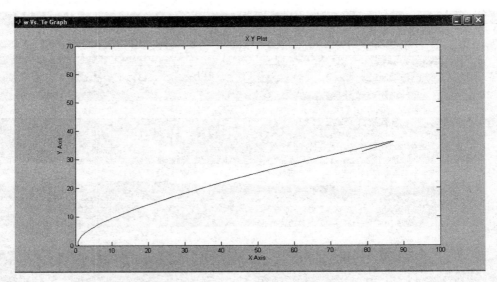

Fig. 9.45 Speed versus electromagnetic torque

9.5.2 DC Series Motor

In a DC motor, the field winding is connected in series with the armature as shown in Fig. 9.46. In this motor, the armature current (I_a), load current (I), and the field current (I_f) are equal. Thus, for this motor, the back EMF $E_b = V - I_a(R_a + R_{se})$, where R_a is the armature resistance, R_{se} is the resistance connected in series with the armature, and V is the supply voltage. The field winding of a DC series machine has few turns and is made up of heavy gauge wire so that it can carry the load current. The amount of current that passes through the winding determines the amount of torque the motor shaft can produce. It is difficult to control the speed of a series motor as it is totally dependent on the size of the load in comparison to the size of the motor. For heavy loads, the rotor speed is lower and for light loads the rotor speed is higher. These motors are not stated without load as runaway can occur.

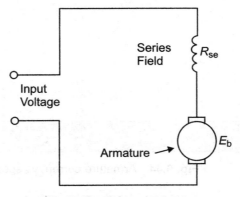

Fig. 9.46 DC series motor

The model for speed control of DC series motor by armature resistance control is shown in Fig. 9.47. The block parameters for this motor are $R_a = 11.2\ \Omega$, $L_a = 0.1215$ H, $R_f = 2.813\ \Omega$, $L_f = 56$ H, $L_{af} = 1.976$ H, total inertia = 0.02215 kg m^2, viscous friction = 0.002953 N m, coulomb friction torque = 0.5161 N m, and initial speed = 3 rad/s. The subsystem block contains series resistances as shown in Fig. 9.48. The field winding is connected in series of the motor so that it acts as a series motor as can be seen from Fig. 9.47. The voltage at the armature terminal rises gradually as shown in Fig. 9.49. Figure 9.50 shows the motor speed in rad/s, armature current in A, field current in A, and electromagnetic torque of the motor in N m. The armature current versus speed and armature current versus electromagnetic torque graphs are shown in Figs 9.51 and 9.52 respectively.

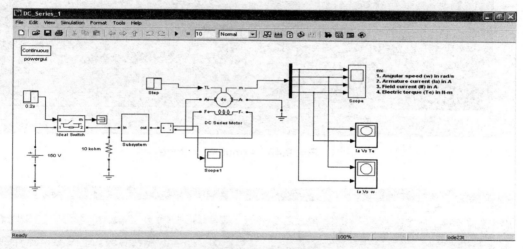

Fig. 9.47   DC series motor

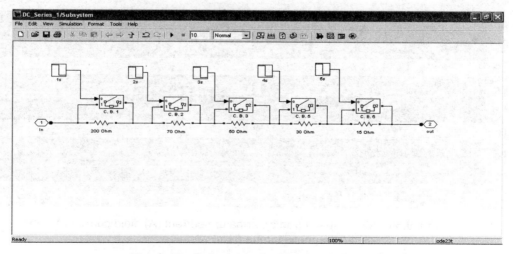

Fig. 9.48   Subsystem for variable resistance

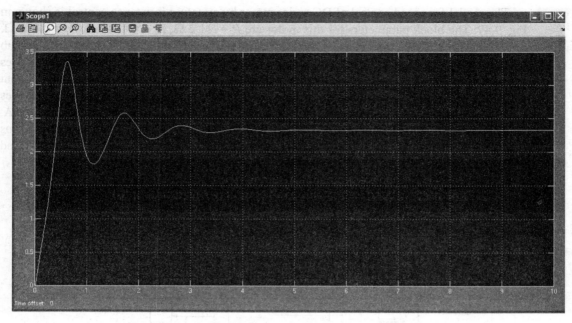

Fig. 9.49 Armature voltage

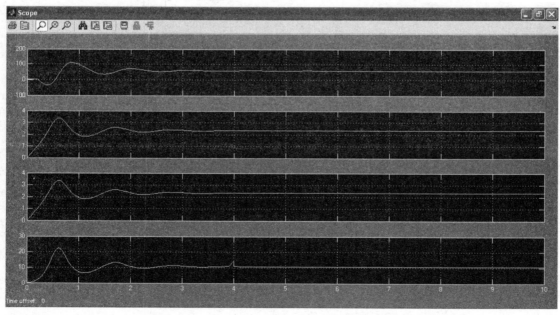

Fig. 9.50 Motor speed (rad/s), armature current (A), field current (A), and electromagnetic torque (N m)

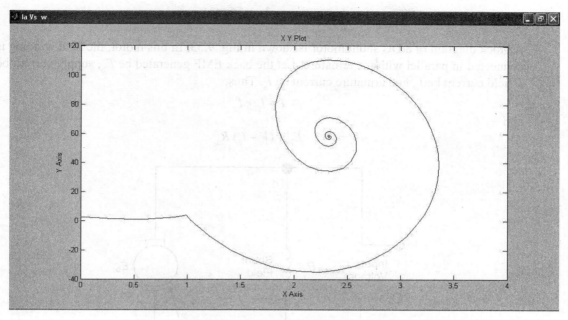

Fig. 9.51 Armature current vs speed

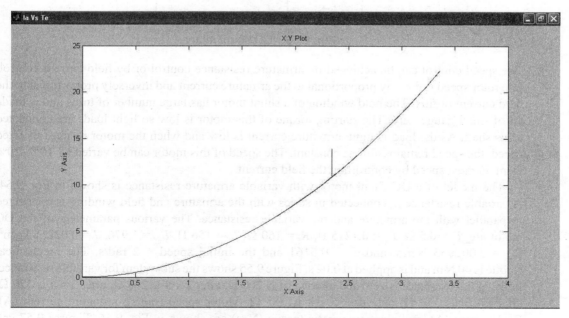

Fig. 9.52 Armature current vs electromagnetic torque

9.5.3 DC Shunt Motor

A block diagram of a DC shunt motor is shown in Fig. 9.53. In this motor, the field winding is connected in parallel with the armature. Let the back EMF generated be E_b, supply current be I, field current be I_f, and armature current be I_a. Thus,

$$I = I_f + I_a$$

and

$$E_b = (V - I_a)\, R_a$$

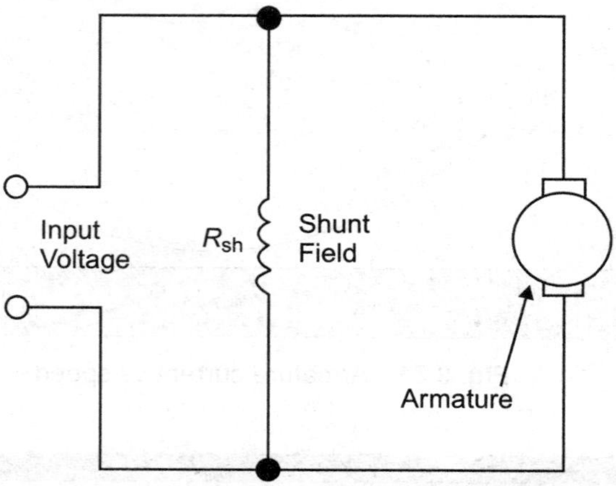

Fig. 9.53 DC shunt motor

The speed control can be achieved by armature resistance control or by field current control. The rotor speed is directly proportional to the armature current and inversely proportional to the field current or flux. The field winding of a shunt motor has large number of turns and is made up of small gauge wire. The starting torque of this motor is low so light loads are connected to its shaft. As the load is light, armature current is low and when the motor reaches its rated speed, the speed remains almost constant. The speed of this motor can be varied by 10%–20% from its rated speed by controlling the field current.

The model of a DC shunt motor with variable armature resistance is shown in Fig. 9.54. A variable resistance is connected in series with the armature and field winding is connected in parallel with the armature and the variable resistance. The various parameters of the DC motor are $R_a = 0.5\ \Omega$, $L_a = 0.1215\ \text{H}$, $R_f = 160\ \Omega$, $L_f = 156\ \text{H}$, $L_{af} = 1.976$, $J = 0.02215\ \text{kg}\,\text{m}^2$, $B_m = 0.002953\ \text{N m s}$, and $T_f = 0.5161$ and the initial speed $= 2$ rad/s. The mechanical torque is 40 N m and is applied at 0.02 s. Figure 9.55 shows the subsystem for variable resistance. Initially, the resistance is zero and after 1 s it is 200 Ω, after 2 s it is 270 Ω, after 3 s it is 320 Ω, after 4 s it is 350 Ω, and after 5 s it is 365 Ω. Motor speed (rad/s), armature current (A), field current (A), and electromagnetic torque (N m) are shown in Fig. 9.56. Figures 9.57 and 9.58 show the armature current versus speed and armature current versus electromagnetic graphs respectively.

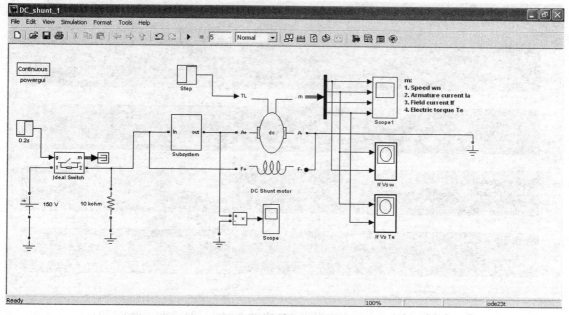

Fig. 9.54 Speed control of DC shunt motor

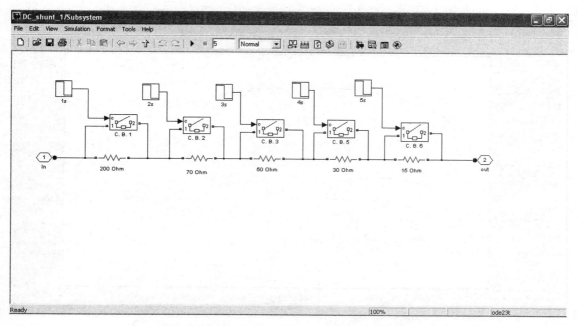

Fig. 9.55 Subsystem for variable resistance

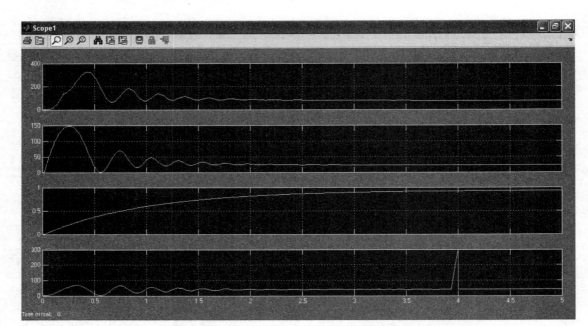

Fig. 9.56 Motor speed (rad/s), armature current (A), field current (A), and electromagnetic torque (N m)

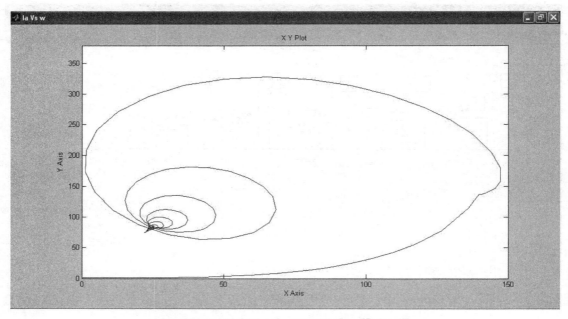

Fig. 9.57 Armature current vs speed curve

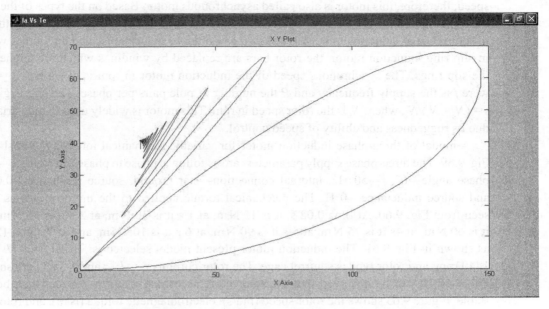

Fig. 9.58 Armature current versus electromagnetic torque curve

9.6 AC MOTORS

An AC motor is driven by an AC supple. It has two main parts—stator and rotor. The stator of the motor produces rotating magnetic field and the rotor provides the necessary torque required to drive the mechanical load. Primarily, there are two types of AC motors—induction motor and synchronous motor—depending on the type of rotor used. In this section, the dynamics of both these motors are analyzed.

9.6.1 Three-phase Induction Motor

A three-phase induction motor is a singly excited AC machine. The stator winding of this motor is connected to the supply whereas the rotor winding receives energy from the stator by induction. An induction motor can be considered as an electric transformer whose magnetic circuit is separated by an air gap between two relatively movable parts, one carrying the primary winding (stator) and the other the secondary winding (rotor). When the AC supply is connected to the primary winding, it induces opposing currents in the secondary winding if it is closed by external impedance. The relative motion is produced between the primary and the secondary windings by the electromagnetic force. This machine can be distinguished from other types of machines because its secondary currents are produced solely by induction as in a transformer instead of being supplied by a DC exciter or other external sources as in the case of synchronous or DC machines. The speed of the rotor is always less than the synchronous speed; therefore, this motor is also called asynchronous motor. Based on the types of the rotors, the induction motor is mainly of two types—squirrel cage and slip ring. The bars of the rotor in squirrel cage induction motor have skewing in order to reduce noise and harmonics whereas in slip ring induction motor, the rotor bars are replaced by windings which are connected to the slip rings. The synchronous speed of the induction motor in rpm is given by $N_s = 60f/P$, were f is the supply frequency and P the number of pole pairs per phase. The slip is given by $s = (N_s - N_r)/N_s$, where N_r is the rotor speed in rpm. This motor is widely used in industrial drive due its ruggedness and ability of speed control.

A model of three-phase induction motor for variable mechanical torque (T_m) is shown in Fig. 9.59. The three-phase supply parameters are as follows: phase to phase rms voltage—400 V, phase angle—0°, f—50 Hz, internal connection—star ground, source resistance—0.034 Ω, and source inductance—0 H. The mechanical torque applied to the motor varies as can be seen from Fig. 9.60. At time 0.02 s, it is 15 Nm, at 1 s it is 30 Nm, at 2 s it is 45 Nm, at 3 s it is 60 Nm, at 4s it is 75 Nm, at 5 s it is 90 Nm, at 6 s it is 105 Nm, and at 7 s it is 120 Nm as shown in Fig. 9.61. The induction motor present model selected is: 10 HP, 400 V, 50 Hz, 1440 rpm, and rotor type is squirrel cage. The rotor currents I_{r_a}, I_{r_b}, and I_{r_c} (A) are shown in Fig. 9.62. It can be observed that the rotor currents increases up to 7 s and then becomes stable. Figure 9.63 shows the rotor speed (rad/s), electromagnetic torque (Nm), and rotor angle (rad). The rotor speed decreases slightly and electromagnetic torque increases slightly with increase in mechanical torque. The rotor speed versus electromagnetic torque curve is shown in Fig. 9.64.

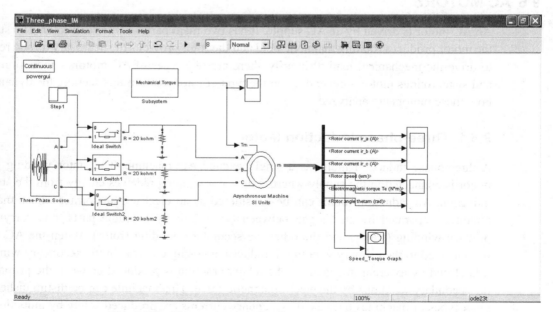

Fig. 9.59   Three-phase induction motor with variable torque

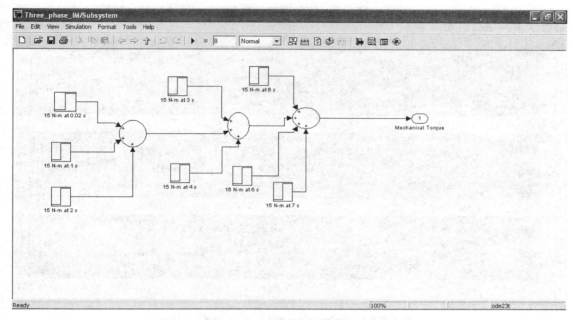

Fig. 9.60   Subsystem for variable mechanical torque

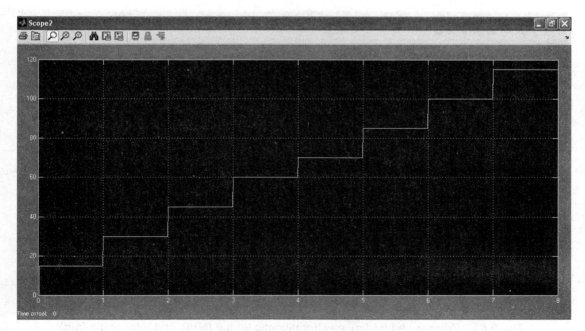

Fig. 9.61   Mechanical torque (N m)

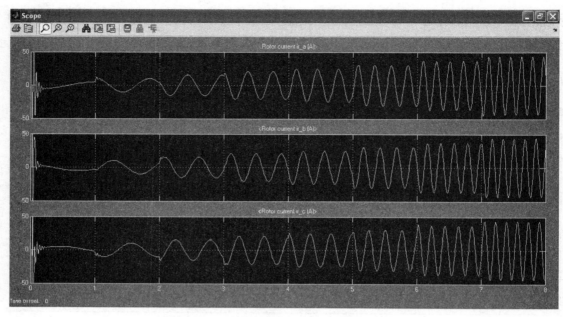

Fig. 9.62 Rotor current I_{r_a}, I_{r_b}, and I_{r_c} (A)

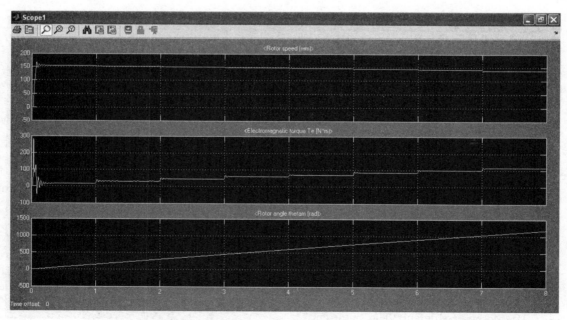

Fig. 9.63 Rotor speed (rad/s), electromagnetic torque (N m), and rotor angle (rad)

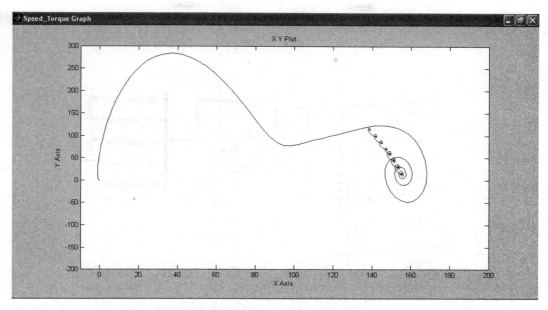

Fig. 9.64 Speed vs torque curve

9.6.2 Three-phase Synchronous Motor

A synchronous motor operates synchronously with the line frequency. In a synchronous motor, the three-phase AC supply to the stator produces the rotating magnetic field around the rotor. The rotor is energized with a DC supply and therefore acts as a bar magnet. The rotating magnetic field attracts the rotor magnetic field and the rotor rotates. A drawback of synchronous motor is that it is not self-starting. It has torque only when it is running at synchronous speed. A squirrel cage winding is added to its rotor which causes it to start. The speed of a synchronous motor is given by $\omega = 120f/n$, where ω is the rotor speed in rpm, f is the supply frequency in Hz, and n is the number of magnetic poles. These motors are used for power factor correction as synchronous condensers, electric clocks, and in applications where constant speed is required.

The model for a three-phase synchronous motor is shown in Fig. 9.65. In this model, two three-phase voltage sources are taken. The first voltage source is of frequency 50 Hz and the second (bottom one) is of frequency 100 Hz. The other parameters of these sources are same and are as follows: phase-to-phase rms voltage—400 V, phase angle of phase A—0°, internal connection—star neutral, source resistance—0.034 Ω, and source inductance—0 H. First source supplies power to the motor from 0 to 1 s, and the second source supplies power from 1 to 2 s (simulation stop time). The mechanical power supplied to the motor is –100 W and phase-to-phase rms internal voltage is 300 V. The parameters of the three-phase synchronous motor are by default. The stator currents are shown in Fig. 9.66, and rotor speed (rad/s) and electrical power (W) are shown in Fig. 9.67.

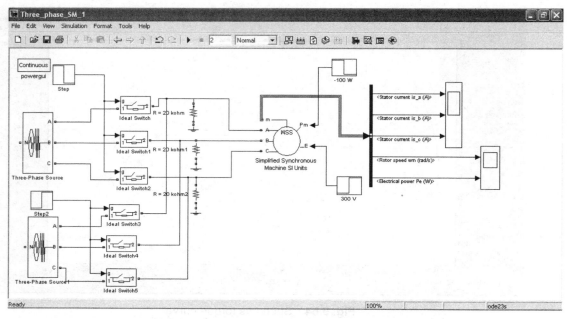

Fig. 9.65   Three-phase synchronous motor

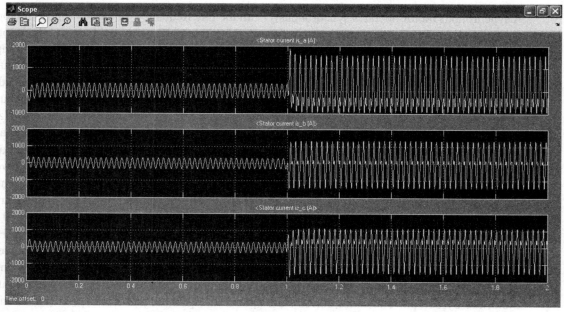

Fig. 9.66   Stator currents I_{r_a}, I_{r_b}, and I_{r_c} (A)

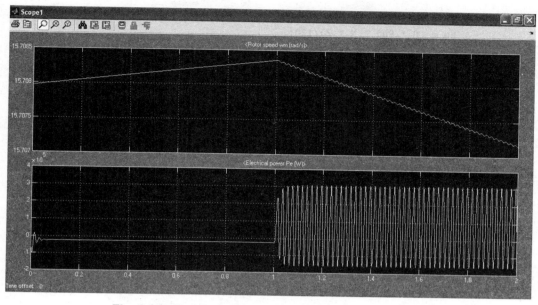

Fig. 9.67 Rotor speed (rad/s) and electrical power (W)

9.7 SIMULATION PROJECTS

Project 1

MATLAB program for computing back EMF and armature current for a DC motor

```
%%%%%%%%%%%%%%%%%%%%%%%%%%%%%%%%%%%%%%%%%%%%%%%%%%%%%%%%%%%%%%%
% MATLAB program to compute the back EMF and armature current %
%                       for a DC motor                        %
%%%%%%%%%%%%%%%%%%%%%%%%%%%%%%%%%%%%%%%%%%%%%%%%%%%%%%%%%%%%%%%
clc;
clear all;

P = input('Number of poles are: ');
Phi = input('Flux per pole in Wb is: ');
Z = input('Total number of armature conductors are: ');
N = input('Armature speed in rpm is: ');
V = input('Supply voltage in volts is: ');
Ra = input('Armature resistance in Ohms is: ');
W = input('1 for lap wining and 2 for wave winding: ');
if W == 1
    Eb = (Phi * N * Z) / 60;
end
```

```
if W == 2
    Eb = (P * Phi * N * Z)/120;
end

Ia = (V - Eb)/ Ra;

disp('Back EMF developed in the motor in volts is:'), Eb
disp('Armature current in A is: '); Ia
```

The output of the program for a set of parameters is as follows:

```
Number of poles are: 4
Flux per pole in Wb is: 0.025
Total number of armature conductors are: 100
Armature speed in rpm is: 700
Supply voltage in volts is: 230
Armature resistance in Ohms is: 12
1 for lap wining and 2 for wave winding: 2
Back EMF developed in the motor in volts is:

Eb =

   58.3333

Armature current in A is:

Ia =

   14.3056
```

Project 2

MATLAB program for converting transfer function to state space and vice versa

```
%%%%%%%%%%%%%%%%%%%%%%%%%%%%%%%%%%%%%%%%%%%%%%%%%%%%%%%%%%%%%%%%%%
% MATLAB program to convert transfer function to state space and  %
%                        vice versa                               %
%%%%%%%%%%%%%%%%%%%%%%%%%%%%%%%%%%%%%%%%%%%%%%%%%%%%%%%%%%%%%%%%%%

clc;
clear all;

%%%%%%%%%%%%%%%%%%%%%%%%%%%%%%%%%%%%%%%%%%%%%%%%%%%%%%%%%%%%%%%%%%
% T. F. = num / den = (26.02 s^2 + 37 s + 1)/             %
%          (s^4 + 6.023 s^3 + 7.02 s^2 + 3.01 s + 10.31)  %
%%%%%%%%%%%%%%%%%%%%%%%%%%%%%%%%%%%%%%%%%%%%%%%%%%%%%%%%%%%%%%%%%%

num = [26.02 37 1];
den = [1 6.023 7.02 3.01 10.31];
```

```
%%%%%%%%%%%%%%%%%%%%%%%%%%%%%%%%%%%%%%%%%%%%
%    State space x' = A x + Bu; y = C x + D u    %
%%%%%%%%%%%%%%%%%%%%%%%%%%%%%%%%%%%%%%%%%%%%

[A B C D] = tf2ss(num,den) % Transfer function to state space

[num den] = ss2tf(A, B, C, D) % State space to transfer function
```

The output is as follows:

```
A =

    -6.0230    -7.0200    -3.0100    -10.3100
     1.0000          0          0           0
          0     1.0000          0           0
          0          0     1.0000           0

B =

     1
     0
     0
     0

C =

     0    26.0200    37.0000     1.0000

D =

     0

num =

     0     0.0000    26.0200    37.0000     1.0000

den =

     1.0000     6.0230     7.0200     3.0100     10.3100
```

SUMMARY

The main points of this chapter can be summarized as follows:

- Control system engineering is based on the foundations of feedback theory and linear and nonlinear system analysis, and it integrates the concepts of network theory and communication engineering.
- Depending upon the presence of feedback, the control systems may be classified as open-loop-control and closed-loop control systems.

- The time response of a control system includes its transient response and steady-state response. The commonly used test signals for time response analysis are step signal, impulse signal, ramp signal, exponential signal, and sinusoidal signal.
- Time response specifications for a control system are delay time, rise time, peak time, peak overshoot, settling time, and steady-state error.
- A system is undamped if $\xi = 0$, under-damped if $0 < \xi < 1$, critically damped if $\xi = 1$, and over-damped if $\xi > 1$.
- Root locus is a graphical technique used for examining how the roots of a system vary with the gain of the system.
- A Bode plot is a combination of two plots, one of log of frequency versus magnitude in dB and second of log of frequency versus phase angle.
- Nyquist stability criteria states that if the Nyquist plot of system encircles the $(-1 + j0)$ point in the counterclockwise direction as many times as the number of right-hand side poles $G(s) H(s)$ in the s-plane, the system is stable.
- Nichols chart is similar to Nyquist plot but has the gain on log scale and phase on linear scale with origin at $(0 \text{ dB}, -180°)$ and is symmetrical about the $-180°$ axis.
- DC machines can be classified based on the method of excitation as separately excited and self-excited DC machine. The self-excited DC machines can further be divided based on the connection of field winding as series, shunt, and compound machines.
- An induction motor can be considered as an electrical transformer whose magnetic circuit is separated by an air gap between two relatively movable parts, one carrying the primary winding, i.e., stator, and the other the secondary, i.e., rotor winding.
- The rotor of a synchronous machine rotates synchronously with the line frequency.

REVIEW QUESTIONS

1. What do you mean by an electrical machine? Differentiate between an electric motor and an electric generator.
2. Differentiate between self inductance and mutual inductance.
3. On what factors do Eddy current and hysteresis losses depend in a transformer?
4. What are the advantages of the star-connected transformer over the delta-connected transformer?
5. How does the principle of motor differ from that of motor?
6. What is the significance of back EMF in DC motors?
7. Classify the different types of DC motors.
8. Explain armature reaction in a DC motor and its effects.
9. Draw the speed–torque characteristics of DC series, shunt, and compound motors.
10. Define a physical system.
11. What do you mean by a control system?
12. What is a mathematical model of a system?

13. Define a linear time invariant system model.
14. Classify different types of control systems.
15. What are the advantages of feedback control system?
16. What do you mean by negative and positive feedback?
17. Why is negative feedback preferred in the control system?
18. Define open-loop and closed-loop systems.
19. Distinguish between linear and nonlinear control systems.
20. Define continuous time and discrete time control systems.
21. What is the effect of feedback on overall gain, stability, external disturbance, and sensitivity of the system?
22. What do you mean by multivariable control system?
23. Define a transfer function.
24. What do you mean by time response of a control system?
25. Mention the standard test signals used in control systems.
26. What is the significance of characteristic equation of a system?
27. Define impulse response, step response, and ramp response of a system.
28. Define the term damping coefficient of a control system. How is control system classified depending on the damping coefficient?
29. Define undamped natural frequency and damped frequency of a control system.
30. Define delay time, rise time, peak time, peak overshoot, settling time, and steady-state error.
31. Distinguish between type and order of a control system.
32. What is a PID controller and where is it used?
33. Define a root locus and state its applications.
34. Mention the advantages of root locus technique.
35. What are the conditions to be satisfied for the root locus to exist at any point in the s-plane?
36. What is the effect of adding poles and zeros to the open-loop function on the root locus?
37. Define frequency response of a system. What are the advantages of frequency response analysis?
38. What do you mean by phase margin and gain margin?
39. How will you determine the gain margin and phase margin of a system from its Bode plot?
40. What are the drawbacks of Nyquist stability criterion? Mention the advantages of Nyquist plot.
41. What are constant M and N circles?
42. What is a Nichols chart and what are its advantages?
43. Distinguish between conventional and modern control theory.
44. Define state-space equations.
45. Define observability and controllability of a system.

EXERCISES

1. Write a MATLAB program to obtain the unit impulse, unit step, and unit ramp response of the system whose transfer function is as follows:
$C(s)/R(s) = 100/(s^2 + s + 0.5)$

2. Obtain the unit step, unit impulse, and unit ramp response of the following state space system by using MATLAB: $A = [-2 \ -1; \ 3 \ 0]$, $B = [0.3; \ 0]$, $C = [1 \ 0]$, and $D = [0]$.

3. Construct a Simulink model for a second-order undamped system, under-damped system, critically damped system, and over-damped system, and view their unit step and unit impulse response on the Scope.

4. The open-loop transfer function of a unity feedback control system is $16/(s^2 + 5s)$. By using derivative control, the damping coefficient is made 0.85. Determine the value of delay time, rise time, peak time, and maximum overshoot with and without derivative control by using MATLAB. Also plot the response of both the systems for unit step signal.

5. Plot the root locus of the system $G(s) \ H(s) = K(s + 3)/s(s + 1)$ by MATLAB and also compute the range of values for the system to be under-damped.

6. Plot the root locus of the control system whose open-loop transfer function is $G(s)H(s) = Ke^{-0.5s}/s(s + 1)$ by using MATLAB.

7. Construct the Nyquist plot for the closed-loop system whose open-loop transfer function is $G(s)H(s) = K(s + 3)/s^2(s + 1)$ and determine the range of values of K for which the system is stable.

8. Construct the Nyquist plot for the closed-loop system whose open-loop transfer function is $G(s)H(s) = K(s - 4)/(s + 1)^2$ by using MATLAB and determine whether the system is stable or not.

9. Construct a MATLAB program to plot Nichols chart of the system $G(s)H(s) = s/(s + 1)(s + 2)$ and determine the gain margin and phase margin of the system.

10. Determine the EMF generated by a 6 pole wave wound armature having 50 slots with 18 conductors per slot driven at 1,000 rpm by using MATLAB. The flux per pole is 20 mWb.

11. Determine the speed in rpm of a DC generator for 220 V EMF, 80 slots, 10 conductors per slot, lap wound using MATLAB. If the EMF required is 400 V, compute the speed again in rpm.

12. Construct a model of a DC shunt motor for field control of the speed. The mechanical torque on the motor is 100 Nm and the input supply voltage is 300 V. Plot the graph of field current versus speed.

13. Develop a model for a three-phase induction motor of 100 HP, 575 V, 60 Hz, and 1780 rpm. The mechanical torque (T_m) on the motor varies from 10 to 100 Nm in 10 s. View and study its rotor currents, speed, and electromagnetic torque on the Scope.

14. Develop a model for a three-phase synchronous motor of 50 Hz , 400 V, 8.1 kVA, and 15,000 rpm. The mechanical power to the machine is constant and is 100 W. The field winding supply voltage varies from 10 to 110 V in 11 s. View the stator currents, rotor speed, electrical power, and electromagnetic torque on the Scope.

SUGGESTED READING

For Control System Engineering

Anderson, B.D.O. and J.B. Moore, *Optimal Control*, Prentice-Hall, New Jersey, 1989.

Athans, M., and P.L. Falb, *Optimal Control: An Introduction to the Theory and Its Applications*, McGraw-Hill, New York, 1965.

Bayliss, L.E., *Living Control Systems*, English Universities Press Limited, London, 1966.

Bode, H.W., *Network Analysis and Feedback Design*, Van Nostrand Reinhold, New York, 1945.

Choudhury, D. Roy, *Modern Control Engineering*, Prentice-Hall of India, New Delhi, 2005.

Drof, R.C., *Modern Control Systems*, 6th Ed., Addison-Wesley Publishing Company, Reading, MA, US, 1992.

Friedland, B., *Control System Design*, McGraw-Hill, New York, 1986.

Hahn, W., *Theory and Application of Liapunov's Direct Method*, Prentice Hall, New Jersey, 1963.

Kuo, B.C., *Automatic Control Systems*, 3rd Ed., Prentice Hall, New Jersey, 1975.

Nagrath, I.J. and M. Gopal, *Control Systems Engineering*, 4th Ed., New Age International Publishers, New Delhi, 2005.

Ogata, K., *Discrete-Time Control Systems*, 2nd Ed., Prentice Hall, New Jersey, 1995.

Ogata, K., *Modern Control Engineering*, 3rd Ed., Prentice Hall, New Jersey, 1997.

Ogata, K., *Solving Control Engineering Problems with MATLAB*, Prentice Hall, New Jersey, 1994.

Raven, F.H., *Automatic Control Engineering*, 4th Ed., McGraw-Hill, Singapore, 1987.

For Electrical Machines

Alger, P.L., *Induction Machines*, 2nd Ed., Gordon and Breach, New York, 1970.

Bimbhra, P.S., *Electrical Machinery*, 5th Ed., Khanna Publishers, Delhi, 2000.

Chapman, S.J., *Electric Machinery Fundamentals*, 3rd Ed., McGraw-Hill, Singapore, 1999.

Draper, A., *Electric Machines*, 2nd Ed., Longman, London, 1967.

Guru, B.S. and H.R. Hiziroglu, *Electric Machinery and Transformer*, 3rd Ed., Oxford University Press, New York, 2001.

Kosow, I.L., *Electric Machinery and Transformers*, 2nd Ed., Prentice Hall, New Jersey, 1991.

Kothari, D.P. and I.J. Nagrath, *Electric Machines*, 3rd Ed., McGraw-Hill, New Delhi, 2006.

Lansdrof, E.H., *Theory of Alternating Current Machinery*, McGraw-Hill, New York, 1955.

Leonhard, W., *Control of Electric Drives*, Springer-Verlag, Berlin, Heidelberg, 1985.

McPherson, G., *An Introduction to Electrical Machines and Transformers*, John Wiley, New York, 1981.

Nasar, S.A. and L.E. Unnewehr, *Electromechanics and Electric Machines*, John Wiley, New York, 1979.

Richardson, D.V., *Rotating Electric Machinery and Transformer Technology*, Reston Publishing Co., London, 1978.

Ryff, P.T., D. Platnick, J.A. Karnas, *Electric Machines and Transformers: Principles and Applications*, Prentice Hall, New Jersey, 1987.

Sarma, M.S., *Synchronous Machines*, Gordon & Breach Science Publishers, New York, 1979.

Say, M.G. and E.O. Taylor, *Direct Current Machines*, Addison Wesley, Reading, 1980.

Sen, P.C., *Principles of Electric Machines & Power Electronics*, 2nd Ed., John Wiley & Sons, New York, 1997.

Shanmugasundaram, A., G. Gangadharan, and R. Palani, *Electrical Machine Design Data Book*, Wiley Eastern, New Delhi, 1979.

Veinott, C., *Theory and Design of Small Induction Motors*, McGraw-Hill, New York, 1959.

10

MISCELLANEOUS APPLICATIONS

10.1 INTRODUCTION TO COMMUNICATION SYSTEMS

Communication is the process of conveying data (audio, video, or symbolic) over a distance. The communication systems using the electrical and electronic technology have a significant impact on modern society. This system transmits and receives information over significant distances for the purpose of communication. Telecommunication is defined as the transmission and reception of any messages, signals, or signs through an electromagnetic medium called channel. In earlier days, communication was by audio signals such as coded drumbeats, lung blown horns, loud whistles, or by video signals such as smoke, beacons, signal flags, and optical heliographs over a short distance of few miles. Radios and wireless were used for several years for communication. Alexander Graham Bell patented the telephone in 1876. Rapid developments in long distance radio telephony were made possible by the invention of diode in 1904 (Fleming) and of triode in 1906 (Lee De Forest). The transistor (BJT) by John Bardeen, Walter Houser Brattain, and William Bradford Shockley, which led to the development of integrated circuits, paved the way for compact electronic systems. The rapid development of mobile and personal communication systems was made feasible by the continuous advances in microelectronic circuits. Mobile communication is the widely used technology for message communication.

A basic communication system comprises the following primary units that are present in some form or the other:

1. A transmitter that takes information and modulates it to a transmission signal.
2. A transmission medium which is a physical channel that carries the transmission signal.
3. A receiver that receives the transmission signal and demodulates it to get the usable information.

In a radio broadcasting station, for instance, the transmitter is the power amplifier used to modulate the audio signal. The transmission antenna is the interface between the transmitter and the transmission channel which is air or free space. The receiver's antenna, i.e., antenna of the radio set, is the interface between the channel and the receiver. The radio receiver demodulates the signal and it will be available as sound to the listener.

Modulation is a process of varying one or more parameters, i.e., amplitude, frequency, or phase of a high frequency periodic wave, called as carrier signal in accordance with the instantaneous value of the message signal. So, by modulation we vary the characteristics of the carrier signal. Modulation is done so as to superimpose the low frequency message on a high frequency carrier. Design of receiver and transmitter amplifier circuit becomes easier at higher frequencies. Also, the sizes of the transmitter and the receiver antenna are reduced at higher frequencies. The antenna radiates effectively when its size is of the order of the wavelength of the signal. Modulation is also helpful in multiplexing the channel as signals will not interfere with each other after modulation. The three parameters—amplitude, phase, and frequency—of a carrier signal can be varied. The modulation process in which the amplitude of the carrier wave is varied in accordance with the instantaneous value of the message or modulating signal is called as amplitude modulation (AM). Similarly, the modulating process in which the phase or frequency of the carrier wave is varied in accordance with the instantaneous value of the message or modulating signal is called as phase modulation (PM) or frequency modulation (FM) respectively. On the other hand, demodulation is the process of extracting the original message bearing signal from the modulated carrier wave. There are several ways of demodulation depending on the way the parameters, such as amplitude, phase, or frequency, of the message signal are transmitted in the carrier signal. For instance, an amplitude modulated signal may be demodulated by a synchronous detector.

The bandwidth required for a given transmission should ideally depend on the bandwidth occupied by the message signal. However, the transmitted bandwidth is not exactly the same as the bandwidth of the message signal. If the message signals are sinusoidal in nature, the modulated signal bandwidth is simply the frequency range between the highest and the lowest sine-wave signal. If the message signals are non-sinusoidal, as in any real-life situation, complexity arises while estimating the bandwidth of the modulated signal. While selecting a carrier frequency for a particular application, the most important factor is the frequency band covered by the message signal. Table 10.1 shows different types of radio signals and their typical carrier frequencies.

Table 10.1 Types of applications and their typical carrier frequencies

S. No.	Name of the Radio Signal	Transmission Bandwidth	Typical Carrier Frequency Range
1.	Telegraphy	80 Hz to 2 kHz	18 kHz to 30 MHz
2.	Telephony (AM)	10 kHz	500 kHz to 30 MHz
3.	Telephony Signals (FM)	150 kHz	88 MHz to 108 MHz
4.	Facsimile Signals	6 kHz	500 kHz to 30 MHz

(Continued)

Table 10.1 (*Continued*)

S. No.	Name of the Radio Signal	Transmission Bandwidth	Typical Carrier Frequency Range
5.	Television Signals	6 kHz	54 MHz to 216 MHz
6.	Radar Signals	2 MHz to 10 MHz	200 MHz to 30,000 MHz

10.2 MODULATION AND DEMODULATION TECHNIQUES

10.2.1 Amplitude Modulation and Demodulation

The amplitude modulation is a technique which is used mostly for transmitting information by a radio station. In radio communication, a continuous wave radio frequency signal is amplitude modulated by an audio waveform before transmitting. Suppose the message signal is given by $a_m = A_m \cos(\omega_m t)$, where A_m is the amplitude and ω_m is the angular frequency of the message signal, and the carrier signal is given by $a_c = A_c \cos(\omega_c t)$, where A_c is the amplitude and ω_c is the angular frequency of the carrier signal, also $\omega_c \gg \omega_m$. Now, the amplitude modulated signal $m(t)$ can be obtained mathematically by multiplying the two signals as follows:

$$m(t) = A_c \cos(\omega_c t) \times A_m \cos(\omega_m t)$$

$$= \frac{1}{2} A_c A_m \{\cos(\omega_c + \omega_m)t + \cos(\omega_c - \omega_m)t\}$$

The first term, i.e., $\cos(\omega_c + \omega_m) t$, is called as upper sideband (USB) in frequency domain and the second term, i.e., $\cos(\omega_c - \omega_m) t$ is called as lower sideband (LSB) in frequency domain. Now for demodulation, we can again multiply the modulated signal $m(t)$ by the carrier wave, i.e., for demodulation,

$$m(t) \times A_c \cos(\omega_c t) = A_c^2 \cos^2(\omega_c t) \times A_m \cos(\omega_m t)$$

$$= A_c^2 A_m \frac{1}{2} \{1 + \cos(2\omega_c t)\} \times \cos(\omega_m t)$$

$$= \frac{1}{2} A_c^2 A_m \{\cos(\omega_m t) + \cos(2\omega_c t) \cos(\omega_m t)\}$$

The first term, i.e., $\cos(\omega_m t)$, is the original message signal. Thus, the message signal can be obtained by filtering the above signal through a low pass filter which can pass the signals of frequency up to ω_m. Amplitudes of the components of the signal are not taken into account as their role is insignificant in the process modulation and demodulation. However, they play an important role in determining the efficiency of a communication system. The modulation index $\mu = A_m/A_c$. If $\mu > 1$, it is over-modulation and if $\mu < 1$, it is under-modulation. Figure 10.1 shows the model for amplitude modulation and demodulation. The modulation is done by multiplying the message signal with the carrier signal. For demodulation, the modulated signal is again multiplied by the carrier signal and is passed through a low pass filter as shown in Fig. 10.1. Waveforms of the message signal, the modulated signal, and the demodulated signal are shown in Fig. 10.2.

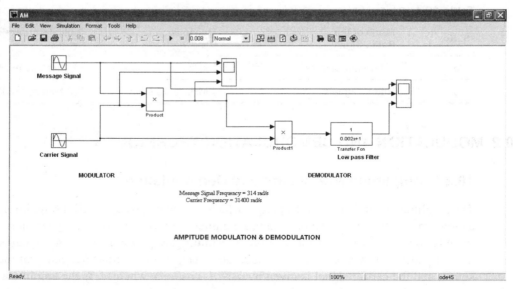

Fig. 10.1 Amplitude modulation and demodulation

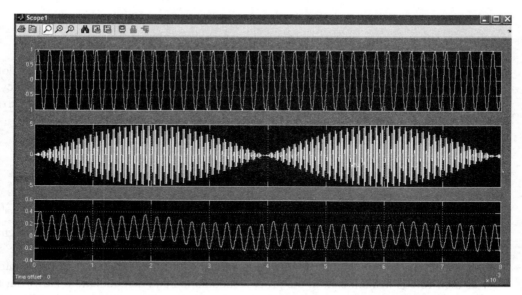

Fig. 10.2 Message signal, amplitude modulated wave, and demodulated signal

In the frequency domain, the amplitude modulation produces a signal with the power concentrated at the carrier frequency and in two adjacent sidebands. Each sideband of the modulated signal is of equal bandwidth and is like a mirror image of the other. The AM which contains two sidebands and the carrier signal is called double sideband amplitude modulation (DSB-AM). This modulation technique is not efficient as two-thirds of the power is concentrated in the carrier signal which contains no useful information. In order to increase transmitter efficiency, the carrier signal can be removed or suppressed. This carrier suppressed amplitude modulated signal is called double sideband suppressed carrier (DSBSC) signal. This scheme is three times more efficient than DSB-AM. Also, the bandwidth of the modulated signal can be reduced at the expense of increased transmitter and receiver complexity by completely suppressing both the carrier and one of the sidebands (LSB or USB). This type of amplitude modulation is called single sideband suppressed carrier (SSBSC). Thus, depending on the carrier and the sidebands, the following are the five different types of AM techniques:

1. **Double sideband (DSB)** In this technique of AM, the carrier signal, along with both the sidebands (LSB and USB), are transmitted. This type of AM is also called as conventional amplitude modulation.

2. **Double sideband suppressed carrier (DSBSC)** In order to increase the efficiency of the transmitter, the carrier signal may be suppressed from the DSB signal. This technique in which the carrier signal is suppressed is called double sideband suppressed carrier.

3. **Single sideband (SSB)** The technique of AM in which only one sideband (LSB or USB) is transmitted is called single sideband modulation. This type of modulation technique is used in shortwave radio or shortwave broadcasting.

4. **Single sideband suppressed carrier (SSBSC)** In this technique of AM, the carrier signal and one of the sideband (LSB or USB) are suppressed before transmitting. This type of modulation is used in amateur radio.

5. **Vestigial sideband (VSB)** In this modulation technique, only one part of the LSB or USB is suppressed along with the carrier signal. In other words, one sideband, say LSB, and the some part of the other sideband, i.e., USB, is transmitted. This type of modulation is used in television broadcasting.

Three types of AM techniques, i.e., SSB-AM, DSB-AM, and DSBSC-AM, are used in the model as shown in Fig. 10.3. The SSB-AM demodulator is used to demodulate the DSB-AM signal. These block sets are available in Communications Blockset of Simulink. The Message Signal is a sine wave of amplitude 5 and frequency 200 rad/s. For SSB-AM block, the USB is selected for modulation and the carrier frequency is taken as 3 kHz. The same carrier frequency, i.e., 3 kHz, is taken for DSB-AM and DSBSC-AM. For demodulation by DSB-AM, the carrier frequency is taken as 3 kHz and Butterworth method for low pass filter design is selected with 'filter order' of 4 and 'cutoff frequency' of 300 Hz. The modulated waves are shown in Fig. 10.4 and the signal obtained after demodulation is shown in Fig. 10.5. It can be observed that the demodulated signal is almost same as the message signal. In the time domain, it is difficult to differentiate between DSB-AM, SSB-AM, and DSBSC-AM signals as can be observed from Fig. 10.4.

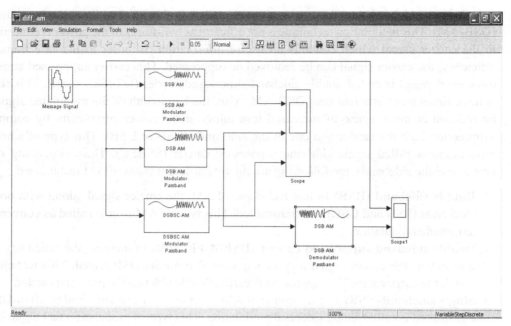

Fig. 10.3 SSB, DSB, and DSBSC amplitude modulation and DSB demodulation

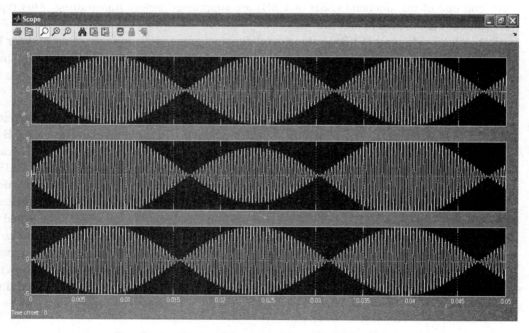

Fig. 10.4 SSB, DSB, and DSBSC modulated waves

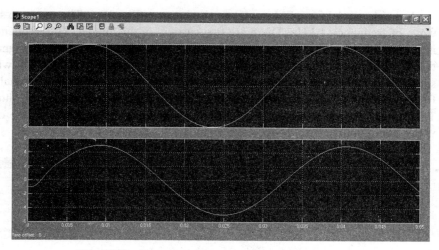

Fig. 10.5 Message signal and demodulated signal

10.2.2 Frequency Modulation and Demodulation

The frequency modulation conveys information through a carrier wave by varying its instantaneous frequency. Suppose the message signal to be transmitted is $m(t)$ and the carrier is a sinusoidal signal $c(t) = A_c \cos(2\pi f_c t)$, where f_c is the carrier base frequency and A_c is the carrier amplitude. The frequency modulator combines the carrier with the baseband message signal to get the transmitted signal:

$$y(t) = A_c \cos\left(2\pi \int_0^t f(\tau)d\tau\right)$$

$$= A_c \cos\left(2\pi \int_0^t [f_c - \Delta f m(\tau)]d\tau\right)$$

$$= A_c \cos\left(2\pi f_c t + 2\pi \Delta f \int_0^t m(\tau)d\tau\right)$$

where $f(t)$ is the instantaneous frequency of the oscillator and Δf is the frequency deviation. The instantaneous frequency of FM signal varies with time. The maximum change in instantaneous frequency from the average frequency, i.e., f_c, is called as frequency deviation. The frequency deviation can be expressed as

$$\Delta f = |K_f f(t)|_{\max}$$

The constant K_f represents the frequency sensitivity of the modulator expressed in Hz/V. The frequency deviation is a useful parameter for determining the bandwidth of the FM signals. If the value of K_f is small, the bandwidth of FM signal is narrow and is called narrowband FM; whereas, if K_f is very large, the FM signal has a wide bandwidth and is called wideband FM. The estimated bandwidth of FM signal is given by

$$BW_{FM} = 2(\Delta f + 2f_m)$$

A simple model for a FM modulator is shown in Fig. 10.6. The message signal is a square wave of frequency 1 kHz. There are two carriers in this model. The frequency of 'Carrier 1' is 12,560 rad/s and that of 'Carrier 2' is 15,700 rad/s. When the message signal is high, i.e., 1, the 'switch' passes 'Carrier 1' signal, else it passes 'Carrier 2' signal. So, when the message signal is '1' the carrier frequency is 12,560 rad/s and when the message signal is '0' the carrier frequency is 15,700 rad/s. The FM waveform and the message signal are shown in Fig. 10.7. This model intends to describe the FM in a simple manner otherwise the carrier frequency varies continuously with the message signal.

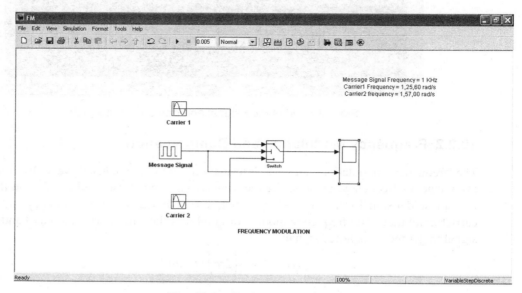

Fig. 10.6   Frequency modulation

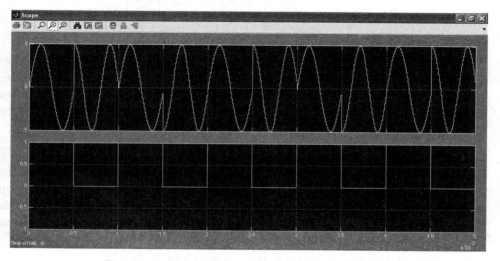

Fig. 10.7   Frequency modulated wave and message signal

The FM generation circuits can be grouped into two categories: direct FM generation and indirect FM generation. The FM signal can be generated directly by varying the frequency of an oscillator (direct generation). A capacitor microphone or a varactor diode can be used in the oscillator circuit. As the capacitance of the microphone or the varactor diode varies with the audio signal, the frequency of the oscillator also varies ($f = 1/2\pi\sqrt{(LC)}$). A reactance modulator circuit can also be used for FM generation. In a reactance modulator, a transistor is made to act like a variable reactance. When the reactance modulator is placed across the LC circuit of the oscillator, its reactance varies with the audio signal and the frequency of the oscillator varies as well. A voltage controlled oscillator (VCO) can also be used for FM generation. The output frequency of a VCO is proportional to the input voltage applied to it. All these methods of direct frequency generation suffer from the serious problem of frequency drift. The second method for FM generation is indirect generation or Armstrong modulation. In Armstrong modulator, first a PM signal is generated and is converted into a FM signal later. Mathematically, it can be shown that PM using the integral of the message signal is identical to FM using the audio signal itself and thus a FM signal is generated.

The following MATLAB program performs frequency modulation and demodulation. In this MATLAB program, the frequency of the message signal is taken as 1 kHz and that of the carrier signal taken as 100 MHz. The frequency deviation is of 8 MHz and the initial phase of the frequency modulated signal is considered to be zero. The frequency modulation is done by the MATLAB function fmmod and demodulation is done by the MATLAB function fmdemod. The plots of the message signal, FM signal, and the demodulated signal are shown in Figs 10.8, 10.9 and 10.10 respectively.

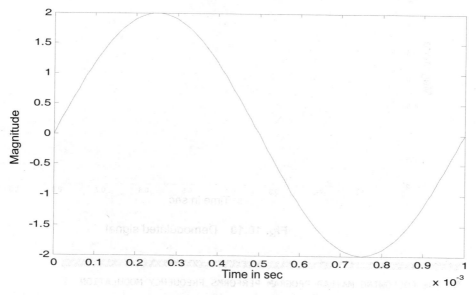

Fig. 10.8 Message signal

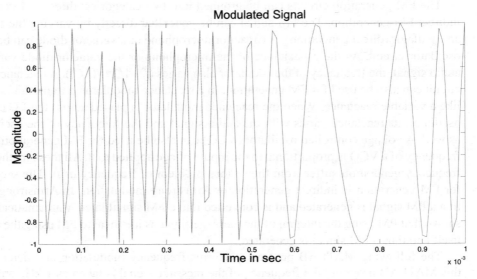

Fig. 10.9 FM signal

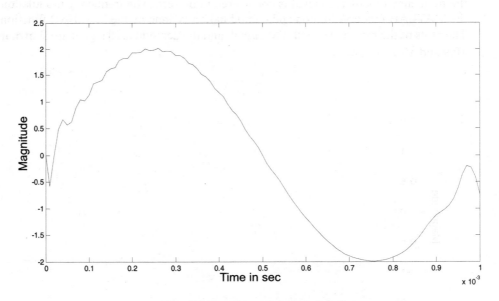

Fig. 10.10 Demodulated signal

```
%%%%%%%%%%%%%%%%%%%%%%%%%%%%%%%%%%%%%%%%%%%%%%%%%%%%%%%%%%%%
%   THE FOLLOWING MATLAB PROGRAM PERFORMS FREQUENCY MODULATION   %
%                  AND DEMODULATION                              %
%%%%%%%%%%%%%%%%%%%%%%%%%%%%%%%%%%%%%%%%%%%%%%%%%%%%%%%%%%%%
```

```
 clc;
clear all;

t = 0:10e-6:0.001;

%%%%%%%%%%%%%%%%%%%%%%%%%%%%%%%%%%%%%%%%%%
%  MESSAGE AND CARRIER SIGNAL PARAMETERS  %
%%%%%%%%%%%%%%%%%%%%%%%%%%%%%%%%%%%%%%%%%%
fm = 1e3; % Frequency of message signal in Hz
fc = 20e6; % Frequency of carrier signal in Hz
fs = 100e6; % Sampling frequency in Hz fs > 2fc
df = 8e6;   % Frequency deviation in Hz
ini_phase = 0; % Initial phase angle of the frequency modulated signal

x = 2*sin(2*pi*fm*t); % Message Signal

FM = fmmod(x,fc,fs,df,ini_phase); % Frequency modulated signal

figure(1)
plot(t,x)
title('Message Signal')
xlabel('Time in sec')
ylabel('Magnitude')

figure(2)
plot(t,FM)
title('Modulated Signal')
xlabel('Time in sec')
ylabel('Magnitude')

FDM = fmdemod(FM,fc,fs,df,ini_phase); % Frequency demodulated Signal

figure(3)
plot(t,FDM)
title('Demodulated Signal')
xlabel('Time in sec')
ylabel('Magnitude')
```

10.2.3 Phase Modulation and Demodulation

In phase modulation, the phase angle of the carrier signal is varied in accordance with the instantaneous value of the message signal. The phase modulation (PM) changes the phase angle of the complex envelope in direct proportion to the message signal. Let the modulating or message signal be $m(t)$ and the carrier signal be given by

$$c(t) = A_c \sin(\omega_c t + \varphi_c)$$

Now, the phase modulated signal will be

$$y(t) = A_c \sin(\omega_c t + \{m(t) + \varphi_c\})$$

So, the message signal $m(t)$ modulates the phase, i.e., $\varphi = m(t) + \varphi_c$ of the carrier signal. Also, it may be noted that $\omega = d\varphi/dt$, i.e., angular frequency is the rate of change of phase angle.

Frequency modulation (FM) requires the oscillator frequency to deviate both above and below the carrier frequency. During the process of frequency modulation, the peaks of each successive cycle in the modulated waveform occur at times other than they would, if the carrier was unmodulated. This is actually an incidental phase shift that takes place along with the frequency shift in FM. In phase modulation, it's just the opposite. In phase modulation, the time period of each successive cycle varies in the modulated wave according to the message signal variation. As frequency is the inverse of time period ($f = 1/T$), a phase shift in the carrier will cause its frequency to change. The change in carrier frequency in FM is vital, but it is merely incidental in PM. This change in frequency has nothing to do with the resultant wave shape in PM.

A simple model of a phase modulation is shown in Fig. 10.11. The message signal is a square wave of frequency 100 Hz. The frequency of both the carriers, i.e., 'Carrier 1' and 'Carrier 2', is equal and is 3,140 rad/s. The 'Carrier 2' signal has a phase delay of $\pi/2$ rad. If the amplitude of the message signal is greater than zero, 'Switch' passes 'Carrier 1', else 'Carrier 2' signal. Thus, the output modulated wave has a phase difference of 90° when the message signal changes the value of its amplitude. The PM signal and message signal are shown in Fig. 10.12. It may also be noted that the phase angle of the carrier wave is not varying continuously. This model is built only to demonstrate the basic concept of phase modulation.

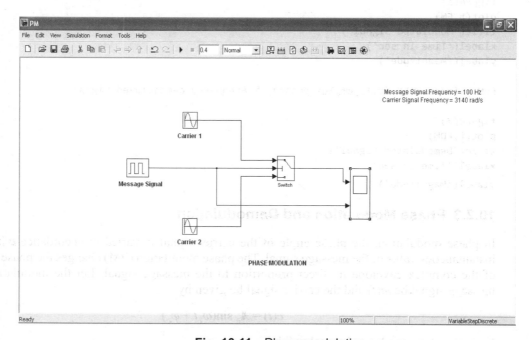

Fig. 10.11 Phase modulation

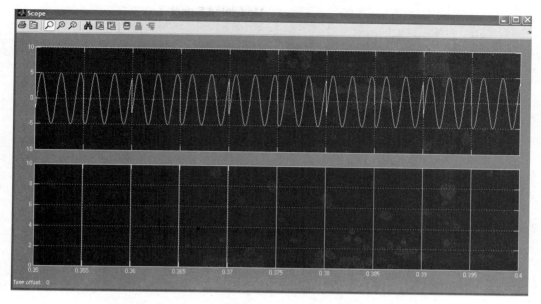

Fig. 10.12   Phase modulated wave and message signal

The following MATLAB program performs phase modulation and demodulation. The frequency of the message signal and the carrier signal is 1 kHz and 20 MHz respectively. The sampling frequency is taken as 100 MHz and the phase deviation considered is $\pi/2$. The initial phase angle of the PM signal is taken as zero. The MATLAB function pmmod is used for phase modulation and function pmdemod is used for phase demodulation. The plots of the message signal, phase modulated signal, and the demodulated signal are shown in Figs 10.13, 10.14 and 10.15 respectively.

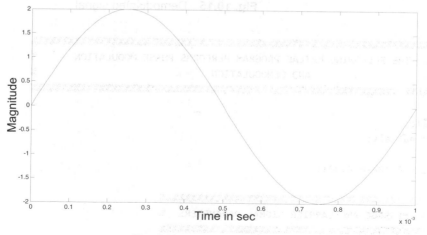

Fig. 10.13   Message signal

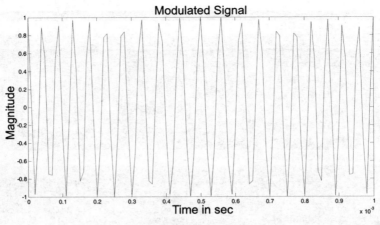

Fig. 10.14 PM signal

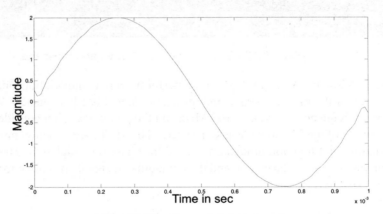

Fig. 10.15 Demodulated signal

```
%%%%%%%%%%%%%%%%%%%%%%%%%%%%%%%%%%%%%%%%%%%%%%%%%%%%%%%%%%%%%%%%%
%   THE FOLLOWING MATLAB PROGRAM PERFORMS PHASE MODULATION      %
%                   AND DEMODULATION                            %
%%%%%%%%%%%%%%%%%%%%%%%%%%%%%%%%%%%%%%%%%%%%%%%%%%%%%%%%%%%%%%%%%

clc;
clear all;

t = 0:10e-6:0.001;

%%%%%%%%%%%%%%%%%%%%%%%%%%%%%%%%%%%%%%%%%%%%
%  MESSAGE AND CARRIER SIGNAL PARAMETERS  %
%%%%%%%%%%%%%%%%%%%%%%%%%%%%%%%%%%%%%%%%%%%%
```

```
fm = 1e3; % Frequency of message signal in Hz   Y =
PMMOD(X,Fc,Fs,PHASEDEV)
fc = 20e6; % Frequency of carrier signal in Hz
fs = 100e6; % Sampling frequency in Hz fs > 2fc
dph = pi/2;   % Phase deviation in rad
ini_phase = 0; % Initial phase angle of the phase modulated signal

x = 2*sin(2*pi*fm*t); % Message Signal

PM = pmmod(x,fc,fs,dph,ini_phase); % Phase modulated signal

figure(1)
plot(t,x)
title('Message Signal')
xlabel('Time in sec')
ylabel('Magnitude')

figure(2)
plot(t,PM)
title('Modulated Signal')
xlabel('Time in sec')
ylabel('Magnitude')

PDM = pmdemod(PM,fc,fs,dph,ini_phase); % Phase demodulated Signal

figure(3)
plot(t,PDM)
title('Demodulated Signal')
xlabel('Time in sec')
ylabel('Magnitude')
```

10.2.4 Digital Modulation Techniques

In digital modulation, the analog carrier wave is modulated by a digital message signal. A digital message signal consists of bit streams of '0' and '1'. The following are the three basic types of digital modulation techniques described in this section:

1. Amplitude shift keying (ASK)
2. Frequency shift keying (FSK)
3. Phase shift keying (PSK)

In ASK, a finite number of amplitudes of the carrier wave are used for modulation and other parameters, i.e., phase and frequency of the carrier remains unchanged. The bit '1' of the digital message signal is transmitted by a carrier of particular amplitude and the bit '0' is transmitted by a carrier of other (changed) amplitude. Thus, the amplitude of the carrier changes keeping the frequency and phase constant. ON-OFF keying (OOK) is a special case of ASK where the amplitude of one of the carriers is taken as zero.

A model for ASK is shown in Fig. 10.16. In this model, a square wave signal is taken as a digital message signal. If the amplitude of the square wave is 1, it represents bit '1' and if the amplitude is –1, it represents bit '0' of the digital message. The amplitude of the carrier signal is one for bit '1' and zero for bit '0'. So, this is an OOK. The frequency of both the carriers, i.e., 'Carrier 1' and 'Carrier 2', is taken as 31,400 rad/s. The ASK signal and the message signal are shown in Fig. 10.17.

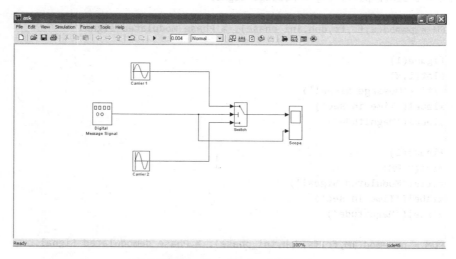

Fig. 10.16 Amplitude shift keying

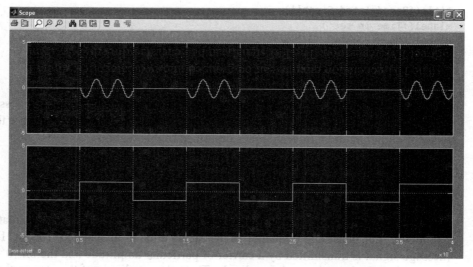

Fig. 10.17 ASK signal and the digital message signal

In FSK, a finite number of frequencies of the carrier signal are used and the frequency of the carrier signal is changed keeping the phase and amplitude constant. The bit '1' of the digital message signal is transmitted by carrier of frequency f_1 and bit '0' is transmitted by a carrier of frequency f_2. The model for FSK is shown in Fig. 10.18. The digital message signal is a square wave. If amplitude of the square wave is 1, it represents bit '1' and if the amplitude is –1, it represents bit '0'. The frequency of 'Carrier 1' is 31,400 rad/s and that of 'Carrier 2' is 3,140 rad/s. The amplitude of the carrier signals is 1 and their phase angle is 0. If the message is bit '1', it is transmitted by 'Carrier 1' and if it is bit '0', it is transmitted by 'Carrier 2'. The FSK signal and the message signal are shown in Fig. 10.19.

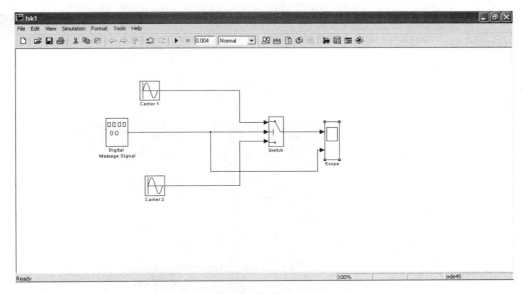

Fig. 10.18 Frequency shift keying

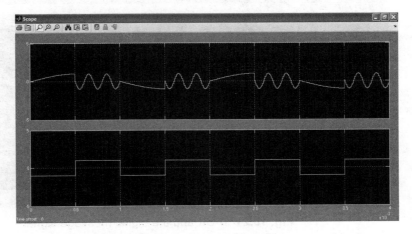

Fig. 10.19 FSK signal and the digital message signal

In PSK, a finite number of phases of the carrier signal are used and the phase of the carrier signal is changed keeping the amplitude and frequency constant. The bit '1' of the digital message signal is transmitted by a carrier phase φ_1 and bit '0' is transmitted by a carrier of phase φ_2. The phase shift represents the information contained in the modulated signal. The model for PSK is shown in Fig. 10.20. The digital message signal is a square wave. If the amplitude of the square wave is 1, it represents bit '1' and if the amplitude is –1, it represents bit '0'. The phase of 'Carrier 1' is 0 rad and of 'Carrier 2' is $\pi/2$ rad. The amplitude of the carrier signals is one and their frequency is 31,400 rad/s. If the message is bit '1' it is transmitted by 'Carrier 1' and if it is bit '0', it is transmitted by 'Carrier 2'. The PSK signal and the message signal are shown in Fig. 10.21.

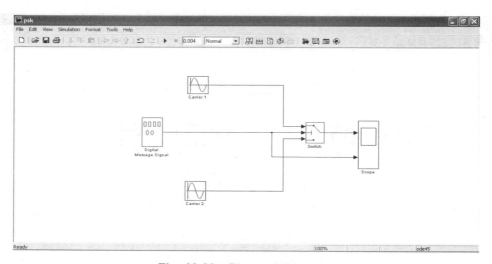

Fig. 10.20 Phase shift keying

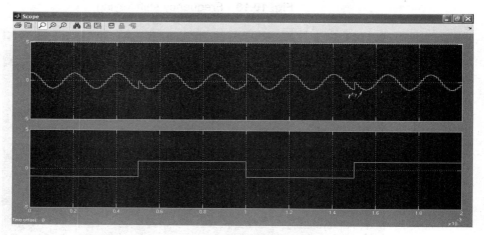

Fig. 10.21 PSK and the digital message signal

10.3 INTRODUCTION TO COMPUTER SCIENCE

Computer science is the study of the theoretical foundations of information and computation and their applicability in computer systems. It is the systematic study of the feasibility, structure, expression, and algorithms that emphasizes the acquisition, representation, processing, storage, communication, and access to information. The information that computer scientists uncover, process, store, and communicate is often encoded in computer memory in bits and bytes. Bits (0 and 1) aid in transferring information between the digital machines. Bytes or kilo-bytes are the unit of information measurement. Computer science can be used to explore the transfer of this information.

10.4 APPLICATIONS IN COMPUTER SCIENCE

This section discusses some of the applications of MATLAB in the field of computer science. MATLAB is widely being used for animations, fuzzy logic, and artificial neural networks programming these days.

10.4.1 Animation

The rapid display of a sequence of images in 2D or 3D so as to create an illusion that they are moving is called animation. The object appears to be moving due to the optical illusion of motion which is caused by the persistence of human vision, and can be created and demonstrated by several ways. The animation can generally be seen in motion pictures or video games. Most of the twentieth-century films were animated. The individual frames of the traditionally animated film are photographs of drawings which are first drawn on papers where each drawing differs slightly from the previous one in order to create an illusion of movement. The animators' drawings are traced or photocopied onto transparent acetate sheets called cels which are filled in with paints in assigned colors or tones on the side opposite the line drawings. The completed characters cels are photographed one-by-one onto motion pictures film against a painted background by a rostrum camera.

Nowadays, the animated drawings and the backgrounds are either scanned or drawn directly into a computer system. Various software programs are used to color the drawings and simulate camera movement and effects. The final animated piece is the output to one of the several delivery media, like a computer screen or digital video. Animated sequences with MATLAB graphics can be created by creating a number of different pictures that can be saved and then played as a movie, or by objects that can be erased and redrawn on the screen by making incremental changes in each redraw or by redefining xdata, ydata, zdata or cdata plot objects properties and updating them continuously. The following MATLAB programs in Examples 10.1 and 10.2 create animated objects on screen.

Example 10.1

This program creates an animation of stretching a rope so as to make it straight. The `drawnow` function is used to update the events on the figure window. The current figure window is updated after the `drawnow` function. The initial picture displayed on the figure window is shown in Fig. 10.22. The animation stops when it becomes a straight line as shown in Fig. 10.23.

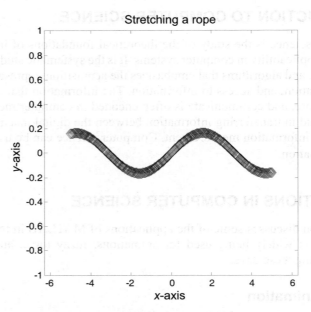

Fig. 10.22   Initial position of the rope

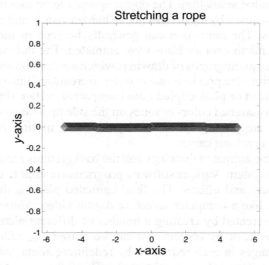

Fig. 10.23   Final position of the rope

```
%%%%%%%%%%%%%%%%%%%%%%%%%%%%%%%%%%%%%%%%%%%%%%%%%%%%%%%%%%%%%%%%%
%  This program creates an animation of stretching a rope      %
%           so that it becomes straight                        %
%%%%%%%%%%%%%%%%%%%%%%%%%%%%%%%%%%%%%%%%%%%%%%%%%%%%%%%%%%%%%%%%%
clc;
clear all;
clf;
```

```
    theta = -5:0.02:5;
    f = sin(2 * theta);
    line = plot(theta, f,'kd-');
    title('\bfStretching a rope')
    xlabel('\bfx-axis')
    ylabel('\bfy-axis')
    axis([-2*pi 2*pi -1 1])
    axis square
    grid off

    set(line,'EraseMode','xor','MarkerSize',10)

    for t = 0:0.1:7
        drawnow
        f = sin(theta) * exp(-t);
        set(line,'YData',f)
        pause(0.1)
    end
```

Example 10.2

This program creates an animation of a rotating flowerpot with vertical and horizontal rotations. The function drawnow is used to update the figure window. The initial and the final positions of the flowerpot are shown in Figs 10.24 and 10.25 respectively. The animation of the rotating flowerpot can better be viewed by running this program on MATLAB.

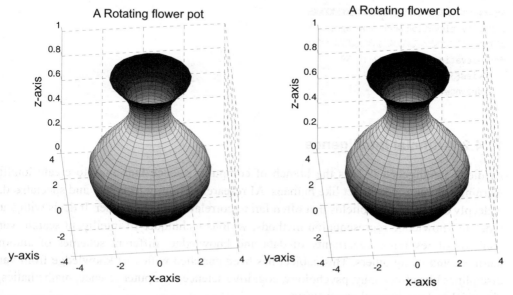

Fig. 10.24 Initial position of the flower pot   **Fig. 10.25** Final position of the flower pot

```
%%%%%%%%%%%%%%%%%%%%%%%%%%%%%%%%%%%%%%%%%%%%%%%%%%%%%%%%%%%%%%%%%%%%%%%%%%%
% This program creates a rotating flowerpot with vertical and horizontal   %
% rotations. The rotation of the flowerpot can be viewed on the screen     %
%%%%%%%%%%%%%%%%%%%%%%%%%%%%%%%%%%%%%%%%%%%%%%%%%%%%%%%%%%%%%%%%%%%%%%%%%%%

clc;
clear all;
clf;

t = 0:pi/20:2*pi;
[X,Y,Z] = cylinder(2+sin(t));
surf(X,Y,Z)
axis square
title('\bfA Rotating flowerpot')
xlabel('\bfx-axis');
ylabel('\bfy-axis');
zlabel('\bfz-ais');
%%%%%%%%%%%%%%%%%%%%%%%%%%%%%%%%%%%%%%%%%%%%
% Modify the z-axis and draw again  %
%%%%%%%%%%%%%%%%%%%%%%%%%%%%%%%%%%%%%%%%%%%%
for zrotate = -20 : .2 : 40;
    view(zrotate,40)
    drawnow
end


%%%%%%%%%%%%%%%%%%%%%%%%%%%%%%%%%%%%%%%%%%%%
% Modify elevation and draw again  %
%%%%%%%%%%%%%%%%%%%%%%%%%%%%%%%%%%%%%%%%%%%%
for elevation = 50 : -.2 : -50
    view(30, elevation)
    drawnow
end
```

10.4.2 Artificial Intelligence

Artificial intelligence (AI) is the branch of computer science that aims to create intelligent learning machines which act like humans. AI research is highly technical and specialized, and is deeply divided into subfields that often fail to correlate with each other. It deals with various types of knowledge representation methods, various techniques of intelligent search, various methods of resolving uncertainty of data and knowledge, different schemes of automated learning, and many others. This subject has been enriched with vast knowledge from various disciplines like philosophy, psychology, cognitive science, computer science, mathematics, and electrical and mechanical engineering.

The human brain is a natural source of intelligence. Our brain processes incomplete information obtained by perception at an incredibly rapid rate. Although the nerve cells function much slower (10^6 times slower) than the electronic computer, the visual and auditory information is processed much faster in our brain. Inspired by this system, many researchers have been exploring artificial neural networks for information processing. They model the brain as a continuous time dynamic system consisting of processing units. This system consists of a set of massive interconnected neurons along with certain weights. It does not need any critical decision flow in its algorithm. The MATLAB toolbox, Neural Network Toolbox, contains the block sets Control Systems, Net Input Functions, Transfer Functions, and Weight Functions. Control System blockset contains controllers used to train the neural network for controlling a process, Net Input blockset contains Net Product and Net Sum blocks used for multiplication/ division or addition/subtraction, Transfer Function blockset contains different types of transfer functions, and Weight Functions blockset contains various types of weight functions.

Example 10.3

In this Simulink model of Fig. 10.26, the same power system is considered as discussed in Chapter 8. Artificial neural networks are unutilized to analyze the harmonics of a power system. Figure 10.27 shows the subsystem of the power system present in Fig. 10.26. The ANN architecture of the subsystem is shown in Fig. 10.28. Figure 10.29 shows the line currents of the three phases, d-q components of the line current, d-q components of the fundamental frequency, and the harmonic output current. This model is simulated for all the fault conditions as mentioned in Chapter 8. This model provides the harmonic output also for unknown loads.

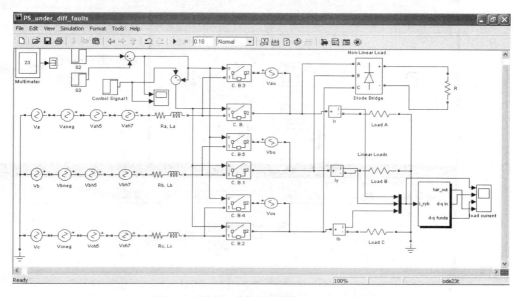

Fig. 10.26 Harmonic analysis by ANNs

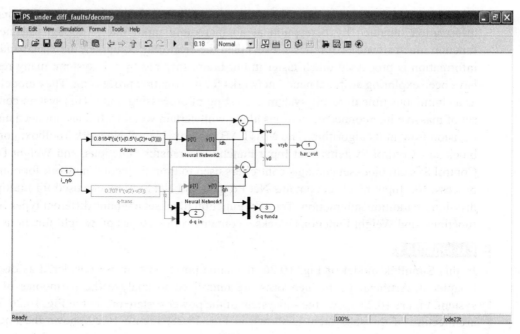

Fig. 10.27 Subsystem

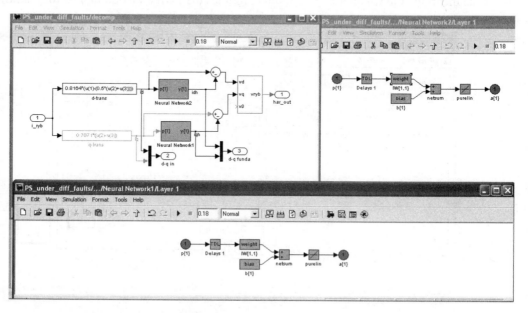

Fig. 10.28 ANNs architecture

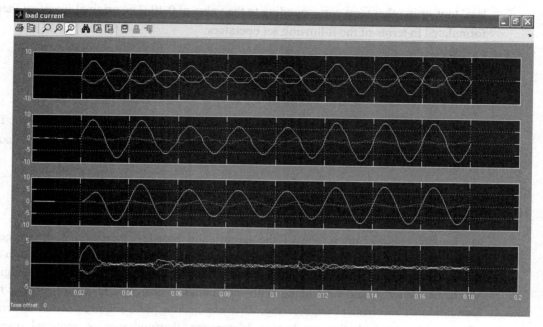

Fig. 10.29 Line current, d-q components of line current and fundamental frequency, and the harmonic output current

Fuzzy logic is a multi-valued logic that allows the intermediate values to be defined between conventional dichotomous evaluations like 0 (false) and 1 (true). The very basic notion of a fuzzy system is a fuzzy set. You might be familiar with the crisp sets studied in classical mathematics. The characteristic function of a crisp set assigns value of 1 or 0 to each member of the universal set, thereby discriminating between the members and nonmembers of the set. If A is a member of set X the value assigned to A is '1' and if it is not a member of set X, the value assigned to A is 0. This concept is sufficient for many areas of applications but it can be observed that this lacks flexibility for some applications like classification of remotely sensed data analysis. In fuzzy set, instead of assigning 0 and 1 values to the members, we assign them values lying between 0 and 1 (including 0 and 1) and call them membership function (μ). The membership function is a graphical representation of the magnitude of participation of each member or input. It associates a weighting with each of the inputs that are processed, define functional overlap between inputs and ultimately determines the output response. It is important to distinguish between fuzzy logic and probability as both operate over the same numeric range of 0 and 1. However, the probabilistic approach yields the statement like, 'There is a 40% chance that C is low', while fuzzy terminology corresponds to, 'C's degree of membership within the set of low interferometric coherence of 0.40'. The semantic difference is significant as the first view supposes that C is or is not low; it is just that we only have 40% chance of knowing which set it is in. On the other hand, fuzzy terminology supposes that C is 'more or less' low, or in other term corresponding to the value of 0.40. The original fuzzy set theory was formulated in terms of the following set operators:

1. Fuzzy complement: $\mu_A(x) = 1 - \mu_A(x)$
2. Fuzzy union: $\mu_{A \cup B}(x) = \max[\mu_A(x), \mu_B(x)]$
3. Fuzzy intersection: $\mu_{A \cap B} = \min[\mu_A(x), \mu_B(x)]$

Fuzzy logic can be realized in MATLAB by programming or by using 'Fuzzy Logic Toolbox' available in Simulink. This toolbox contains various membership functions and fuzzy logic controller (FLC) as well.

10.5 MECHANICAL ENGINEERING APPLICATIONS

Mechanical engineering is one of the largest and broadest engineering disciplines. Civil engineering is the oldest branch of engineering. The people from Germany wanted to have the military techniques applied to civilian use and the engineering was thus called civil engineering. Mechanics, energy, and heat along with mathematics, design, and manufacturing form the foundation of mechanical engineering. This engineering field requires a vast understanding of core concepts including mechanics, kinematics, thermodynamics, material science, and structural analysis.

Example 10.4

A mechanical system consisting of a mass, dashpot, and a spring is shown in Fig. 10.30. This mechanical system can be described by the following second-order equation:

$$m \frac{d^2}{dt^2} x(t) + c \frac{d}{dt} x(t) + k x(t) = F(t)$$

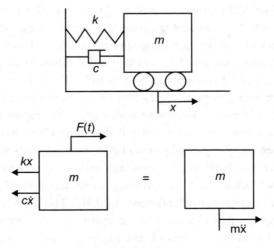

Fig. 10.30 Mechanical system with a mass, spring, and dashpot

where m represents the mass of the block, c is a positive constant of proportionality of the force that dashpot exerts on the block, and k is a positive constant of proportionality of the force that a spring exerts on the block. The mass of the spring and dashpot, and the friction are neglected in this system. The force $F(t)$ is the input and the output is the resultant displacement $x(t)$ of the body. Consider the numeric constant as $F(t) = 10 \sin(t) \, u(t)$, $m = 1$, $k = 7$, and $c = 4$, and the initial conditions as $x(t) = 4$ and $dx(t)/dt = 0$.

Now the state variables of the system are defined as follows:

$$x_1(t) = x(t)$$

and

$$x_2(t) = dx_1(t)/dt$$

Thus, the state-space equations of the system are as follows:

$$\frac{dx_1(t)}{dt} = x_2(t)$$

and

$$\frac{dx_2(t)}{dt} = -7x_1(t) - 4x_1(t) + 10 \sin(t) \, u(t)$$

In the matrix form, the equations can be written as follows:

$$
\begin{array}{c}
\dfrac{dx_1(t)}{dt} \\[2mm]
\dfrac{dx_2(t)}{dt}
\end{array}
=
\begin{array}{cc}
0 & 1 \\
-7 & -4
\end{array}
\begin{array}{c}
x_1(t) \\
x_2(t)
\end{array}
+
\begin{array}{c}
0 \\
10 \sin(t)
\end{array}
\, u(t)
$$

These state-space equations can be modeled in the State-Space block of the Simulink as shown in Fig. 10.31. The amplitude of the sine wave is 10 and the frequency is 1 Hz. The output waveform of the model is shown in Fig. 10.32.

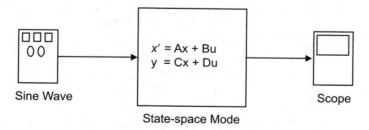

Sine Wave   x' = Ax + Bu
                   y = Cx + Du   Scope

State-space Mode

Fig. 10.31   State-space model of the mechanical system

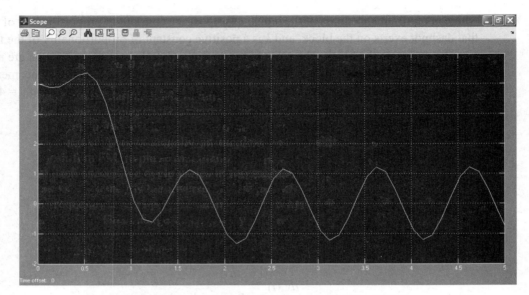

Fig. 10.32   Waveform for the model of Fig. 10.30

The same model of Fig. 10.30 is modeled in the following MATLAB program for exponential input function e^{ix}. The plot of displacement $x(t)$ and velocity $v(t)$ of the system with respect to time is shown in Fig. 10.33.

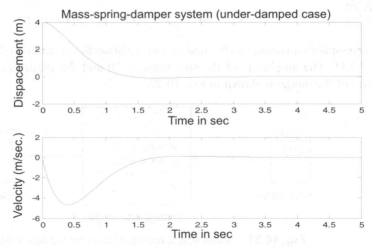

Fig. 10.33   Displacement and velocity of the system

```
clc;
clear all;

m = 1;                      %  Mass of the block in kg
```

```
k = 7;                          %  stiffness of the spring (N/m)
c = 4;                          %  Dashpot constant N-s/m
wn = sqrt(k/m);                 %  Natural frequency (rad/sec)
zeta = c/(2*wn*m);              %  Damping factor

tf = 2;                         %  Calculation Time
n = 10e5;                       %  Number of data points

s1 = (-zeta+sqrt(zeta^2-1))*wn;
s2 = (-zeta-sqrt(zeta^2-1))*wn;

%%%%%%%%%%%%%%%%%%%%%%%%%%%%%
%    Initial conditions     %
%%%%%%%%%%%%%%%%%%%%%%%%%%%%%

x_initial = 4;                  %  Initial displacement in m
v_initial = 0;                  %  Initial velocity in m/s
x0 = [x_initial,v_initial];

t_final = 5;
t = 0:t_final/n:t_final;

if zeta ~= 1                          % Under-damped and over-damped cases

  A = [[1,1];[s1,s2]];
  X = A^(-1)*x0';

  X_1 = X(1);
  X_2 = X(2);

  x = real(X_1*exp(s1*t)+X_2*(exp(s2*t)))+...
      imag(X_1*exp(s1*t)+X_2*(exp(s2*t)));        % Displacement x(t)
  v = real(s1*X_1*exp(s1*t)+s2*X_2*(exp(s2*t)))+...
        imag(s1*X_1*exp(s1*t)+s2*X_2*(exp(s2*t)));  % Velocity v(t)

elseif zeta == 1                             % Critically damped case

  X_1 = x_initial;
  X_2 = v_initial+X_1*wn;
  x = (X_1+X_2.*t).*exp(-wn*t);               % Displacement x(t)
  v = (-wn*(X_1+X_2.*t)+X_2).*exp(-wn*t);     % Velocity v(t)

end

figure(1);
clf;
orient landscape;
subplot(2,1,1),plot(t,x);
title('\bfMass-Spring-Damper system (under-damped case)')
xlabel('\bfTime in sec');
ylabel('\bfDisplacement (m)');

subplot(2,1,2),plot(t,v);
xlabel('\bfTime in sec');
ylabel('\bfVelocity (m/sec.)');
```

10.6 SIMULATION PROJECTS

Project 1

MATLAB program for amplitude modulation and demodulation

```
%%%%%%%%%%%%%%%%%%%%%%%%%%%%%%%%%%%%%%%%%%%%%%%%%%%%%%%%%%%%%%%%
%  THE FOLLOWING MATLAB PROGRAM PERFORMS AMPLITUDE MODULATION  %
%                   AND DEMODULATION                           %
%%%%%%%%%%%%%%%%%%%%%%%%%%%%%%%%%%%%%%%%%%%%%%%%%%%%%%%%%%%%%%%%

clc;
clear all;

t = 0:10e-6:0.004;

%%%%%%%%%%%%%%%%%%%%%%%%%%%%%%%%%%%%%%%%%%%%%%%%
%  MESSAGE AND CARRIER SIGNAL PARAMETERS  %
%%%%%%%%%%%%%%%%%%%%%%%%%%%%%%%%%%%%%%%%%%%%%%%

fm = 1e3;    % Frequency of message signal in Hz
fc = 2e6;    % Frequency of carrier signal in Hz
fs = 10e6;   % Sampling frequency in Hz fs > 2fc
ini_phase = 0; % Initial phase angle of the amplitude modulated signal

x = 2*sin(2*pi*fm*t); % Message Signal

AM = ammod(x,fc,fs,ini_phase); % Amplitude modulated signal

subplot(3,1,1)
plot(t,x);
title('Message Signal')
xlabel('Time in sec')
ylabel('Magnitude')

subplot(3,1,2)
plot(t,AM);
title('Modulated Signal')
xlabel('Time in sec')
ylabel('Magnitude')

ADM = amdemod(AM,fc,fs,ini_phase); % Amplitude demodulated Signal

subplot(3,1,3)
plot(t,ADM);
title('Demodulated Signal')
xlabel('Time in sec')
ylabel('Magnitude')
```

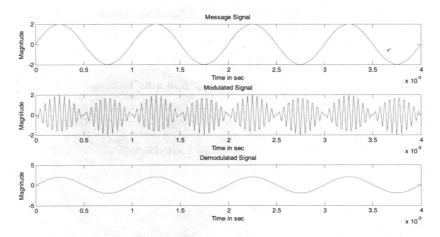

Fig. 10.34   Message signal, modulated signal, and demodulated signal

Project 2

MATLAB program to animate a revolving sphere

```
%%%%%%%%%%%%%%%%%%%%%%%%%%%%%%%%%%%%%%%%%%%%%%%%%%%%%%%%%%%%%%%%%%%%
% This program creates an rotating sphere with vertical and horizontal  %
%   rotation. The rotation of the sphere can be viewed on the screen     %
%%%%%%%%%%%%%%%%%%%%%%%%%%%%%%%%%%%%%%%%%%%%%%%%%%%%%%%%%%%%%%%%%%%%
clc;
clear all;
clf;

sphere(45)
axis('vis3d')
title('\bfA Rotating Sphere')
xlabel('\bfx-axis');
ylabel('\bfy-axis');
zlabel('\bfz-ais');

%%%%%%%%%%%%%%%%%%%%%%%%%%%%%%%%%%%%%%%%%
% Modify the z-axis and draw again  %
%%%%%%%%%%%%%%%%%%%%%%%%%%%%%%%%%%%%%%%%%
for zrotate = -20 : .2 : 40;
    view(zrotate,40)
    drawnow
end

%%%%%%%%%%%%%%%%%%%%%%%%%%%%%%%%%%%%%%%%%
% Modify elevation and draw again  %
%%%%%%%%%%%%%%%%%%%%%%%%%%%%%%%%%%%%%%%%%
for elevation = 50 : -.2 : -50
    view(30, elevation)
    drawnow
end
```

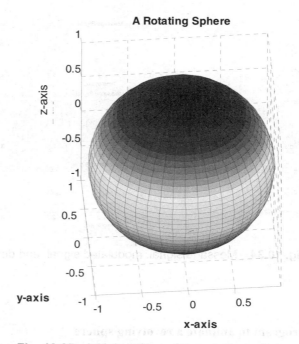

Fig. 10.35 Initial position of the rotating sphere

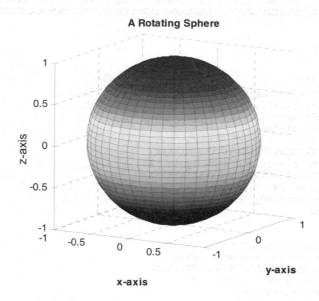

Fig. 10.36 Another position of the rotating sphere

Project 3

Animating the motion of a mechanical spring

```
%%%%%%%%%%%%%%%%%%%%%%%%%%%%%%%%%%%%%%%%%%%%%%%%%%%%%%%%%%%%%%%%%%
%    This program animates the motion of a mechanical spring    %
%%%%%%%%%%%%%%%%%%%%%%%%%%%%%%%%%%%%%%%%%%%%%%%%%%%%%%%%%%%%%%%%%%

clc;
clear all;
clf;

theta = 0 : pi/40 : 64*pi;
A = 1;
B = 0.25;
w = 2*pi/15;

M = moviein(16);
for t = 1 : 16;
    x= A * cos(theta);
    y = B * sin(theta);
    z = (1 + A*cos(w*(t-1))* theta);

    plot3(x,y,z,'k-');
    title '\bfMotion of a Mechnical spring'
    xlabel 'x-axis'
    ylabel 'y-axis'
    zlabel 'z-axis'
    axis([-1 1 -1 1 -40*pi 40*pi]);
    M(:,t) = getframe;
end

movie(M,25)
```

Motion of a Mechanical Spring

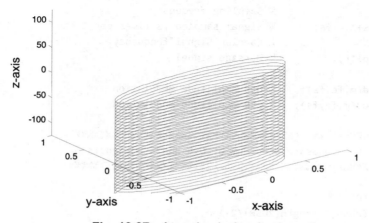

Fig. 10.37 A mechanical spring in motion

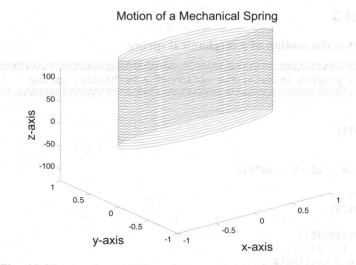

Fig. 10.38 Another position of the mechanical spring in motion

Project 4

To compare SSB and DSB amplitude modulation

```
%%%%%%%%%%%%%%%%%%%%%%%%%%%%%%%%%%%%%%%%%%%%%%%%%%%%%%%%%%%%%%%%%%
% This program compares Single Sideband and Double Sideband    %
%                          Amplitude Modulation                %
%%%%%%%%%%%%%%%%%%%%%%%%%%%%%%%%%%%%%%%%%%%%%%%%%%%%%%%%%%%%%%%%%%

clc;
clear all;

fs = 100;                % Sampling frequency
t = [0:2*fs+1]'/fs;      % Signal sampled fs times per 2 sec
fc = 25;                 % Carrier Signal frequency
m = sin(2*pi*t);         % Message signal

DSB = ammod(m,fc,fs);    % DSB amplitude modulation
SSB = ssbmod(m,fc,fs);   % SSB amplitude modulation

%%%%%%%%%%%%%%%%%%%%%%%%%%%%%%%%%%%%%%%%%%%%%%%%%%%%%%%%%
% Compute the frequency spectrum of DSB and SSB signals %
%%%%%%%%%%%%%%%%%%%%%%%%%%%%%%%%%%%%%%%%%%%%%%%%%%%%%%%%%

zDSB = fft(DSB);
zDSB = abs(zDSB(1:length(zDSB)/2+1));
frqdouble = [0:length(zDSB)-1]*fs/length(zDSB)/2;
```

```
zSSB = fft(SSB);
zSSB = abs(zSSB(1:length(zSSB)/2+1));
frqsingle = [0:length(zSSB)-1]*fs/length(zSSB)/2;

%%%%%%%%%%%%%%%%%%%%%%%%%%%%%%%%%%%%%%%%%%%%%%%%%%%%%%%%%%%
%  Plot of frequency spectra of DSB and SSB signals %
%%%%%%%%%%%%%%%%%%%%%%%%%%%%%%%%%%%%%%%%%%%%%%%%%%%%%%%%%%%

orient landscape;
figure(1);
subplot(2,1,1); plot(frqdouble,zDSB);
title('\bfFrequency spectrum of double-sideband signal');
xlabel('\bfFrequency in Hz');
ylabel('\bfAmplitude')
subplot(2,1,2); plot(frqsingle,zSSB);
title('\bfFrequency spectrum of single-sideband signal');
xlabel('\bfFrequency in Hz');
ylabel('\bfAmplitude')
```

nd

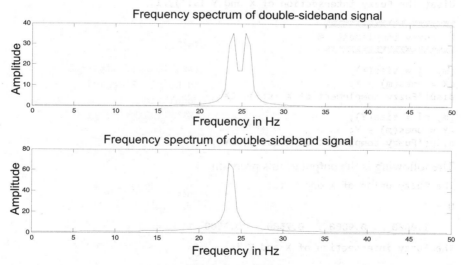

Fig. 10.39   Frequency spectrum of double-sideband and single-sideband AM signal

Project 5

To compute fuzzy union, fuzzy intersection, and fuzzy complement of two fuzz sets

This MATLAB program computes fuzzy union, fuzzy intersection, and fuzzy complement to two fuzzy sets X and Y given by

$$X = \{1/1,\ 0.75/2,\ 0.4/3,\ 0.35/4\}$$

$$Y = \{0.5/1, 0.9/2, 0.75/3, 0.21/4\}$$

```
%%%%%%%%%%%%%%%%%%%%%%%%%%%%%%%%%%%%%%%%%%%%%%%%%%%%%%%%%%%%%%%%%%%%%%%%%%
% This program computes Fuzzy union, Fuzzy intersection and Fuzzy    %
%              complement of two Fuzzy sets X & Y                    %
%%%%%%%%%%%%%%%%%%%%%%%%%%%%%%%%%%%%%%%%%%%%%%%%%%%%%%%%%%%%%%%%%%%%%%%%%%

clc;
clear all;

X = [1, 0.75, 0.4, 0.35];
Y = [0.5, 0.9, 0.75, 0.21];

%%%%%%%%%%%%%%%%%%%%
% Fuzzy union %
%%%%%%%%%%%%%%%%%%%%

U = max (X, Y);
disp('The Fuzzy union of X and Y is:'), U

%%%%%%%%%%%%%%%%%%%%%%%%%%%%%
% Fuzzy intersection %
%%%%%%%%%%%%%%%%%%%%%%%%%%%%%

I = min(X, Y);
disp('The Fuzzy intersection of X and Y is:'), I

%%%%%%%%%%%%%%%%%%%%%%%%%%
% Fuzzy complement %
%%%%%%%%%%%%%%%%%%%%%%%%%%

[m, n] = size(X);
CX = ones(m) - X;
disp('Fuzzy complement of X is:'), CX

[m, n] = size(Y);
CY = ones(m) - Y;
disp('Fuzzy complement of Y is:'), CY
```

The following is the output of this program:

The Fuzzy union of X and Y is:

U =

       1.0000   0.9000   0.7500   0.3500

The Fuzzy intersection of X and Y is:

I =

       0.5000   0.7500   0.4000   0.2100

Fuzzy complement of X is:

CX =

            0   0.2500   0.6000   0.6500

Fuzzy complement of Y is:

CY =

       0.5000   0.1000   0.2500   0.7900

SUMMARY

The contents of this chapter can be summarized as follows:

- Telecommunication is defined as the transmission, emission, or reception of electromagnetic messages, signals, or signs.
- A basic communication system consists of a transmitter, a channel, and a receiver in some form or the other.
- Modulation is a process that shifts the range of frequencies in a message signal.
- The process of extracting the message signal from the modulated signal is called as demodulation.
- In the AM, the amplitude of the high frequency carrier is varied in accordance with the instantaneous value of the message signal.
- Depending on the carrier and the sidebands present in the transmitting signal, the AM can be classified as DSB, DSB-SC, SSB, SSB-SC, and VSB.
- The FM conveys information through a carrier wave by varying its instantaneous frequency and the PM conveys by varying its instantaneous phase.
- There are three basic types of digital modulation techniques—ASK, FSK, and PSK.
- Digital modulation provides more information capacity, higher data security, compatibility with digital devices, and better quality communication.
- Computer animation is the art of creating moving images by computer programming. To create the illusion of movement, the image displayed on the screen is repeatedly replaced by a new image slightly different from the previous image but slightly advanced in the time domain.
- Artificial intelligence is the field of computer science which aims at creating machines that can engage on behaviors that humans consider as intelligent.
- Artificial neural network is a mathematical model constructed by studying the functional aspects of biological neural networks. Their utility lies in the fact that these can be made to learn from the observations similar to humans.
- Mechanical engineering is the widest and oldest branch of engineering that encompasses the generation and application of heat and mechanical power in production, design, and manufacturing of machines and tools.
- MATLAB can be used as a tool for solving various mechanical engineering problems.

REVIEW QUESTIONS

1. Mention two ways in which a communication medium can affect a signal.
2. What do you mean by 'noise' in a communication system?
3. Define modulation and demodulation. What is the role of carrier in radio communication?
4. What are the basic features of a communication system?

5. How are voice and audio signal transmitted digitally?
6. What do you mean by under- and over-modulation?
7. Define amplitude modulation, frequency modulation, and phase modulation.
8. Distinguish between DSB-SC, SSB-SC, and VSB.
9. What are the benefits of single sideband suppressed carrier AM transmission?
10. Mention the benefits of multiplexing.
11. Mention various methods of FM generation.
12. Define ASK, FSK, and PSK.
13. Mention the advantages of digital data transmission.
14. What do you mean by animation?
15. Mention some applications of animation.
16. Mention the utility of MATLAB functions soundsc and wvplay in animations.
17. What do you mean by membership function in context to Fuzzy logic systems?
18. Explain fuzzy union and fuzzy intersection.
19. Mention real-time applications of fuzzy logic controller.

EXERCISES

1. Develop a state-space model of a mechanical accelerometer consisting of a block, a dashpot, and a spring in series. Simulate this model in MATLAB for the following system parameters: $m = 1$, $k = 0.3$, and $c = 0.2$. Ignore the friction and the mass of the spring and dashpot. The force applied is $1.2\,u(t)$, where $u(t)$ is a unit step function.
2. Write a MATLAB program to plot a unit step function, unit impulse function, and unit ramp function.
3. Write a MATLAB program to create an animation of a triangle running inside a circle.
4. Develop a MATLAB program to create an animation format folding.
5. Develop a MATLAB program for animating a moving rectangular pulse.
6. Develop an animation of a revolving sphere in MATLAB. As the sphere rotates it should produce sound.
7. Develop a simple MATLAB program to train a neural network.
8. Develop a Simulink model to control a washing machine by using fuzzy logic.
9. Write a MATLAB program to control the temperature of a room by fuzzy logic controller.

10. Develop a MATLAB program that can simulate the mechanical suspension system of a car for a step input, impulse input, and an arbitrary input.
11. Develop a MATLAB program to generate and plot a DSB-SC signal. The message signal is $10\cos(10t)$ and the carrier frequency is 300 kHz. Also plot the carrier and the message signal.
12. Plot the frequency spectra of the DSB-SC signal generated in the previous question.
13. A modulating signal (message signal) is given by
   (a) $\cos(50t)$,
   (b) $\cos(50t) + \cos(100t)$,
   (c) $\cos(50t)\cos(100t)$
   Generate the SSB-SC signal $10\,m(t)\cos(10^4 t)$ for each case. Plot the SSB-SC signals generated.
14. Develop a MATLAB program to generate a FM signal for the following parameters:
   $\Delta f = 75$ kHz, $f_c = 98.2$ MHz, $m(t) = \cos(10t) + \cos(20t)$
15. Construct a MATLAB program for FM, where $\Delta f = 20$ kHz, $f_c = 200$ kHz, $m(t) = \cos(25t)$.
16. Develop a MATLAB program to frequency demodulate the FM signal generated in Q. Nos. 14 and 15.

17. Construct a MATLAB program for phase modulation and demodulation, where $\Delta\varphi = 180°$, $f_c = 30$ MHz, $m(t) = \sin(100t)$. Plot the message signal, carrier signal, PM signal, and the demodulated signal on a figure.

18. Develop a Simulink model for demonstration of ASK, FSK, and PSK. The carrier signal is $20\cos(1000t)$ and the message signal is 1001110011.

19. Consider the two fuzzy sets X and Y given by $X = \{0/2,\ 0.5/4,\ 0.4/6,\ 0.7/8,\ 0.2/10\}$ and $Y = \{0.2/2, 0.4/4, 0.6/6, 0.8/8, 0.7/10\}$.

Write a MATLAB program to compute: $X \cap Y$, $X \cup Y$, $X'\ Y'$, $(X \cap Y)'$, and $(X \cup Y)'$, where (') represents fuzzy complement.

SUGGESTED READING

For Communication Systems

Das, J., S.K. Mullick, and P.K. Chatterjee, *Principles of Digital Communication*, Wiley Eastern, New Delhi, 1986.

Feher, K., *Wireless Digital Communication*, Prentice Hall, NJ, US, 1995.

Hancock, J.C., *Introduction to the Principles of Communication Theory*, McGraw-Hill, New York, 1963.

Haykin, S., *An Introduction to Analog and Digital Communications*, 2nd Ed., Wiley, New Delhi, 2007.

Haykin, S., *Communication Systems*, John Wiley, US, 2001.

Lathi, B.P. and Z. Ding, *Modern Digital and Analog Communication Systems*, 4th Ed., Oxford University Press, US, 2009.

Roden, M.S., *Analog and Digital Communication Systems*, Prentice Hall, New Delhi, 1979.

Schwartz, M., W.R. Bennet, and S. Stein, *Communication Systems and Techniques*, McGraw-Hill, New York, 1966.

Shanmugham, K.S., *Digital and Analog Communication Systems*, John Wiley, Australia, 1979.

Stein, S. and J.J. Jones, *Modern Communication Principles*, McGraw-Hill, New York, 1967.

Taub, H. and D.L. Schilling, *Principles of Communication Systems*, Tata McGraw-Hill, New York, 1991.

Tomasiw, *Advanced Electronic Communication Systems*, Pearson Education, 2004.

Wozencraft, J.M. and I.M. Jacobs, *Principles of Communication Engineering*, John Wiley, New York, 1969.

For Computer Science

Austin, M. and D. Chancogne, *Introduction to Engineering Programming: In C, MATLAB, and JAVA*, John Wiley & Sons, New York, US, 1999.

Gray, M.A., *Introduction to the Simulation of Dynamics Using Simulink*, CRC Press, Boca Raton, US, 2010.

Jang, J.S.R., C.T. Sun, and E. Mizutani, *Neuro-Fuzzy and Soft Computing: A Computational Approach to Learning and Machine Intelligence*, PHI, New Jersey, 1997.

Loan, C.F.V., and K.Y.D. Fan, *Insight Through Computing: A MATLAB Introduction to Computational Science & Engineering*, SIAM, Philadelphia, US, 2010.

Mirzai, A.R., *Artificial Intelligence: Concepts and Engineering*, MIT Press, Cambridge, MA, 1990.

Quin, Q.H. and H. Wang, *MATLAB and C Programming for Trefftz Finite Elements Methods*, CRC Press, Boca Raton, FL, 2009.

Phan, J., *MATLAB-C# for Engineers*, LePhan Publishing, 2010.

Phan, J., *MATLAB-Visual Basic .NET for Engineers*, LePhan Publishing, 2010.

Register, A.H., *A Guide to MATLAB Object-Oriented Programming*, Chapman & Hall/CRC, Boca Raton, US, 2007.

Rich, E. and K. Knight, *Artificial Intelligence*, 2nd Ed., McGraw-Hill, New York, 1991.

Sumathi, S., and P. Surekha, *Computational Intelligence Paradigms: Theory & Applications Using MATLAB*, CRC Press, 2011.

Taylor, W.A., *What Every Engineer Should Know About AL*, MIT Press, Cambridge, MA, 1987.

Winston, P.H., *Artificial Intelligence*, 3rd Ed., Addison-Wesley, Reading, MA, 1992.

For Mechanical Engineering

Baaser, H., *Development and Application of the Finite Element Method Based on MATLAB*, Springer, Berlin, Germany, 2010.

Benaroya, H. and M.L. Nagurka, *Mechanical Vibration: Analysis, Uncertainties, and Control*, 3rd Ed., CRC Press, Boca Raton, US, 2010.

Budynas, R.G. and J.K. Nisbett, *Shigley's Mechanical Engineering Design*, 8th Ed., McGraw-Hill, 2008.

Curtis, H.D., *Orbital Mechanics for Engineering Students*, 2nd Ed., Elsevier Science, UK, 2010.

Doebelin, E.O., *Instrumentation Design Studies*, CRC Press, Boca Raton, FL, US, 2010.

Dukkipati, R.V., *MATLAB for Mechanical Engineers*, New Age Science, 2009.

Figliola, R.S., *Theory and Design for Mechanical Measurements*, 3rd Ed., John Wiley & Sons, 2000.

Gopalkrishnan, S. and M. Mitra, *Wavelet Methods for Dynamical Problems: With Application to Metallic, Composite, and Non-composite Structures*, CRC Press, Boca, US, 2010.

Hasbun, J.E., *Classical Mechanics with MATLAB Applications*, Jones & Bartlett Learning, UK, 2009.

Gardner, J.F., *Simulation of Machines Using MATLAB and Simulink*, Thomson Learning, CA, US, 2001.

Tongue, B.H., *Dynamics: Analysis and Design of Systems in Motion*, 2nd Ed., John Wiley & Sons, New York, US, 2010.

Kattan, P.I. and G.Z. Voyiadjis, *Mechanics of Composite Materials with MATLAB*, Springer, 2005.

Lyshevski, S.E., *Electromechanical Systems, Electric Machines, and Applied Mechatronics*, CRC Press, Boca Raton, US, 2000.

Turcotte, L.H., H.B. Wilson, and D. Halpen, *Advanced Mathematics and Mechanics Applications Using MATLAB*, 3rd Ed., Chapman & Hall/CRC, 2003.

Waldron, K.J. and G.L. Kinzel, *Kinematics, Dynamics, and Design of Machinery*, John Wiley & Sons, Australia, 2000.

11

SIMULATION OF POWER CONVERTERS

11.1 INTRODUCTION

This chapter aims to discuss the recent trends in the area of electrical engineering. The main focus is on matrix converters and pulse width modulation (PWM) rectifiers in terms of their performances and technical issues. The reader need not be an expert on these devices, only elementary knowledge about the power electronic converters is sufficient to go through the chapter. This chapter discusses the design and analysis of these converters by using MATLAB/Simulink, and the results so obtained are further analyzed and elaborated.

11.2 WHAT IS MATRIX CONVERTER?

The first publication on the concept of matrix converter was in 1976 by Guyugi and Pelly in their book titled *Static Power Frequency Changers: Theory, Performance, and Applications*, John Wiley & Sons. This circuit was considered to be a cycloconverter (a device used to change the frequency of AC supply) in which the devices were fully controllable. It was also called as forced commutated cycloconverter or matrix converter. A matrix converter is a device that changes the frequency of AC supply with better control. After the popular first matrix converter paper of Venturine in 1980 titled 'A new sine wave in sine wave out conversion technique which eliminates reactive elements' in *Proceedings of POWERCON7* and the next landmark paper in 1981 titled 'Solid-state power conversion: A Fourier analysis approach to generalized transformer synthesis', in *IEEE Transactions on Circuits and Systems*, which put the matrix converter control algorithms on a strong mathematical foundation, a lot in research interest evoked. At that time, the industrial interest in the converter was very limited due to the number of switching devices required, complex current commutation techniques, and complex control algorithms involved. With the introduction of new devices, these problems got sorted and the focus shifted to the protection and application part.

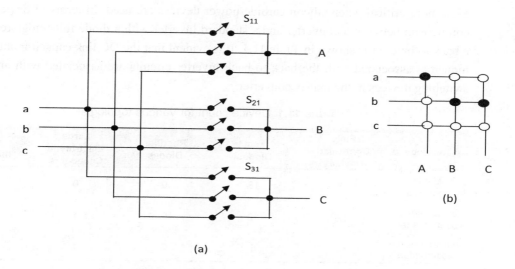

Fig. 11.1 Basic topology: (a) electric scheme and (b) symbol

Recently, there has been a considerable interest in the potential benefits of matrix converter technology, especially in applications where the size, weight, and long-term reliability are important factors. The nine bidirectional switches of the matrix converter allow any of the output to be connected to any of the input phases for a given length of time. This can be observed from Fig. 11.1. The input terminals of the converter are connected to a three-phase voltage-fed system, while the output is connected to a three-phase current-fed system. These nine bidirectional switches are then controlled to achieve the desired output. The output waveform is then created by using a suitable PWM modulation pattern similar to that of a normal inverter, except that the input is a three-phase voltage instead of a fixed DC voltage. This approach eliminates the need for the large reactive energy storage components used in the conventional inverter-based converter systems. An input line filter is included to eliminate the high frequency switching harmonics from the circuit.

The matrix converter has many advantages over the conventional rectifier–inverter based techniques. It is inherently bidirectional so that the energy can be regenerated back to the supply or grid. It draws sinusoidal input currents from the supply and depending on the modulation technique, unity power factor can also be achieved at the supply side irrespective of the type of load. Also, the overall size and weight of the converter circuit is less compared to the conventional circuit. In other words, the power to weight ratio is more, as there are no large capacitors or inductors required for energy storage. It is believed that the matrix converter can have significant advantages over the traditional DC link converter in all applications since the DC link capacitor is eliminated. The elimination of the DC link capacitor may become even more critical when silicon carbide power devices are used. In terms of device count, a comparison between a converter and a standard inverter with a diode full-bridge rectifier and a back to inverter is shown in Table 11.1. It is evident that the DC link capacitor and the input inductors associated with the back-to-back inverter circuits are associated with an extra six switching devices in the matrix converter.

Table 11.1 Device count for various topologies

Topology	Fully Controlled Devices	Fast Diodes	Rectifier Diodes	Large Electrolytic Capacitors	Large Inductors
Matrix converter	18	18	0	0	0
Back-to-back inverter	12	12	0	1	3
Inverter with diode bridge	6	6	6	1	0 or 1

11.3 BASICS OF MATRIX CONVERTER

The matrix converter is a direct AC–AC converter for converting one frequency AC supply to another frequency AC supply without involving an intermediate DC link capacitor. The matrix converter has several advantages over the traditional rectifier–inverter type two-stage power frequency converters. It provides sinusoidal input and output waveforms with a few higher order frequency harmonics and almost no sub-harmonics. Bidirectional power flow is possible and it can generate sinusoidal input currents with controllable power factor. Intensive research is being carried out in matrix converters since mid-1970s, and because of its potential benefits over the other topologies, it has attracted considerable commercial interest over a period of time from industries. Generation of control switching pulses for controlling the device has attracted much attention from scientists and engineers. Two modulation schemes—the Venturini method or direct modulation and the space vector modulation (SVM) or indirect modulation—are well known. The design approaches of these schemes are different and hence they differ in their performances. For proper current commutation in matrix converter, there should always be a continuous path for current to flow in each output phase of the converter. If two switches of a group are ON at any time instant, then a phase-to-phase short circuit occurs. On the other hand, if all the switches of a group are OFF at any time instant, an open circuit occurs. The semiconductor switches used for the implementation of the converter has a certain switching time which is not equal to zero. A number of techniques have been proposed in literature for safe commutation.

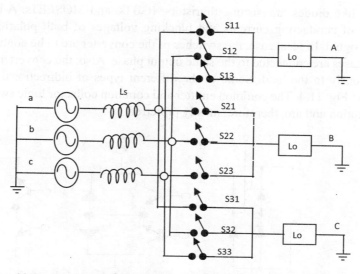

Fig. 11.2 Matrix converter schematic block diagram representation

A block diagram of a three-phase to three-phase matrix converter is shown in Fig. 11.2. It has a three-phase input supply, i.e., V_a, V_b, and V_c. The three-phase output voltages obtained are V_A, V_B, and V_C. There are nine bidirectional switches from S_{11} to S_{33} which represent the nine matrix components mathematically. It can be seen from the circuit shown in Fig. 11.2 that any input phase can be connected to any output phase in order to get the desired output waveform.

The three-phase to three-phase matrix converter converts the three-phase input of a given amplitude (V_i) and frequency (f_i) to three-phase output of a fixed amplitude (V_o) and frequency (f_o). Any desirable output frequency can be achieved by this converter. The three-phase inputs of the converter are given by $V_a = V_i \cos(w_i t)$, $V_b = V_i \cos(w_i t + 2 \times \pi/3)$, and $V_c = V_i \cos(w_i t + 4 \times \pi/3)$. The required output voltages are given by $V_A = V_o \cos(w_o t)$, $V_B = V_o \cos(w_o t + 2 \times \pi/3)$, and $V_C = V_o \cos(w_o t + 4 \times \pi/3)$. The input and output voltages are related to each other according to the following matrix equations shown in Fig. 11.3. Here, M_{11} is the duty cycle of switch S_{11} and so on.

$$
\begin{bmatrix} V_A(t) \\ V_B(t) \\ V_C(t) \end{bmatrix} = \begin{bmatrix} M_{11} & M_{12} & M_{13} \\ M_{21} & M_{22} & M_{23} \\ M_{31} & M_{32} & M_{33} \end{bmatrix} = \begin{bmatrix} V_a(t) \\ V_b(t) \\ V_c(t) \end{bmatrix}
$$

Fig. 11.3 Input and output voltages relation

11.3.1 Bidirectional Switches

The matrix converter requires bidirectional switches for blocking voltages and conducting currents in both the directions. These switches can be constructed by using discrete semiconductor devices like diodes, transistors, thyristors, IGBTs, and MOSFETs. A bidirectional switch is capable of conducting currents and blocking voltages of both polarities, depending on the control signal. In ideal case, the switches of the converter are to be controlled such that no two input phases are connected to the same output phase. Also, the converter should supply current continuously to the load. Some of the different types of bidirectional switches possible are shown in Fig. 11.4. The common emitter and common collector logic switches have low power consumption and are, therefore, mostly preferred.

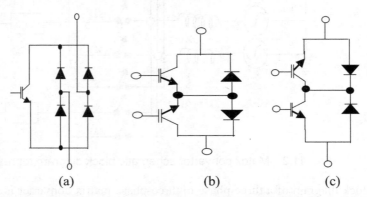

Fig. 11.4 Bidirectional switch topologies using unidirectional switches: (a) diode embedded switch, (b) common emitter logic (CE), and (c) common collector switch (CC)

Figure 11.5 shows the circuit diagram of a two-phase to single-phase matrix converter. V_1 and V_2 are input voltages and S_1 and S_2 are bidirectional switches. S_{1f} and S_{2f} when turned ON by the gate pulse provide a path for current to flow from the input to the output whereas S_{1r} and S_{2r} when turned ON provide a path for current to flow from the output to the input.

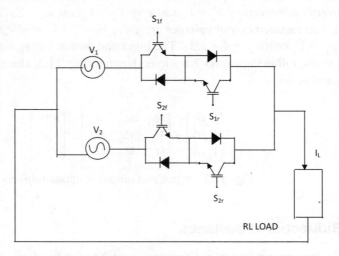

Fig. 11.5 Two-phase to single-phase matrix converter

11.3.2 Commutation Problem

Theoretically, the switches in the converter should be controlled such that no two input lines are connected to the same output line as this would result in input phase-to-phase short circuit. Also, every output line should always be connected to an input line to avoid open circuit at the load terminals. Practically, the power semiconductor devices do not switch instantaneously (i.e., $t_{ON} > 0$ and $t_{OFF} > 0$) and perfect synchronization of incoming and outgoing switches is impossible. In other words, a short period of overlap or under-lap is necessary for commutation, unless natural commutation is arranged. The matrix converters have been built using dead times to overcome the shorting of the input lines due to the finite switching times of the devices. If the dead times are used, then some form of voltage clamping or an alternate current path must be provided to avoid an uncontrolled open circuit of the motor (load). There are, however, some technical challenges in the practical implementation of the converter. Traditional inverter topologies such as the voltage source inverter include natural freewheel paths, which allow straightforward commutation of current from one device to another. There are, however, no problem-causing freewheel paths in the matrix converter when commutating between the switches.

Therefore, for commutation between two switches, the following two basic rules are to be followed for safe operation of the converter:

- Do not connect two different input lines to the same output line (short circuit of the mains, which causes over-currents)
- Do not disconnect the output line circuits (interrupt inductive currents, which causes over-voltages)

Four-step Current Commutation

A number of methods have been proposed for safe commutation of the matrix converters. These methods mostly rely on the knowledge of either the output current sign or relative magnitude of the input voltages. The four-step current commutation technique is more popular as it relies on the direction of the output load current. For understanding the current commutation, consider a two-phase to single-phase matrix converter as shown in Fig. 11.5. All the possible commutation instances in a three-phase to three-phase matrix converter can be analyzed using this circuit.

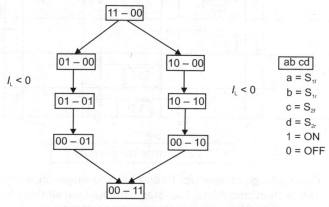

Fig. 11.6 Four-state commutation sequence

In this strategy, three different sequences of switching OFF and switching ON of bidirectional switches are used when a rising edge or a falling edge of a modulation pulse occurs depending on the positive, negative, or zero current. These different sequences are shown in Fig. 11.6. Two of them are shown in the state diagram for positive and negative current. Assume that switches S_{1f} and S_{1r} are ON, and S_{2f} and S_{2r} are OFF and we want this condition to reverse. Assume that the load current is flowing into the load as shown in Fig. 11.5. In such a case, we can turn OFF switch S_{1r} as it not carrying any current, followed by switch S_{2f} gated ON to provide an alternate path for the load current to flow. After this switch, S_{1f} is gated OFF and finally switch S_{2r} is gated ON. In other case, i.e., if the load current is flowing inward, the sequence will be S_{1f} OFF, S_{2r} ON, S_{1r} OFF, and S_{2f} ON. The commutation sequence for both the cases is shown in Fig. 11.6. An alternate scheme is two-step current commutation, in which the switch carrying the current is gated when commutation is required and the first and the last steps are not required. Figure 11.7 illustrates the commutation scheme for a three-phase to single-phase matrix converter for three switches S_1, S_2, and S_3 when the load current is negative. Similarly, for the positive load current, the forward switches will be replaced by reverse switches and vice versa. In case of a three-phase to three-phase matrix converter, the same procedure can be adopted for the other two groups containing six switches.

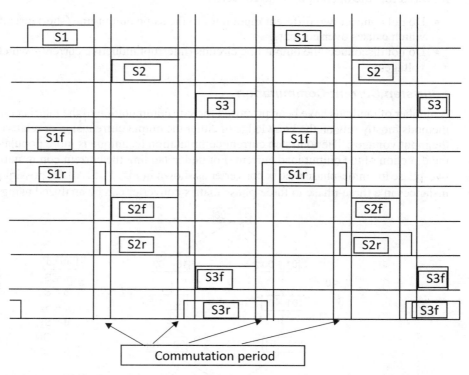

Fig. 11.7 Commutation scheme for three-phase to single-phase converter when load current is less than zero. Here, four-step commutation strategy is used but the time between one switch OFF and next ON is of three steps

In case of zero-load current, two-step commutation scheme is adopted as illustrated for a group of three switches as shown in Fig. 11.8.

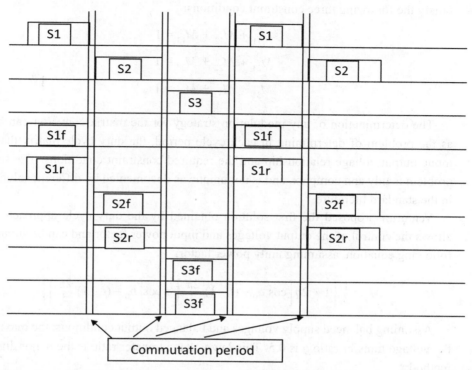

Fig. 11.8 Commutation scheme for three-phase to single-phase converter when load current is zero. Here two-step commutation strategy is used but the time between one switch OFF and next ON is of one step

11.3.3 Modulation Techniques

The output waveforms of the matrix converter are formed by selecting each of the input phases in a particular sequence at different instants of time. The output phase sequence for each output phase remains the same. Various modulation techniques are being proposed for the modulation of matrix converter switches. Venturini method, SVM, and min-mid-max modulation schemes are the popular methods. An optimum modulation strategy is that which minimizes the input current and output voltage harmonic distortion and the switching losses as well and provides almost unity power factor at the supply side.

Venturini Method

The relations between input and output voltages are related to the states of the nine bidirectional switches as illustrated in Fig. 11.3 with the condition that $0 \le M_{ij} \ge 1$, where $i, j = 1, 2, 3$. The variables M_{ij} are the duty cycles of nine switches shown in Fig. 11.1. In order to avoid a short circuit on the input side and to ensure uninterrupted load current flow, these duty cycles must satisfy the following three constraint conditions:

$$M_{11} + M_{12} + M_{13} = 1$$

$$M_{21} + M_{22} + M_{23} = 1$$

$$M_{31} + M_{32} + M_{33} = 1$$

The determination of any modulation strategy for the matrix converter can be formulated as the problem of determining in each cycle period, the duty cycle matrix that satisfies the input–output voltage relationship and the required constraint conditions. The solution to this problem is tidy and complex, and is not unique as documented by different solutions proposed in the standard literature.

Venturini proposed the first solution obtained by the duty cycle approach. This method allows the control of the output voltages and input power factor and can be summarized in the following equation, assuming unity power factor:

$$M_{ij} = \frac{1}{3} \left\{ 1 + 2q \left[\cos a_0 - (i-1) \frac{2\pi}{3} \right] \times \cos \left[b_0 - (j-1) \frac{2\pi}{3} \right] \right\}$$

Assuming balanced supply voltages and balanced output conditions, the maximum value of the voltage transfer ratio q is 0.5. This low voltage transfer ratio is the major drawback of this method.

In order to improve the performance of this method, a second solution known as Venturini optimum method was proposed. In this case, the maximum voltage transfer ratio is 0.866. For this method, the modulation functions can be defined by the following relationships:

$$M_{ij} = \frac{1}{3} \left\{ 1 + 2q \left[\cos b_h - (j-1) \frac{2\pi}{3} \right] \times \left[\cos \left(a_0 - (i-1) \frac{2\pi}{3} \right) - 1/6 \cos(3a_0) \right. \right.$$
$$\left. \left. + \left(\frac{1}{2}\sqrt{3} \right) \cos(3b_h) \right] - \left(\frac{2}{3}\sqrt{3} \right) q \cos \left[\left(4b_h - (j-1) \frac{2\pi}{3} \right) - \cos \left(2b_h + (j-1) \frac{2\pi}{3} \right) \right] \right\}$$

Space Vector Modulation

This method is based on the instantaneous space-vector representation of input and output voltages and currents. There are 27 switching combinations possible with nine bidirectional switches in matrix converter. Bearing in mind the conditions and modulation constraints, only 21 switching combinations from these 27 can be usefully employed to SVM and are shown in Table 11.2.

Table 11.2 Space vector modulation switching configurations

Switching configuration list	Switches ON			V_o	a_0	i_i	b_i
+1	S_{11}	S_{22}	S_{32}	$2/3\ v_{12i}$	0	$2/\sqrt{3}\ i_{01}$	$-\pi/6$
−1	S_{12}	S_{21}	S_{31}	$-2/3\ v_{12i}$	0	$-2/\sqrt{3}\ i_{01}$	$-\pi/6$
+2	S_{12}	S_{23}	S_{33}	$2/3\ v_{23i}$	0	$2/\sqrt{3}\ i_{01}$	$\pi/2$
−2	S_{13}	S_{22}	S_{32}	$-2/3\ v_{23i}$	0	$-2/\sqrt{3}\ i_{01}$	$\pi/2$
+3	S_{13}	S_{21}	S_{31}	$2/3\ v_{31i}$	0	$2/\sqrt{3}\ i_{01}$	$7\pi/6$
−3	S_{11}	S_{23}	S_{33}	$-2/3\ v_{12i}$	0	$-2/\sqrt{3}i_{01}$	$7\pi/6$
+4	S_{12}	S_{21}	S_{32}	$2/3\ v_{12i}$	$2\pi/3$	$2/\sqrt{3}\ i_{02}$	$-\pi/6$
−4	S_{11}	S_{22}	S_{31}	$-2/3\ v_{12i}$	$2\pi/3$	$-2/\sqrt{3}\ i_{02}$	$-\pi/6$
+5	S_{13}	S_{22}	S_{33}	$2/3\ v_{23i}$	$2\pi/3$	$2/\sqrt{3}\ i_{02}$	$\pi/2$
−5	S_{12}	S_{23}	S_{32}	$-2/3\ v_{23i}$	$2\pi/3$	$-2/\sqrt{3}\ i_{02}$	$\pi/2$
+6	S_{11}	S_{23}	S_{31}	$2/3\ v_{31i}$	$2\pi/3$	$2/\sqrt{3}\ i_{02}$	$7\pi/6$
−6	S_{13}	S_{21}	S_{33}	$-2/3\ v_{31i}$	$2\pi/3$	$-2/\sqrt{3}\ i_{02}$	$7\pi/6$
+7	S_{12}	S_{22}	S_{31}	$-2/3\ v_{12i}$	$4\pi/3$	$2/\sqrt{3}\ i_{03}$	$-\pi/6$
−7	S_{11}	S_{21}	S_{32}	$-2/3\ v_{12i}$	$4\pi/3$	$-2/\sqrt{3}\ i_{03}$	$-\pi/6$
+8	S_{13}	S_{23}	S_{32}	$2/3\ v_{23i}$	$4\pi/3$	$2/\sqrt{3}\ i_{03}$	$\pi/2$
−8	S_{12}	S_{22}	S_{33}	$-2/3\ v_{23i}$	$4\pi/3$	$-2/\sqrt{3}\ i_{03}$	$\pi/2$
+9	S_{11}	S_{21}	S_{33}	$2/3\ v_{31i}$	$4\pi/3$	$2/3\ i_{03}$	$7\pi/6$
−9	S_{13}	S_{23}	S_{31}	$2/3\ v_{31i}$	$4\pi/3$	$-2/\sqrt{3}\ i_{03}$	$7\pi/6$
0_1	S_{11}	S_{21}	S_{31}	0	—	0	—
0_2	S_{12}	S_{22}	S_{32}	0	—	0	—
0_3	S_{13}	S_{23}	S_{33}	0	—	0	—

The first 18 switching configurations determine an output voltage vector V_o and an input current vector i_i. The magnitude of these vectors depends upon the instantaneous values of the input line-to-line voltages and output currents respectively and are known as 'active configurations'. The last three switching configurations determine zero input current and output voltage vectors and are known as 'zero configurations'. The remaining six configurations have each output phase connected to a different input phase. In this case, the output voltage and input current vectors have variable directions and cannot usefully synthesize the reference vectors. The SVM algorithm has the capability to achieve the full control of both the output voltage vector and the instantaneous input current displacement angle.

MIN-MID-MAX Modulation

This modulation strategy labels the input line-to-neutral voltages as MIN, MID, and MAX depending on the voltage situation, i.e., whether the voltage value is most negative or most positive, or lies between these two values. The middle value of the input sinusoidal current reference is used to determine whether the MIN or MAX input line-to-neutral voltage can be used as a base value. One of the output phases is connected to the base voltage, the output phase with the lowest reference line-to-neutral voltage for a MIN base voltage or that with the highest reference line-to-neutral voltage for a MAX base voltage. The other two output phases are then connected to the input phases according to the switching sequence MIN-MID-MAX-MAX-MID-MIN for a MIN base voltage value or MAX-MID-MIN-MIN-MID-MAX for a MAX base voltage value. The determination of modulation timings are such that the target line-to-line output voltages and the target input current ratios are achievable. In this method, the commutation losses are minimized as minimum commutations per switching cycle are employed and also the voltages changes at each commutation are minimized. For further literature on this modulation strategy, refer to Imayavramban et al. (2004), Hosseini and Babaei (2003), and Ecklebe et al. (2009).

11.4 PROGRAMMING AND SIMULATIONS OF MATRIX CONVERTERS

Example 11.1

A three-phase to three-phase matrix converter as shown in Fig. 11.2 can also be programmed in MATLAB in order to understand its working. Consider a three-phase input voltage of 50 Hz, 230 V rms. Assuming the input voltages to be cosine waves, they can be defined as follows:

$$V_{i1} = 230 \times 1.414 \times \cos(2\pi \times 50 \times t)$$

$$V_{i2} = 230 \times 1.414 \times \cos\left(2\pi \times 50 \times t + \frac{(2 \times \pi)}{3}\right)$$

$$V_{i3} = 230 \times 1.414 \times \cos\left(2\pi \times 50 \times t + \frac{(4 \times \pi)}{3}\right)$$

Let the output line voltage frequency and amplitude be f_o and V_{out}. Now, based on these parameters, the modulating signals of all the nine bidirectional switches can be calculated by using Venturini direct method. Once these modulating pulses are known, the output line voltages can be estimated by the following equations:

$$V_{out1} = m_{11} \times V_{i1} + m_{12} \times V_{i2} + m_{13} \times V_{i3}$$
$$V_{out2} = m_{21} \times V_{i1} + m_{22} \times V_{i2} + m_{23} \times V_{i3}$$
$$V_{out3} = m_{31} \times V_{i1} + m_{32} \times V_{i2} + m_{33} \times V_{i3}$$

In order to estimate the practical switching pulses from these modulating signals, current commutation logic can be utilized. The switching pulses obtained can be fed to the bidirectional switches in order to obtain the desired output line voltages.

The following set of MATLAB instructions would program a three-phase to three-phase matrix converter assuming all the bidirectional switches to be ideal. Figure 11.9 shows the three-phase input voltages fed to the matrix converter. For three-phase output voltage of 100 Hz and 110 V rms, the modulating signals of the first group of switches obtained are shown in Fig. 11.10. The output line voltages obtained by these modulating pulses are shown in Fig. 11.11, and the switching pulses of the first group are shown in Fig. 11.12. It can be observed from these figures that the modulating signals vary from 0 to 1, and the output voltages are of 110 V rms, 100 Hz as desired by the user.

```
%%%%%%%%%%%%%%%%%%%%%%%%%%%%%%%%%%%%%%%%%%%%%%%%%%%%%%%%%%%%%%%%%%%%%%%%%
%   Program to illustrate a three phase to three phase matrix converter   %
%%%%%%%%%%%%%%%%%%%%%%%%%%%%%%%%%%%%%%%%%%%%%%%%%%%%%%%%%%%%%%%%%%%%%%%%%
clc;
clear all;
clf;
Vi = 230*1.414; % Peak value of the input phase voltage
Vout = input('RMS value of the output phase voltage required in volts: ');
Vo = Vout*1.414;  % Peak value of the output line voltage desired

wi = 50*2*pi; % Input phase frequency in rad/sec
fo = input('Enter output phase frequency required in Hz: ');
wo = 2*pi*fo; % Output line frequency in rad/sec

t = 0:5e-6:0.08; % Time array

%%%%%%%%%%%%%%%%%%%%%%%%%%%%%%%%%%%%%%%%%%%%%%%%%%%%%%%%%%%%%%%%%%%
%                      Input three phase voltages                    %
%%%%%%%%%%%%%%%%%%%%%%%%%%%%%%%%%%%%%%%%%%%%%%%%%%%%%%%%%%%%%%%%%%%

Vi1 = Vi*cos(wi*t);
Vi2 = Vi*cos(wi*t + (2*pi)/3);
Vi3 = Vi*cos(wi*t + (4*pi)/3);
al = Vo/(3*Vi);

%%%%%%%%%%%%%%%%%%%%%%%%%%%%%%%%%%%%%%%%%%%%%%%%%%%%%%%%%%%%%%%%%%%%%
%                   Plot of input three phase voltages                  %
%%%%%%%%%%%%%%%%%%%%%%%%%%%%%%%%%%%%%%%%%%%%%%%%%%%%%%%%%%%%%%%%%%%%%

figure(1)
subplot(3,1,1),plot(t,Vi1,'k-')
Axis([0 0.04 -Vi Vi])
xlabel('\bfVi1');
ylabel('\bfAmplitude');
subplot(3,1,2),plot(t,Vi2,'k-')
Axis([0 0.04 -Vi Vi])
```

```
xlabel('\bfVi2 ');
ylabel('\bfAmplitude');
subplot(3,1,3),plot(t,Vi3,'k-')
Axis([0 0.04 -Vi Vi])
xlabel('\bVi3');
ylabel('\bfAmplitude');


%%%%%%%%%%%%%%%%%%%%%%%%%%%%%%%%%%%%%%%%%%%%%%%%%%%%%%%%%%%%%%%%%%%%%%%%%
%            Modulating pulses for Nine bidirectional switches        %
%%%%%%%%%%%%%%%%%%%%%%%%%%%%%%%%%%%%%%%%%%%%%%%%%%%%%%%%%%%%%%%%%%%%%%%%%


m11 = 1/3 + al*cos((wo-wi)*t) + al*cos(-(wo+wi)*t);
m12 = 1/3 + al*cos((wo-wi)*t -(2*pi)/3) + al*cos(-(wo+wi)*t -(2*pi)/3);
m13 = 1/3 + al*cos((wo-wi)*t -(4*pi)/3) + al*cos(-(wo+wi)*t -(4*pi)/3);
m21 = 1/3 + al*cos((wo-wi)*t -(4*pi)/3) + al*cos(-(wo+wi)*t -(2*pi)/3);
m22 = 1/3 + al*cos((wo-wi)*t) + al*cos(-(wo+wi)*t -(4*pi)/3);
m23 = 1/3 + al*cos((wo-wi)*t -(2*pi)/3) + al*cos(-(wo+wi)*t);
m31 = 1/3 + al*cos((wo-wi)*t -(2*pi)/3) + al*cos(-(wo+wi)*t -(4*pi)/3);
m32 = 1/3 + al*cos((wo-wi)*t -(4*pi)/3) + al*cos(-(wo+wi)*t);
m33 = 1/3 + al*cos((wo-wi)*t) + al*cos(-(wo+wi)*t -(2*pi)/3);


%%%%%%%%%%%%%%%%%%%%%%%%%%%%%%%%%%%%%%%%%%%%%%%%%%%%%%%%%%%%%%%%%%%%%%%%%
%                    plot of modulating signals                      %
%%%%%%%%%%%%%%%%%%%%%%%%%%%%%%%%%%%%%%%%%%%%%%%%%%%%%%%%%%%%%%%%%%%%%%%%%


figure(2)
subplot(3,1,1),plot(t,m11,'k-')
Axis([0 0.04 0 1])
xlabel('\bfm11');
ylabel('\bfMagnitude');
subplot(3,1,2),plot(t,m12,'k-')
Axis([0 0.04 0 1])
xlabel('\bfm12');
ylabel('\bfMagnitude');
subplot(3,1,3),plot(t,m13,'k-')
Axis([0 0.04 0 1])
xlabel('\bfm13');
ylabel('\bfMagnitude');


%%%%%%%%%%%%%%%%%%%%%%%%%%%%%%%%%%%%%%%%%%%%%%%%%%%%%%%%%%%%%%%%%%%%%%%%%%%
%                    Output three phase line voltages                  %
%%%%%%%%%%%%%%%%%%%%%%%%%%%%%%%%%%%%%%%%%%%%%%%%%%%%%%%%%%%%%%%%%%%%%%%%%%%


Vout1 = m11.*Vi1 + m12.*Vi2 + m13.*Vi3;
Vout2 = m21.*Vi1 + m22.*Vi2 + m23.*Vi3;
Vout3 = m31.*Vi1 + m32.*Vi2 + m33.*Vi3;
```

```
%%%%%%%%%%%%%%%%%%%%%%%%%%%%%%%%%%%%%%%%%%%%%%%
%                Plot output voltages            %
%%%%%%%%%%%%%%%%%%%%%%%%%%%%%%%%%%%%%%%%%%%%%%%

figure(3)
subplot(3,1,1),plot(t,Vout1,'k-')
Axis([0 0.04 -Vo Vo])
xlabel('\bfVout1');
ylabel('\bfAmplitude');
subplot(3,1,2),plot(t,Vout2,'k-')
Axis([0 0.04 -Vo Vo])
xlabel('\bfVout2');
ylabel('\bfAmplitude');
subplot(3,1,3),plot(t,Vout3,'k-')
Axis([0 0.04 -Vo Vo])
xlabel('\bfVout3');
ylabel('\bfAmplitude');


%%%%%%%%%%%%%%%%%%%%%%%%%%%%%%%%%%%%%%%%%%%%%%%%%%%%%%%%%%%%%%%%%%
%                Current Commutation logic                        %
%%%%%%%%%%%%%%%%%%%%%%%%%%%%%%%%%%%%%%%%%%%%%%%%%%%%%%%%%%%%%%%%%%

x(1) = 0;
y(1) = 0;
j = 1:1:16001
for i = 2:16001
    x(i) = x(i-1) + 0.1;
    z(i) = x(i) - rem(x(i),1);
    if mod(z(i),2)== 0
        y(i) = mod(x(i),1);
    else y(i) = 1 - mod(x(i),1);
    end
end
for l = 1:16001
    if m11(l)> y(l)
        p(l) = 1;
    else p(l) = 0;
    end
end
for l = 1:16001
    if m12(l)> y(l)
        q(l) = 1;
    else q(l) = 0;
```

```
        end
    end
    for l = 1:16001
        if m13(l)> y(l)
            r(l) = 1;
        else r(l) = 0;
        end
        u(l) = p(l)+q(l)+r(l);
    end
    for d = 1:800
       sum = 0;
      for k = 1:20
            sum   = sum + u((d-1)*10 + k);
        end
       s(d) = sum/20;
    end


%%%%%%%%%%%%%%%%%%%%%%%%%%%%%%%%%%%%%%%%%%%%%%%%%%%%%%%%%%%%%%%%%%%%%%%%
%                       Plot of switching pulses                      %
%%%%%%%%%%%%%%%%%%%%%%%%%%%%%%%%%%%%%%%%%%%%%%%%%%%%%%%%%%%%%%%%%%%%%%%%

figure(4)
subplot(3,1,1),plot(t,q,'k-')
Axis([0.001 0.002 0 3])
xlabel('\bfTime in sec');
ylabel('\bfAmplitude ');
subplot(3,1,2),plot(t,r,'k-')
Axis([0.001 0.002 0 3])
xlabel('\bfTime in sec');
ylabel('\bfAmplitude ');
subplot(3,1,3),plot(t,p,'k-')
Axis([0.001 0.002 0 3])
xlabel('\bfTime in sec');
ylabel('\bfAmplitude ');


%%%%%%%%%%%%%%%%%%%%%%%%%%%%%%%%%%%%%%%%%%%%%%%%%%%%%%%%%%%%%%%%%%%%%%%%
%                         End of the program                         %
%%%%%%%%%%%%%%%%%%%%%%%%%%%%%%%%%%%%%%%%%%%%%%%%%%%%%%%%%%%%%%%%%%%%%%%%
```

While executing the program, it will ask for two input parameters on the command window as follows:

```
RMS value of the output phase voltage required in volts: 110
Enter output phase frequency required in Hz: 100
```

The values entered by the user are 110 V and 50 Hz.

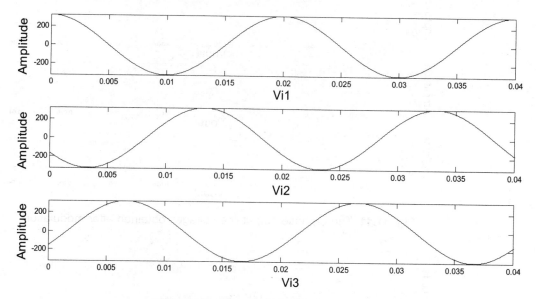

Fig. 11.9   Three-phase input voltages

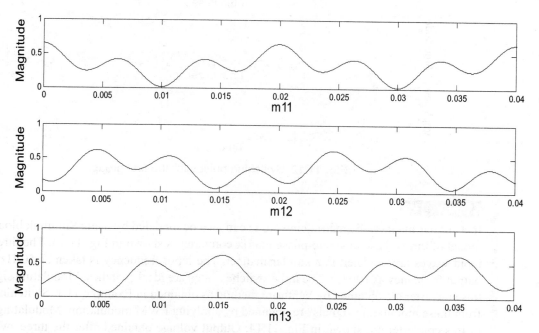

Fig. 11.10   Modulating signals of the first group, i.e., m_{11}, m_{12}, and m_{13}

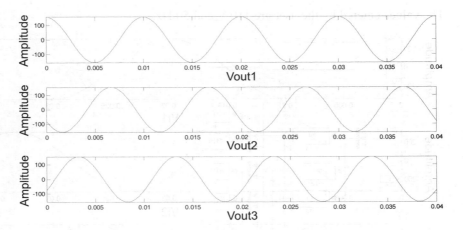

Fig. 11.11 Three-phase output line voltages obtained after modulation

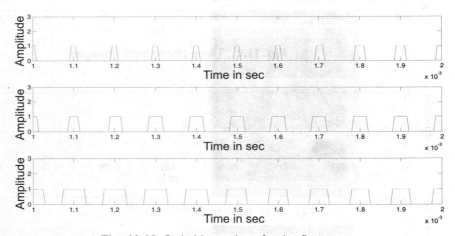

Fig. 11.12 Switching pulses for the first group

Example 11.2

In a similar manner, the matrix converters can also be modeled by using Simulink blocksets. A model of three-phase to single-phase matrix converter is shown in Fig. 11.13. The three-phase input waves are modeled in a similar manner. The input frequency is taken as 50 Hz and the output frequency as 100 Hz. The three switches taken are ideal switches. Modulating signals for this converter are obtained by Venturini method as obtained in Example 11.1. Switching pulses from these modulating signals are obtained by applying PWM modulation. Modulating signals of this converter are shown in Fig. 11.14. Output voltage obtained after the three switches is shown in Fig. 11.15, and Fig. 11.16 shows the combined output voltage from time 0 to 0.04 s. Switching pulses of the switches are shown in Fig. 11.17.

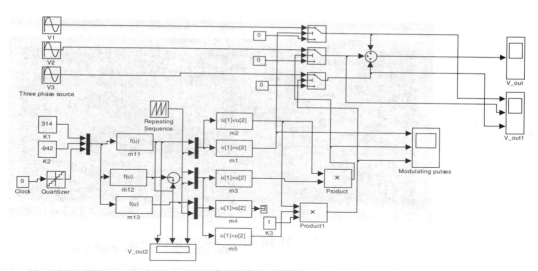

Fig. 11.13 Simulink model of a three-phase to single-phase matrix converter

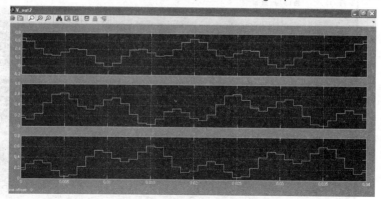

Fig. 11.14 Modulating signals of the three switches

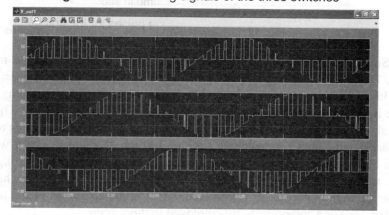

Fig. 11.15 Modulated output voltage in terms of three-phase input voltages

Fig. 11.16 Chopped output voltage

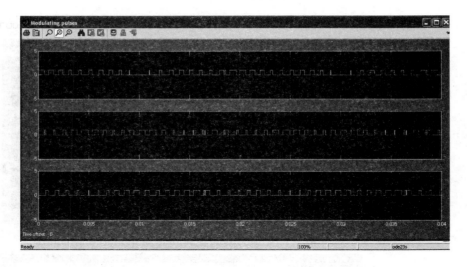

Fig. 11.17 Switching pulses of the three switches

Similarly, a three-phase to three-phase matrix converter can be modeled as shown in Fig. 11.18. The switching frequency of the PWM pulse should be at least greater than 20 times the highest input/output frequency. This model is executed for switching frequency of 10 kHz. The input and output frequencies are 50 Hz and 100 Hz respectively. The three input voltages are shown in Fig 11.19. The output frequency can also be varied in this model, if desired. This model is simulated for switching time of 100 μs. Modulated output waveform is shown in Fig. 11.20 which contains a fundamental cosine wave of 100 Hz plus higher order harmonics. R-L passive filter with $R = 7$ Ω and $L = 11$ mH is used to filter out the higher order harmonics from the output waveform. The filtered output is shown in Fig. 11.21. It can be observed that no large passive components are employed in this circuit.

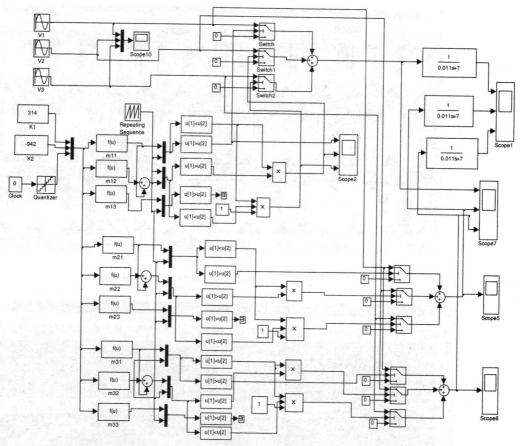

Fig. 11.18   Simulink model of a three-phase to three-phase matrix converter

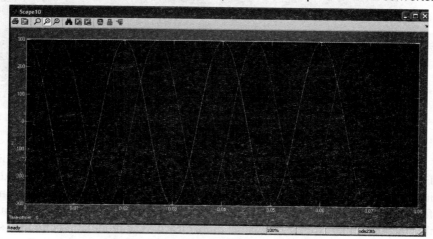

Fig. 11.19   Three-phase input voltage

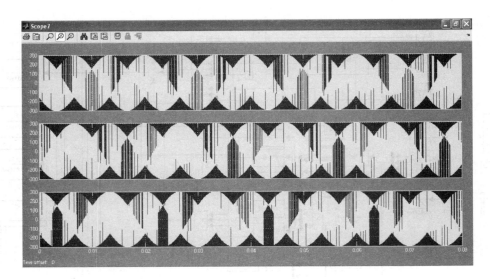

Fig. 11.20 Three-phase output voltage

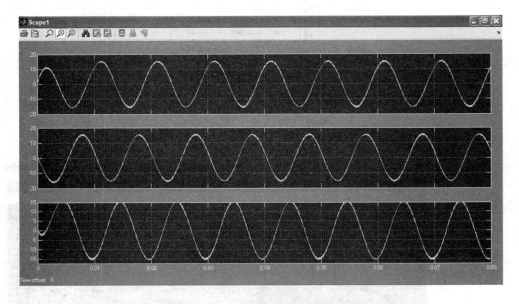

Fig. 11.21 Three-phase output voltage after filtering

This model of a three-phase to three-phase matrix converter is again modified as shown in Fig. 11.22 for proper commutation. Four-step current commutation logic is used for commutation in this model. 'Subsystem1', 'Subsystem2', and 'Subsystem3' are the subsystems of the circuitry for the modulating pulses of the three phases. They are kept in a subsystem in Fig. 11.22 so that the remaining circuitry is visible more clearly. The modulating circuit is same as shown in Fig. 11.18. The output waveforms obtained after filtering are shown in Fig. 11.23.

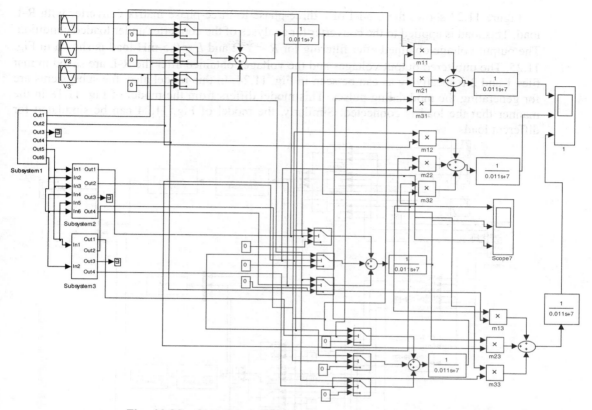

Fig. 11.22 Simulink model of a current commutated matrix converter

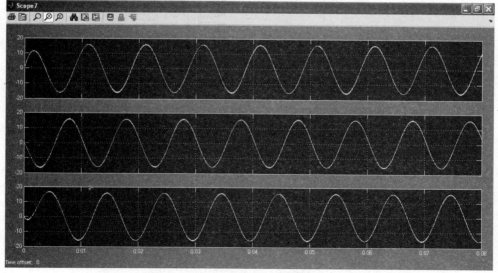

Fig. 11.23 Three-phase output voltages

Figure 11.24 shows the model of a three-phase to three-phase matrix converter with R-L load. This load is applied to the converter for analyses of the converter under loaded condition. The output voltage obtained after filtering for $R = 1\ \Omega$ and $L = 0.5$ mH load is shown in Fig. 11.25. The unfiltered output voltages and the voltages obtained after the R-L are saved in mat files 1 and 2 respectively as can be seen in Fig. 11.24. In this model also, the subsystems are for generating the modulating pulses. This model differs from the model of Fig. 11.18 in the manner that the load is connected. Similarly, the model of Fig. 11.24 can be simulated for different loads.

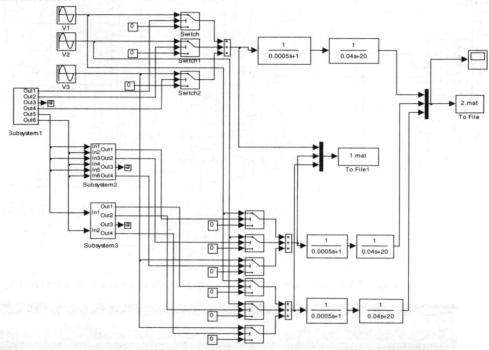

Fig. 11.24 Simulink model of a matrix converter feeding a R-L load

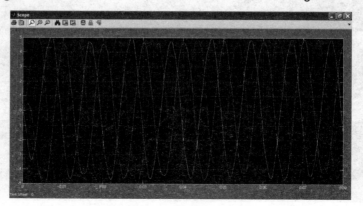

Fig. 11.25 Three-phase output voltages in case of R-L load

Similarly, a three-phase to three-phase sine wave matrix converter can also be modeled as shown in Fig. 11.26. A 'sine wave' matrix converter has sinusoidal input supply waveforms. The input waveforms given to this converter are as follows:

$$V_{i1} = 220 \times 1.414 \times \sin(2\pi \times 50 \times t)$$

$$V_{i2} = 220 \times 1.414 \times \sin\left(\frac{2\pi \times 50 \times t + (2 \times \pi)}{3}\right)$$

$$V_{i3} = 220 \times 1.414 \times \sin\left(\frac{2\pi \times 50 \times t + (4 \times \pi)}{3}\right)$$

The three-phase output voltage waveforms required are as follows:

$$V_{o1} = V_o \sin(w_o \times t)$$

$$V_{o2} = V_o \sin\left(w_o \times t + 2 \times \frac{\pi}{3}\right)$$

$$V_{o3} = V_o \sin\left(w_o \times t + 4 \times \frac{\pi}{3}\right)$$

where $w_o = 2 \times \pi \times f_o$, f_o is the output voltage frequency in Hz.

In a similar manner, modulating signals for the three-phase to three-phase sine wave converter will be as follows:

$$m_{11} = \frac{1}{3} + a_1 \times \sin((w_o - w_i) \times t) + a_1 \times \sin(-(w_o + w_i) \times t)$$

$$m_{12} = \frac{1}{3} + a_1 \times \sin\left(\frac{(w_o - w_i) \times t - (2 \times \pi)}{3}\right) + a_1 \times \sin\left(\frac{-(w_o + w_i) \times t - (2 \times \pi)}{3}\right)$$

$$m_{13} = \frac{1}{3} + a_1 \times \sin\left(\frac{(w_o - w_i) \times t - (4 \times \pi)}{3}\right) + a_1 \times \sin\left(\frac{-(w_o + w_i) \times t - (4 \times \pi)}{3}\right)$$

$$m_{21} = \frac{1}{3} + a_1 \times \sin\left(\frac{(w_o - w_i) \times t - (4 \times \pi)}{3}\right) + a_1 \times \sin\left(\frac{-(w_o + w_i) \times t - (2 \times \pi)}{3}\right)$$

$$m_{22} = \frac{1}{3} + a_1 \times \sin((w_o - w_i) \times t) + a_1 \times \sin\left(\frac{-(w_o + w_i) \times t - (4 \times \pi)}{3}\right)$$

$$m_{23} = \frac{1}{3} + a_1 \times \sin\left(\frac{(w_o - w_i) \times t - (2 \times \pi)}{3}\right) + a_1 \times \sin(-(w_o + w_i) \times t)$$

$$m_{31} = \frac{1}{3} + a_1 \times \sin\left(\frac{(w_o - w_i) \times t - (2 \times \pi)}{3}\right) + a_1 \times \sin(-(w_o + w_i) \times t (4 \times \pi)/3)$$

$$m_{32} = \frac{1}{3} + a_1 \times \sin\left(\frac{(w_o - w_i) \times t - (4 \times \pi)}{3}\right) + a_1 \times \sin(-(w_o + w_i) \times t)$$

$$m_{33} = \frac{1}{3} + a_1 \times \sin((w_o - w_i) \times t) + a_1 \times \sin\left(\frac{-(w_o + w_i) \times t - (2 \times \pi)}{3}\right)$$

where w_i is the input angular frequency, w_o is the output wave desired angular frequency, and $a_1 = V_o/3V_i$. The Simulink model of this converter is shown in Fig. 11.26. For input waveform of 50 Hz and the desired output of 100 Hz, output waveforms obtained after modulation are shown in Fig. 11.27. After filtering the high frequency switching harmonics, we obtain the output as shown in Fig. 11.28. For switching time of 100 μs, switching pulses for the first group of switches (i.e., S_{11}, S_{12}, S_{13}) are shown in Fig. 11.29, and for the second group of switches (i.e., S_{21}, S_{22}, S_{23}) are shown in Fig. 11.30.

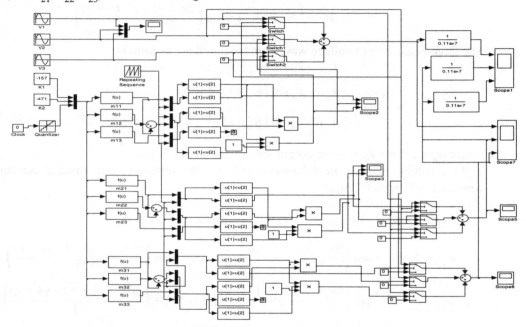

Fig. 11.26 Simulink model of a sine wave matrix converter

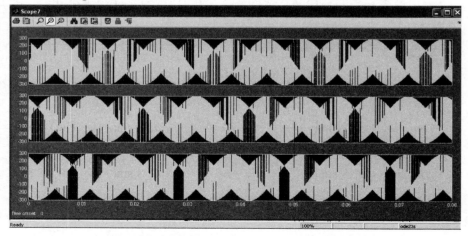

Fig. 11.27 Chopped sine wave output voltages

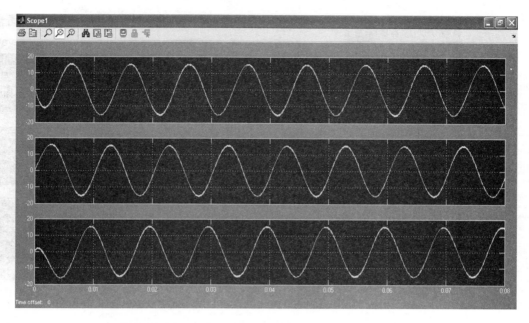

Fig. 11.28   Three-phase sine wave output voltages

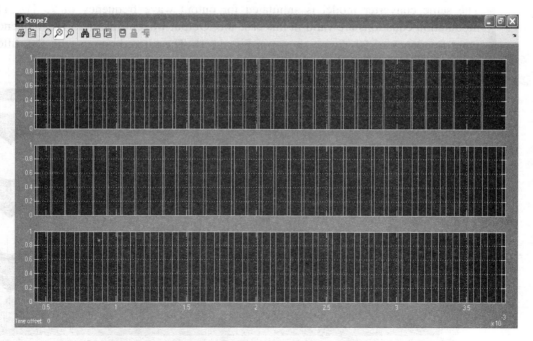

Fig. 11.29   Switching pulses of the first group of switches

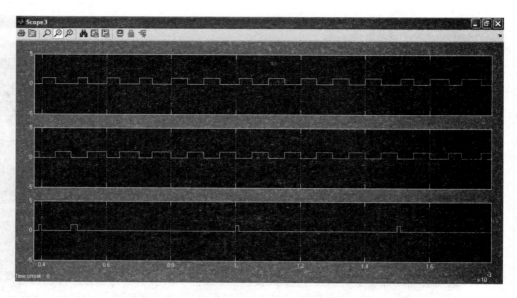

Fig. 11.30 Switching pulses of the second group of switches

Example 11.3

The same converter model is simulated for output wave frequency of 25 Hz. The same modulation equations are valid in this case too. Output waveforms of 25 Hz frequency for an input frequency of 50 Hz are shown in Fig. 11.31. The filtered output with R-L filter taking $R = 7$ and $L = 0.11$ H is shown in Fig. 11.32.

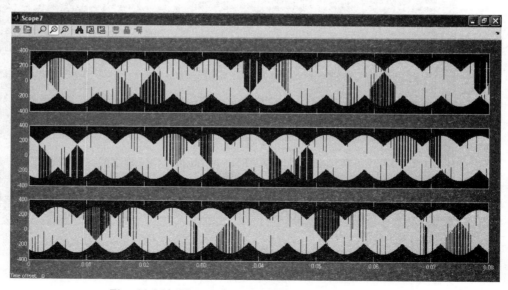

Fig. 11.31 Three-phase output waveform of 25 Hz

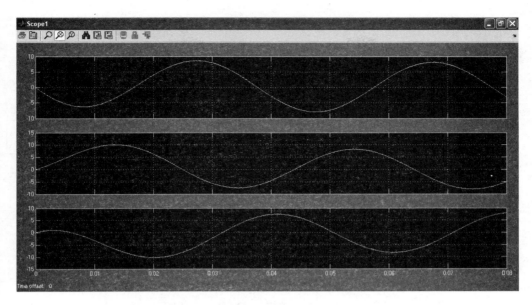

Fig. 11.32 Three-phase filtered output waveform of 25 Hz

11.5 INTRODUCTION TO PWM RECTIFIER

Nowadays, voltage and current source inverters are widely being used in electrical motor drives. In these drives, the DC voltage or the current required are usually obtained by using PWM controlled rectifiers. The major drawbacks of classical rectifiers are lower order harmonics generation on AC line, lagging displacement factor to the grid utility, unidirectional power flow, and requirement of a large DC link filter element. The rapid growth of AC adjustable speed drives (ASDs) in industries is exacerbating the problem of harmonics in power system, caused by using nonlinear devices and elements. Although active or passive filters can be employed for filtering, the best solution could be obtained by using PWM rectifiers. A PWM rectifier has an added advantage of bidirectional flow of electrical power. The circuit diagram of a three-phase PWM rectifier is shown in Fig. 11.33 Also, new limits has been put forward by standards like IEEE519-1992 and IEC 61000-3-2/IEC 6100-3-4 which indicate the currents limits for the electrical converters. The PWM rectifiers can overcome these problems because of its well-known capabilities such as power regeneration, low harmonic contents, sinusoidal input current, high power factor, controlled DC link voltage, small filtering elements, and possibility of four-quadrant operation. These rectifiers are becoming more and more popular in the industry. Since these converters have the ability to provide sinusoidal input current, the unity power factor operation can be easily performed by regulating the currents in phase with the power source voltages.

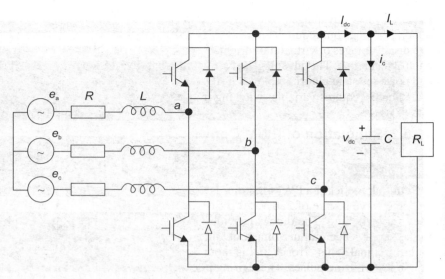

Fig. 11.33 Circuit of a three-phase PWM rectifier

High power, low rpm gearless DC machines used in elevators are of large size. These DC machines work at power levels where no AC machine can be employed. Thyristor-based DC drives which are inherently bidirectional are capable of operating the largest elevator motors although they have drawbacks of variable power factor and power frequency harmonics. These harmonics can be removed by using appropriate filters but variable power factor issue remains unsolvable by this technique. The AC drives, on the other hand, can provide good power factor with similar harmonic contents; but in low cost AC drives, using the conventional diode bridge rectifier at the front end regeneration is not possible. In order to achieve regeneration, low harmonic content and high power factor, high frequency with an AC drive, the conventional diode bridge rectifier should be replaced with the three-phase PWM rectifier as shown in Fig. 11.33, where high frequency PWM technique can be used. With PWM, we can also control the output power of the motor effectively.

11.5.1 Control Techniques

Various control methodologies have been proposed for PWM rectifiers. It is well known that in order to obtain better AC supply power quality and high performance of these converters, it is preferable to directly control the magnitude and phase angle of input supply currents. One such technique is hysteresis band current control technique, which forces the supply current to follow the reference current, because of its ease of implementation , fast current control response, and inherent peak current limiting capability. Broadly, there are four types of control techniques described in literature: voltage-oriented control, voltage-based direct power control, virtual flux-oriented control, and virtual flux-based direct power control. The voltage-oriented control technique has the advantages of fixed switching frequency, cheaper A/D converters, and simple filters but the algorithm required is complex. Voltage-based direct power control technique has advantages of good dynamic performance, lower harmonics, and no current regulation loops but requires high values of sampling frequency and inductance. The advantages and disadvantages of virtual flux-oriented control technique are similar to that of voltage-oriented control technique. The advantages of virtual flux-based direct power control technique are simple and noise resistant algorithm, lower THD, good dynamic performance, and no current regulation loops; but it requires variable switching frequency, fast microprocessors, and A/D converters.

11.5.2 Simulation of PWM Converters

Example 11.4

A Simulink model of a PWM rectifier is shown in Fig. 11.34. In this model, a three-phase input AC sine wave source of 220 V, 50 Hz is constructed. Six thyristors in parallel with diodes are used for switching. The diodes provide a path for the current to flow from the inductive load to the supply. The output voltage obtained is compared with the reference output to generate an error signal. This error signal is modulated by a high frequency sawtooth wave in order to obtain PWM firing pulses. These firing pulses are then fed to the six SCRs. The widths of these firing pulses vary with the error signal so as to obtain the desired output voltage. Figure 11.35 shows the input current of the three phases. The output current and voltage waveforms for a load of $R = 100$ Ω, $L = 10$ mH are shown in Fig. 11.36, and Fig. 11.37 shows the PWM pulses and error signal obtained.

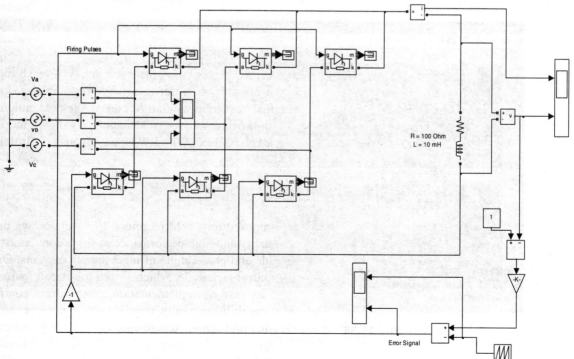

Fig. 11.34   Simulink model of a three-phase PWM rectifier

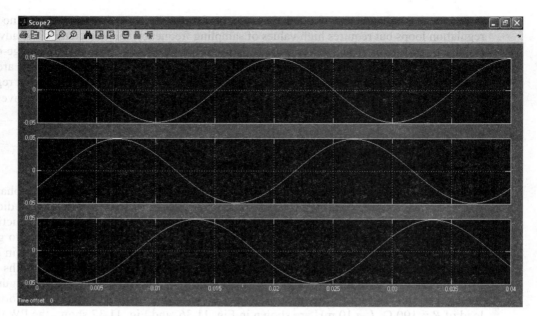

Fig. 11.35 Input current of phase 1, phase 2, and phase 3

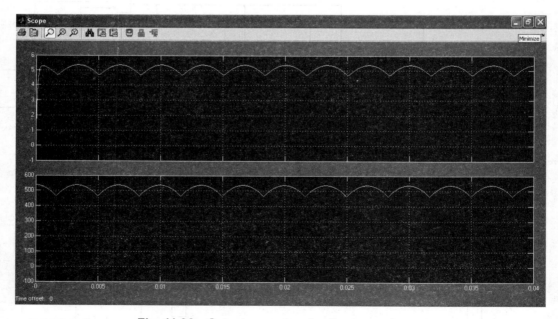

Fig. 11.36 Output current and voltage waveforms

Fig. 11.37 Firing pulses and error signal

The PWM converters as discussed earlier are employed for converting DC voltage into AC voltage of desired magnitude and frequency. This is helpful in getting the AC supply from a DC source like battery. Figure 11.38 shows the Simulink discrete model of a three-phase PWM converter. This model takes 440 V DC voltages as input for the two converters and produces three-phase AC output by PWM control. For switching, three-level bridge block is utilized. A discrete three-phase PWM generator is used for the generation of PWM pulses. The output of the converter is fed to a 12-terminal three-phase linear transformer for isolation of output from input. The three-phase output of the transformer is then supplied to the three-phase load. Figure 11.39 shows the output voltages of the converter.

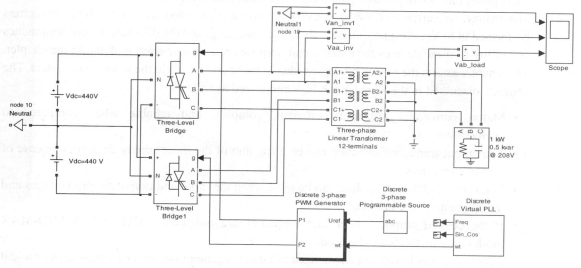

Fig. 11.38 Simulink model of a discrete PWM inverter cum rectifier

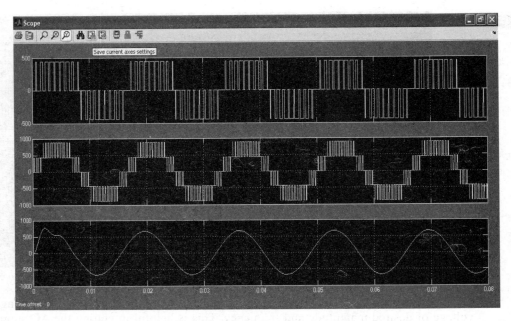

Fig. 11.39 Output voltages of inverter (phase voltage), inverter line-to-line voltage, and line-to-line voltage at the load

SUMMARY

After going through the chapter, the reader will be familiarized with some advanced converters like matrix converter and PWM converters. The analysis and design of these converters performed in MATLAB gives an idea about their working, and the different control approaches and utility of these devices for commercial purpose are also discussed in detail in the chapter. This chapter assists the reader get a clear idea of the recent trends in the field of converters. The contents discussed in this chapter can be summarized as follows:

- Matrix converters are more efficient, compact, and reliable when compared to cycloconverters.
- The output frequency obtained can be a fraction of the input supply frequency in case of matrix converters.
- Venturini modulation method can be used when the input and output supply voltages and the frequencies are known.
- If the desired output voltage and frequency are unknown, SVM or MIN-MID-MAX modulation methods can be used.
- PWM rectifiers have such advantages as power regeneration, lower harmonics, sinusoidal input current, small filtering components, and possibility of four-quadrant operation.

REVIEW QUESTIONS

1. What do you mean by AC to AC converters? Mention some applications of these converters.
2. Explain the working of a step-up and step-down cycloconverter along with its limitations.
3. Describe the problems related to the design and control of a cycloconverter.
4. Elaborate the advantages of matrix converters over the conventionally used techniques.
5. Describe the commutation problem of a three-phase to three-phase matrix converter.
6. Compare the different modulation strategies of matrix converters.
7. Mention the advantages of a PWM rectifier over a conventional rectifier.
8. Explain the different control methods for a PWM rectifier.
9. What is meant by a bidirectional switch? Explain.
10. Mention the advantages of a bidirectional power flow.
11. What are the advantages of four-quadrant operation of a PWM rectifier?
12. Explain the current commutation techniques used for matrix converters.
13. Mention some potential industrial applications of matrix converters.

PROGRAMMING EXERCISES

1. Develop a MATLAB program to model a two-phase to single-phase matrix converter. Assume input supply to be 220 V sinusoidal of 50 Hz frequency. Test this model for output frequencies of 10, 40, 81, and 1,000 Hz.
2. Write a program to implement the four-step current commutation logic for the converter in Q. No. 1?
3. Develop a program in MATLAB for a three-phase PWM rectifier. This rectifier takes 440 V, 60 Hz three-phase voltage supply as input and delivers DC voltage of fixed magnitude as output. Frequency of the PWM pulses should be at least 20 times the maximum input/output frequency.

4. Develop a Simulink model for a matrix converter using TRIAC for bidirectional switches. This converter takes three-phase input supply of 340 V, 55 Hz and delivers an output of 247 Hz.
5. Develop a model of a three-phase to three-phase matrix converter and analyze its input current for the following loads:
   (a) Three-phase induction motor
   (b) Three-phase synchronous motor
   (c) Three-phase nonlinear load
6. Develop a PWM rectifier model which supplies an R-L-E load of 25 Ω, 40 mH, and 40 V. The input supply is of 230 V, 50 Hz. Operate this rectifier in all the four quadrants and observe the output voltage and current waveforms on the Scope.

SUGGESTED READING

Analysis of different schemes of matrix converter with maximum voltage conversion ratio

Imayavaramban, M., K. Latha, and G. Uma, MELECON 2004, *Proceedings of the 12th IEEE Mediterranean Electrotechnical Conference*, Vol. 3, pp. 1137–1140, 12–15 May 2004.

Matrix converter-fed ASDs

Cha, H.J. and P.N. Enjeti, *IEEE Industry Applications Magazine*, Vol. 10, No. 4, pp. 33–39, July–August 2004.

New modulation strategy for a matrix converter with a very small mains filter

Muller, S., U. Ammann, and S. Rees, PESC '03, *IEEE 34th Annual Conference on Power Electronics Specialist,* Vol. 3, pp. 1275–1280, 15–19 June 2003.

A novel multi-level matrix converter

Yang, X., Y. Shi, Q. He, and Z. Wang, APEC '04, *19th Annual IEEE Applied Power Electronics Conference and Exposition*, Vol. 2, pp. 832–835, 2004.

A new structure for three-phase to single-phase AC/AC matrix converters

Babaei, E., A. Aghagolzadeh, S.H. Hosseini, and S. Khanmohammadi, ICECS 2003, *Proceedings of the 2003 10th IEEE International Conference on Electronics, Circuits and Systems,* Vol. 1, pp. 36–39, 14–17 December 2003.

A three-phase AC/AC high-frequency link matrix converter for VSCF applications

Cha, H.J., and P.N. Enjeti, PESC '03, *IEEE 34th Annual Conference on Power Electronics Specialist,* Vol. 4, pp. 1971–1976, 15–19 June 2003.

Evaluation of the single-sided matrix converter driven switched reluctance motor

Goodman, A.S., K.J. Bradley, and P.W. Wheeler, *39th IAS Annual Meeting, Conference Record of the 2004 IEEE Industry Applications Conference,* Vol. 3, pp. 1847–1851, 3–7 October 2004.

Power supply loss ride-through and device voltage drop compensation in a matrix converter permanent magnet motor drive for an aircraft actuator

Wheeler, P.W., J.C. Clare, M. Apap, L. Empringham, C. Whitley, and G. Towers, PESC '04 2004, *IEEE 35th Annual Power Electronics Specialists Conference,* Vol. 1, pp. 149–154, 20–25 June 2004.

An approach for matrix converter–based induction motor drive with unity power factor and minimum switching losses

Sangshin, K., and H.A. Toliyat, IECON '03, *The 29th Annual Conference of the IEEE Industrial Electronics Society,* Vol. 3, pp. 2939–2944, 2–6 November 2003.

Combined control of matrix converter–fed induction motor drive system

Sun, K.H., Lipei, K. Matsuse, and T. Ishida, *38th IAS Annual Meeting, Conference Record of the Industry Applications Conference,* Vol. 3, pp. 1723–1729, 12–16 October 2003.

Matrix converters

Wheeler, P.W., J.C. Clare, L. Empringham, M. Bland, and K.G. Kerris, *IEEE Industry Applications Magazine*, Vol. 10, No. 1, pp. 59–65, January–February 2004,

A new control algorithm for matrix converters under distorted and unbalanced conditions

Hosseini, S.H. and E. Babaei, *CCA 2003, Proceedings of 2003 IEEE Conference on Control Applications,* Vol.1, pp. 1088–1093, 23–25 June 2003.

Effects of input voltage measurement on stability of matrix converter drive system

Casadei, D., G. Serra, A. Tani, and L. Zarri, *IEE Proceedings—Electric Power Applications,* Vol. 151, No. 4, pp. 487–497, 7 July 2004.

A 150 kVA vector controlled matrix converter induction motor drive

Podlesak, T.F. and D. Katsis, 39th IAS Annual Meeting, *Conference Record of the 2004 IEEE Industry Applications Conference,* Vol. 3, pp. 1811–1816, 3–7 October 2004.

Control and implementation of a new modular matrix converter

Angkititrakul, S. and R.W. Erickson, APEC '04, *19th Annual IEEE Applied Power Electronics Conference and Exposition,* Vol. 2, pp. 813–819, 2004.

Bidirectional switch commutation for matrix converters

Ecklebe, A., A. Lindermann, S. Schulz, *Power Electronics, IEEE Transaction*, Vol. 24, 5, pp. 1173–1181, May 2009.

Virtual-flux-based direct power control of three-phase PWM rectifiers

Malinowski, M., M.P. Kazmierkowski, S. Hansen, F. Blaabjerg, and G.D. Marques, *IEEE Transactions on Industry Applications*, Vol. 37, No. 4, July/August 2001.

A unity power factor three-phase PWM SCR rectifier for high power applications in the metal industry

Wallace, I., A. Bendre, J. Nord, and G. Venkataramanan, *IEEE IAS Annual Meeting*, Chicago, IL, USA, October 2001.

Dual current control scheme for PWM converter under unbalanced input voltage conditions

Song, H.-S. and K. Nam, *IEEE Transactions on Industrial Electronics*, Vol. 46, No. 5, October 1999.

Single-phase AC/DC integrated PWM converter

Rossetto, L. and S. Buso, *Telecommunications Energy Conf. 2000*, Department of Electronics and Informatics, University of Padova, pp. 411–418, 10–14 Sept. 2000.

Digital repetitive controlled three-phase PWM rectifier Ph.D. Theiss.

Zhou, K. and D. Wang, *IEEE Transactions on Power Electronics*, Vol. 18, No. 1, January 2003, Poland 2001.

Sensor-less control strategies for three-phase PWM rectifiers

Malinowski, M., Ph. D. Thesis, Faculty of Electrical Engineering, Warsaw University of Technology, Poland, 2001.

Review and comparative study of control techniques for three-phase PWM rectifiers

Malinowski, M., M.P. Kazmierkowski, and A. Trzynadlowski, Available at www.elseviermathematics.com.

Fields vector strategy for an induction motor system using PWM rectifier/inverter link

Chen, L., Department of Energy Technology, Aalborg, Denmark, Aalborg University, Forlag, 1994.

Voltage sag response of PWM rectifiers for variable-speed wind turbines
Ottersten, R., A. Petersson, and K. Pietilainen, www. etdeweb.org.

Advanced control scheme for a single-phase PWM rectifier in traction applications
Song, H.-S., R. Keil, and P. Mutschler, *38th IAS Annual meeting,* Industrial App Conf., Vol. 3, pp. 1558–1565, 12–16 Oct 2003.

A fuzzy-controlled active front-end rectifier with current harmonic filtering characteristics and minimum sensing variables
Dixon, J.W., J.M. Contardo, L.A. Moran, Vol. 14, No. 04, pp. 724–729, July 1999.

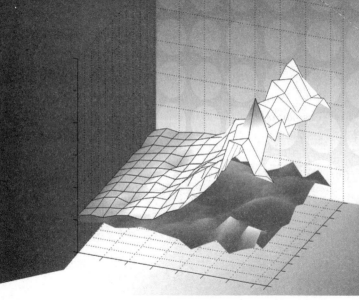

APPENDIX A

FOURIER SERIES AND LAPLACE TRANSFORM

A.1 TRIGONOMETRIC FOURIER SERIES

There are two types of waveforms—periodic and nonperiodic. Waveforms which repeat after a certain time period, i.e., $f(t + T) = f(t)$, are periodic waveforms. For nonperiodic waveforms, $f(t + T) \neq f(t)$. Any periodic waveform can be expressed in terms of sine and cosine waves with the help of Fourier series if it satisfies the following three conditions known as Dirichlet conditions:

1. There should be finite number of discontinuities in the waveform.
2. The average value of the waveform over the time period T should be finite.
3. There should be finite number of positive or negative maxima.

The trigonometric Fourier series of any periodic waveform $f(t) = f(t + T)$ of period T can be expressed as follows:

$$f(t) = a_0 + \sum \{a_n \cos (n\omega t) + b_n \sin (n\omega t)\}$$

where n varies from 1 to ∞.

The Fourier coefficients a's and b's can be obtained for a given waveform by evaluating the following integrals:

$$a_0 = \frac{1}{T} \int_0^T f(\omega t) d\omega t$$

$$a_1 = \frac{2}{T} \int_0^T f(\omega t) \cos \omega t \, d\omega t$$

$$a_2 = \frac{2}{T} \int_0^T f(\omega t) \cos 2\omega t \, d\omega t$$

$$a_n = \frac{2}{T} \int_0^T f(\omega t) \cos n\omega t \, d\omega t$$

$$b_1 = \frac{2}{T} \int_0^T f(\omega t) \sin \omega t \, d\omega t$$

$$b_2 = \frac{2}{T} \int_0^T f(\omega t) \sin 2\omega t \, d\omega t$$

$$b_n = \frac{2}{T} \int_0^T f(\omega t) \sin n\omega t \, d\omega t$$

The Fourier series of some waveforms are determined as follows:

1. A square wave is shown in Fig. A.1. The Fourier series of this wave can be determined as follows:

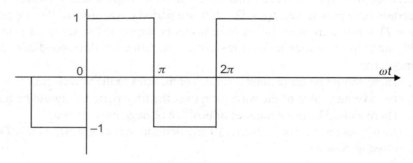

Fig. A.1   Square wave

$$a_0 = \frac{1}{2\pi} \int_0^{2\pi} d\omega t = \frac{1}{2\pi} \left\{ \int_0^{\pi} 1 \, d\omega t + \int_{\pi}^{2\pi} -1 \, d\omega t \right\} = 0$$

$$a_1 = \frac{2}{2\pi} \int_0^{2\pi} f(\omega t) \cos \omega t \, d\omega t = \frac{1}{\pi} \left\{ \int_0^{\pi} \cos \omega t \, d\omega t + \int_{\pi}^{2\pi} -\cos \omega t \, d\omega t \right\} = 0$$

$$a_n = \frac{2}{2\pi} \int_0^{2\pi} f(\omega t) \cos \omega t \, d\omega t = \frac{1}{\pi} \left\{ \int_0^{\pi} \cos n\omega t \, d\omega t + \int_{\pi}^{2\pi} -n\omega t \, d\omega t \right\} = 0$$

$$b_1 = \frac{2}{2\pi} \int_0^{2\pi} f(\omega t) \sin \omega t \, d\omega t = \frac{1}{\pi} \left\{ \int_0^{\pi} \sin \omega t \, d\omega t + \int_{\pi}^{2\pi} -\sin \omega t \, d\omega t \right\} = \frac{4}{\pi}$$

$$b_2 = \frac{2}{2\pi} \int_0^{2\pi} f(\omega t) \sin 2\omega t \, d\omega t = \frac{1}{\pi} \left\{ \int_0^{\pi} \sin 2\omega t \, d\omega t + \int_{\pi}^{2\pi} -\sin 2\omega t \, d\omega t \right\} = 0$$

$$b_3 = \frac{2}{2\pi} \int_0^{2\pi} f(\omega t) \sin 3\omega t \, d\omega t = \frac{1}{\pi} \left\{ \int_0^{\pi} \sin 3\omega t \, d\omega t + \int_{\pi}^{2\pi} -\sin 2\omega t \, d\omega t \right\} = \frac{4}{(3\pi)}$$

$$b_n = \frac{2}{2\pi} \int_0^{2\pi} f(\omega t) \sin n\omega t \, d\omega t = \frac{1}{\pi} \left\{ \int_0^{\pi} \sin n\omega t \, d\omega t + \int_{\pi}^{2\pi} -\sin n\omega t \, d\omega t \right\}$$

$$= \left\{ \frac{2}{(\pi n)} \right\} \times (1 - \cos n\pi)$$

Thus, the Fourier series of this waveform will be as follows:

$$f(\omega t) = 4/\pi \sin \omega t + 4/(3\pi) \sin 3\omega t + 4/(5\pi) \sin 5\omega t + \cdots + \{2/(\pi n)\} \times (1 - \cos n\pi) \sin n\omega t + \cdots$$

2. The half-wave rectified sine wave is shown in Fig. A.2, the Fourier series of which can be determined as follows:

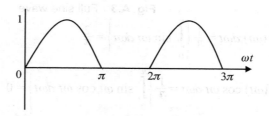

Fig. A.2   Half sine wave

$$a_0 = \frac{1}{2\pi} \int_0^{2\pi} f(\omega t) \, d\omega t = \frac{1}{2\pi} \left\{ \int_0^{\pi} \sin \omega t \, d\omega t + \int_{\pi}^{2\pi} 0 \, d\omega t \right\} = 1/\pi$$

$$a_1 = \frac{2}{2\pi} \int_0^{2\pi} f(\omega t) \cos \omega t \, d\omega t = \frac{1}{\pi} \left\{ \int_0^{\pi} \sin \omega t \cos \omega t \, d\omega t + \int_{\pi}^{2\pi} 0 \, d\omega t \right\} = 0$$

$$a_n = \frac{2}{2\pi} \int_0^{2\pi} f(\omega t) \cos \omega t \, d\omega t = \frac{1}{\pi} \left\{ \int_0^{\pi} \sin \omega t \cos n\omega t \, d\omega t + \int_{\pi}^{2\pi} 0 \, d\omega t \right\}$$

$$= \frac{(1 + \cos n\pi)}{\{\pi(1 - n^2)\}}$$

$$b_1 = \frac{2}{2\pi} \int_0^{2\pi} f(\omega t) \sin \omega t \, d\omega t = \frac{1}{\pi} \left\{ \int_0^{\pi} \sin \omega t \sin \omega t \, d\omega t + \int_{\pi}^{2\pi} 0 \, d\omega t \right\} = \frac{1}{2}$$

$$b_2 = \frac{2}{2\pi} \int_0^{2\pi} f(\omega t) \sin 2\omega t \, d\omega t = 1/\pi \left\{ \int_0^{\pi} \sin \omega t \sin 2\omega t \, d\omega t + \int_{\pi}^{2\pi} 0 \, d\omega t \right\} = 0$$

$$b_n = \frac{2}{2\pi} \int_0^{2\pi} f(\omega t) \sin n\omega t \, d\omega t = \frac{1}{\pi} \left\{ \int_0^{\pi} \sin \omega t \sin n\omega t \, d\omega t + \int_{\pi}^{2\pi} 0 \, d\omega t \right\} = 0$$

Thus, the Fourier series of this waveform will be as follows:

$$f(\omega t) = \frac{1}{\pi} + \frac{1}{2} \sin \omega t - \frac{2}{3\pi} \cos 2\omega t - \frac{2}{15} \cos 4\omega t - \cdots$$

3. A full sine wave is shown in Fig. A.3, the Fourier series of which can be determined as follows:

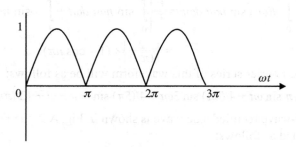

Fig. A.3 Full sine wave

$$a_0 = \frac{1}{\pi} \int_0^\pi f(\omega t) \, d\omega t = \frac{1}{\pi} \left\{ \int_0^\pi \sin \omega t \, d\omega t \right\} = \frac{2}{\pi}$$

$$a_1 = \frac{2}{\pi} \int_0^\pi f(\omega t) \cos \omega t \, d\omega t = \frac{1}{\pi} \left\{ \int_0^\pi \sin \omega t \cos \omega t \, d\omega t \right\} = 0$$

$$a_n = \frac{2}{2\pi} \int_0^{2\pi} f(\omega t) \cos \omega t \, d\omega t = \frac{1}{\pi} \left\{ \int_0^\pi \sin \omega t \cos n\omega t \, d\omega t + \int_0^{2\pi} d\omega t \right\}$$

$$= \frac{(1 + \cos n\pi)}{\{\pi(1 - n^2)\}}$$

$$b_1 = \frac{2}{\pi} \int_0^\pi f(\omega t) \sin \omega t \, d\omega t = \frac{1}{\pi} \left\{ \int_0^\pi \sin \omega t \sin \omega t \, d\omega t \right\} = 0$$

$$b_n = \frac{2}{\pi} \int_0^\pi f(\omega t) \sin n\omega t \, d\omega t = \frac{1}{\pi} \left\{ \int_0^\pi \sin \omega t \sin n\omega t \, d\omega t \right\} = 0$$

Thus, the Fourier series of this waveform will be as follows:

$$f(\omega t) = \frac{2}{\pi} - \frac{4}{3\pi} \cos 2\omega t - \frac{4}{15\pi} \cos 4\omega t - \frac{4}{35\pi} \cos 6\omega t$$

A.2 EXPONENTIAL FOURIER SERIES

The exponential Fourier series of a periodic waveform $f(t)$ can be obtained as follows:

$$f(t) = \sum_{n=\infty}^{-\infty} C_n e^{in\omega t}$$

where C_n is given by

$$C_n = \frac{1}{2\pi} \int_0^{2\pi} f(t)\, e^{-jn\omega t}\, d\omega t$$

Also, the exponential Fourier series of a periodic waveform can be obtained by replacing the sin $n\omega t$ and cos $n\omega t$ terms by the exponential terms as follows:

$$e^{jn\omega t} = \cos n\omega t + j \sin n\omega t \qquad e^{-jn\omega t} = \cos n\omega t - j \sin n\omega t$$

$$\sin n\omega t = \frac{(e^{jn\omega t} - e^{-jn\omega t})}{2j} \qquad \cos n\omega t = \frac{(e^{jn\omega t} - e^{-jn\omega t})}{2}$$

A.3 LAPLACE TRANSFORM

The Laplace transform $F(s)$ of a function $f(t)$ is given by

$$F(s) = \int_0^\infty f(t)\, e^{-st}\, dt$$

The Laplace transforms of some selected functions are as follows:

f(t)	F(s)
1	$1/s$
t	$1/s^2$
e^{-at}	$1/(s + a)$
te^{-at}	$1/(s + a)^2$
$\sin \omega t$	$\omega/(s^2 + \omega^2)$
$\cos \omega t$	$s/(s^2 + \omega^2)$
$e^{-at} \sin \omega t$	$\omega/\{(s + a)^2 + \omega^2\}$
$e^{-at} \cos \omega t$	$(s + a)/\{(s + a)^2 + \omega^2\}$

B.1 PROOF OF MAXIMUM POWER TRANSFER THEOREM

Any network can be converted into a voltage source (V_{TH}) and a series resistance (R_{TH}) by Thevenin's theorem as shown in Fig. B.1.

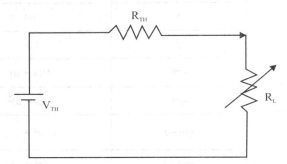

Fig. B.1 Thevenin's equivalent network

Now, the value of load resistance (R_L) has to be computed such that the power dissipated in R_L is maximum. For this circuit,

$$I_L = \frac{V_{TH}}{R_{TH} + R_L}$$

$$P_L = I_L^2 R_L = \left\{\frac{V_{TH}}{(R_{TH} + R_L)}\right\}^2 \times R_L = \left\{\frac{V_{TH}^2}{(R_{TH}^2 + R_L^2 + 2 R_{TH} R_L)}\right\} \times R_L$$

Now, for P_L to be maximum $\dfrac{dP_L}{dR_L} = 0$

Thus,

$$\frac{dP_L}{dR_L} = \frac{V_{TH}^2 (R_{TH}^2 - R_L^2)}{(R_{TH}^2 + R_L^2 + 2R_{TH} R_L)^2} = 0$$

As $V_{TH} \neq 0$, so $\qquad\qquad\qquad R_{TH}{}^2 - R_L{}^2 = 0$ or $R_{TH} = R_L$

Thus, it can be stated that the maximum power is transferred from a DC network to the load resistance when the load resistance is equal to the Thevenin's equivalent resistance of the network.

B.2 DERIVATION OF EMF EQUATION OF TRANSFORMER

When a sinusoidal AC voltage is applied to the primary winding of a transformer, an alternating flux is set up in the iron core which links both the windings. Assume that φ_{max} in webers is the maximum flux in the transformer core and f is the supply frequency in hertz. Thus, the time period of the flux is $1/f$ s. The magnetic flux increases from 0 to φ_{max} in one-fourth of the time period, i.e., $1/4f$ s. Thus,

Rate of change of flux $= \dfrac{d\varphi}{dt} = \dfrac{\varphi_{max}}{(1/4f)} = 4f\varphi_{max}$

Now, the average EMF induced per turn is equal to the rate of change of flux, i.e., $4f\varphi_{max}$ V.

We know that form factor for sinusoidal wave is 1.11 (Form factor $= V_{avg}/V_{rms}$). Therefore, the RMS value of the EMF induced per turn $= 1.11 \times 4f\varphi_{max} = 4.44f\varphi_{max}$ V.

Now, assuming that the number of turns on primary windings is N_1 and on secondary windings is N_2, we get,

EMF induced in primary winding (RMS) $E_1 = 4.44\,f\,N_1\,\varphi_{max}$ V
EMF induced in secondary winding (RMS) $E_2 = 4.44\,f\,N_2\,\varphi_{max}$ V
From these two equations, $\qquad\qquad \dfrac{E_1}{E_2} = \dfrac{N_1}{N_2}$

B.3 EMF EQUATION OF A DC MACHINE

For a DC generator, let

p = Number of poles
φ = Flux/pole, webers (Wb)
Z = Total number of armature conductors
$\quad$ = number of slots × number of conductors/slot
N = Rotational speed of armature in rpm
A = Number of parallel paths in armature
E = Generated EMF per parallel path in armature
Average EMF generated per conductor $= d\varphi/dt$ V
Flux cut per conductor in single revolution, i.e., $d\varphi = p\varphi$ Wb
Number of revolutions per second $= N/60$
Time taken by one revolution, i.e., $dt = 60/N$ s

Therefore,

$$\text{EMF generated per conductor} = (p\,\varphi\,N)/60 \text{ V}$$

Now, for a lap-wound generator,

Number of parallel paths, $A = p$

Number of conductors in one path = Z/p

Thus, EMF generated per path = $(Z\,\varphi\,p\,N)/120$ V

For a wave-wound generator,

Number of parallel paths, $A = p$

Number of conductors in one path = $Z/2$

Thus, EMF generated per path = $(Z\,p\,\varphi\,N)/120$ V

BIBLIOGRAPHY

Biran, A.B. and M.M.G. Breiner, *What Every Engineer Should Know About MATLAB and Simulink*, CRC Press, US, 2011.

Brockman, J.B., *Introduction to Engineering: Modelling and Problem Solving*, John Wiley & Sons Inc., 2009.

Chapman, S.J., *Essentials of MATLAB Programming*, 2nd Ed., Cengage Learning, 2009.

Chapman, S.J., *MATLAB Programming for Engineers*, 3rd Ed., Cengage Learning-Asia, 2007.

Davis, T.A., *MATLAB Primer*, 8th Ed., Chapman and Hall/CRC, US, 2011.

Driscoll, T.A., *Learning MATLAB*, SIAM, Philadelphia, 2009.

Gray, M.A., *Introduction to the Simulation of Dynamics Using Simulink*, CRC Press, 2011.

Hahn, B.D. and D.T. Valentine, *Essential MATLAB for Engineers and Scientists*, 4th Ed., Academic Press, 2010.

Haway, S.A., *MATLAB: A Practical Introduction to Programming and Problem Solving*, Elsevier Science, 2009.

Higham, N.J. and D.J. Higham, *MATAB Guide*, 2nd Ed., SIAM, 2005.

Kalechman, M., *Practical MATLAB Applications for Engineers*, CRC Press Inc., 2009.

Lopez, D.B., *MATLAB with Applications in Engineering, Physics, and Finance*, Taylor & Francis, 2010.

Magrab, E.B., S. Azarm, B. Balachandran, J. Duncan, K. Herold, and G. Walsh, *An Engineer's Guide to MATLAB: With Applications from Mechanical, Aerospace, Electrical, and Civil Engineering*, 3rd Ed., Prentice Hall, 2011.

Marchand, P. and O.T. Holland, *Graphics and GUIs with MATLAB*, 3rd Ed., Chapman & Hall/CRC, 2003.

Musto, J.C., W.E. Howard, and R.R. Williams, *Engineering Computations: An Introduction Using MATLAB and Excel*, McGraw-Hill, 2009.

Moore, H., *MATLAB for Engineers*, 2nd Ed., Prentice Hall, 2009.

Nakamura, S., *Numerical Analysis and Graphic Visualization with MATLAB*, 2nd Ed., Prentice Hall, US, 2002.

Okoro, O.I., *The Essential MATLAB and Simulink for Engineers and Scientists*, Juta and Company Ltd., 2010.

Palm-III, W.J., *Introduction to MATLAB for Engineers*, 3rd Ed., McGraw-Hill, New York, 2011.

Pratap, R., *Getting Started with MATLAB: A Quick Introduction for Scientists and Engineers*, Oxford University Press, 2010.

Smith, S.T., *MATLAB: Advanced GUI Development*, Dog Ear Publishing, 2006.

Stenger, H.G. and C.R. Smith, *Introduction to C++, Excel, and MATLAB & Basic Engineering Numerical Methods*, Vol. 1.1, Pearson Education Inc., 2009.

Van Loan, C.F. and K.Y. Daisyfan, *Insight Through Computing: A MATLAB Introduction to Computational Science and Engineering*, SIAM, 2010.

INDEX